Design Rationale

Concepts, Techniques, and Use

Edited by

Thomas P. Moran
Xerox Palo Alto Research Center

John M. Carroll
Virginia Polytechnic Institute and State University

CRC Press

Taylor & Francis Group

Boca Raton   London   New York

CRC Press is an imprint of the
Taylor & Francis Group, an **informa** business

First Published by
Lawrence Erlbaum Associates, Inc., Publishers
10 Industrial Avenue
Mahwah, New Jersey 07430-2262

Transferred to Digital Printing 2009 by CRC Press
6000 Broken Sound Parkway, NW Suite 300, Boca Raton, FL 33487
270 Madison Avenue New York, NY 10016
2 Park Square, Milton Park Abingdon, Oxon OX14 4RN, UK

Cover design by Gail Silverman

Library of Congress Cataloging-in-Publication Data

Design rationale : concepts, techniques, and use / edited by Thomas P.
Moran, John M. Carroll.
    p.   cm.
Includes bibliographical references and index.
ISBN 0-8058-1566-X (c) ISBN 0-8058-1567-8 (p)
    1. System design. 2. Human-computer interaction. I. Moran,
Thomas P. II. Carroll, John M.
QA76.9.S88D474 1996
004'.01'9—dc20   95-24785
                                                         CIP

Publisher's Note
The publisher has gone to great lengths to ensure the quality of this reprint
but points out that some imperfections in the original may be apparent.

Design Rationale

Concepts, Techniques, and Use

COMPUTERS, COGNITION, AND WORK

A series edited by:
Gary M. Olson, Judith S. Olson, and Bill Curtis

Contents

1

Overview of Design Rationale*

Thomas P. Moran
Xerox Palo Alto Research Center

John M. Carroll
Virginia Polytechnic Institute and State University

The primary goal of design is to give shape to an artifact—the *product* of design. This artifact is the result of a complex of activities—the design *process*. But the artifact is a concrete form that does not (except in very subtle ways) manifest this process of creation. It does not give evidence for the motivations that initiated its design, the stated requirements, the conditions that gave rise to its shape, the struggles and deliberations and negotiations, the trials and reflections, the careful balancing and tradeoffs of various factors, the reasons for its particular features, the reasons against features it does not have, and so on. Such background information can be valuable, even critical, to various people who deal with the artifact: not only its users and servicers, but also its builders, marketers, and so on, as well as other designers who want to build on the ideas. This kind of context in which design takes place is what design rationale is about.

Thomas Moran is a cognitive/computer scientist interested in human–computer interaction and collaborative system design, with an emphasis on supporting informal intellectual work practices; he is Principal Scientist and Manager of the Collaborative Systems Area at the Xerox Palo Alto Research Center. **John Carroll** is a cognitive psychologist interested in the analysis of human learning and problem-solving in human–computer interaction contexts and in the design of methods and tools for instruction and design; he is Professor of Computer Science and Psychology and Head of the Computer Science Department at Virginia Polytechnic Institute and State University.

*Note that chapters written in 1995 for this volume (e.g., Buckingham Shum, 1995) are listed 1995 in text and references. The actual copyright date of this volume is 1996.

CONTENTS

Design rationale is the notion that design goes beyond merely accurate descriptions of artifacts, such as specifications, and articulates and represents the reasons and the reasoning processes behind the design and specification of artifacts. Design rationale has recently become a focus of design research in various design domains, such as engineering design and the design of human–computer interaction.

The purpose of this chapter is to give a brief overview of the issues surrounding design rationale research in order to provide context for reading any chapter in this book. Before outlining the specific contents of the chapters in this book (in Section 3), we first consider how rationale fits into the design process (Section 1) and define more specifically what we mean by design rationale (Section 2). We conclude by raising some of the social and cultural aspects of design rationale (Section 4).

1. RATIONALE IN THE DESIGN PROCESS

Design is the process of creating tangible artifacts to meet intangible human needs. The problem of design is to bridge the gap from the realm of needs to the realm of concrete expression. The immense difficulties involved in doing this is what gives the design process its distinctive character and what gives relevance to a broad notion of design rationale. Let us consider the nature of the design process in order to make clear that it is permeated with research issues for design rationale.

Researchers from many different fields have attempted to portray the nature of design with both theoretical and descriptive accounts. A few of these are major influences in current design research. From a designer's point of view, Christopher Alexander has perhaps been the most articulate. His early work, *Notes on the Synthesis of Form* (Alexander, 1964), characterized design as a decomposition and resynthesis problem[1]; and this approach spawned much related research in the 1960s. Herbert Simon's influential book, *The Sciences of the Artificial* (Simon, 1969), gave a cognitive science account of design, involving search in a design space. Horst Rittel (Rittel & Weber, 1973) viewed design as a process of negotiation and deliberation, fundamentally dealing with uncertainty and conflict; and he was the first to propose an explicit representation of rationale. More recently, Donald Schön (1983) characterized design from a social and ethnographic perspective, emphasizing the reflective nature of design.

Like other professionals, designers are well aware that they are coping with elusive problems, searching, analyzing, synthesizing, negotiating, deliberating, and reflecting. Indeed, to be effective designers they must engage in and manage all of these modes of activities.

Research Issue: In their natural modes of practice, do designers spontaneously capture information that we might consider design rationale? How do they do it, and why do they do it?

1.1. The Structural Complexity of Design

The most obvious characteristic of the design process is that it is complex, for the simple reason that there are so many aspects to a design problem. It is impossible to deal with all aspects at once. Design problems have to be broken down into manageable subproblems, which can be addressed separately; but in practice it is not possible to find totally independent subproblems. Simon (1962) called such systems "nearly decomposable"; and the purpose of Alexander's (1964) decomposition method was to identify maximally independent subproblems. One of the most complex aspects of design is to manage the interrelations among subproblems and subsystems.

Research Issue: How can we represent not only the reasons for solutions to subproblems, but also for the tradeoffs and compromises made to adjudicate between the demands of the different subproblems?

It is typical that the needs being addressed in design work are poorly specified—vaguely stated, usually tacit, often latent, and sometimes wrong (people aren't sure what they want). Requirements are often wishfully

---

1. Alexander rejected this mathematical line of thinking and later proposed a view of design as consisting of interlocking patterns (Alexander et al., 1977).

overstated, and it is in the process of design that they are cut back to what is realistic. Defining the problem is as much a part of design as defining the artifact.

Research Issue: *How can we represent the changing problem definition, the reasons for the changes, and the reasons for the design decisions with respect to the state of the problem definition?*

Design problems are more than complex; they are dilemmas—"wicked problems" in Rittel's (Rittel & Weber, 1973) colorful phrase. Such problems cannot be "stated" per se or "solved" in the sense of definitive answers, because the criteria for evaluating goals and outcomes are innumerable, subjective, and conflicting. Any solution will generate "waves of consequences" that interact among themselves and with other problems, changing the problem situation in irreversible and unknown ways. Thus, each wicked problem is merely a symptom of further wicked problems; their solutions cannot even be finally evaluated.

Research Issue: *How can we keep track of the state of the design: the decisions taken and those left pending, the concerns that need to be refined and articulated as decisions, the appeals to various criteria, etc.*[2]

Design usually proceeds under conditions of great uncertainty; the answers to critical questions cannot be had. But designers must continue to move forward, which they do by making assumptions that may pan out or later prove to be unfounded. At any moment the state of a designed artifact is a hypothesis about the problem and the context of its solution.

Research Issue: *How can we keep track of the assumptions made during the design process, many of which are implicit?*

Finally, the design process is open in the sense that, although there are constraints, such as time and budget, the boundaries of the process are not limited. Not only are the goals underdetermined, so are the means to achieve them, the issues and options to be considered, the social and organizational arrangements under which the process is conducted, and so on (see, e.g., Brown & Duguid, 1994). There is not a prescribed set of solutions.[3] Design calls for creativity and ingenuity.

Research Issue: *How can we use an explicit design rationale as a generative aid to uncovering new possibilities for design?*

---

2. We will see later that Rittel proposed a particular representational scheme, called IBIS, to address these issues.

3. This is the difference between a design problem and a decision problem. In the huge literature on decision making, a *decision problem* is defined as one with a fixed set of options to choose among.

1.2. Managing the Design Process

Design projects are almost always too large for one person. Many different technical disciplines are required, as well as management discipline, in addition to creative and integrative skills. Design is a communication-intensive collaborative activity. The communication problem is heightened by the fact that the various stakeholders speak different disciplinary languages, are motivated by different values, see different issues when looking at the same design problem, and have different interests.[4]

> **Research Issue:** *Can an explication of design rationale aid the communication between different stakeholders in the design process?*[5]

The design process is heavily constrained by the practicalities of managing complex problems, especially in large design projects; and design processes have distinct self-imposed structural properties. Designers are historically aware, they build on (i.e., borrow from) other designs; designers cannot afford to start from scratch. Design problem decomposition into loosely connected subproblems is usually along discipline boundaries (and usually handled by different people). Design basically proceeds by iterative refinement; investigation of distinct alternatives is deliberately limited; skilled designers know how to home in on areas of likely solutions, which are incrementally developed. In different domains, design often proceeds in characteristic phases, such as preliminary design, refinement, and detail design (Goel & Pirolli, 1992). How much time and effort to spend in various phases, on the breadth of search, on various parts of the problem, and so on, are critical management decisions.

> **Research Issue:** *Can management rationale, as well as technical rationale, be made explicit? How does this affect the design process?*

Many unresolved issues must be carried along during the design process. The parts of a design are highly interdependent, but each part cannot wait for dependent parts to be resolved. Partial solutions are proposed, and tentative commitments are made and propagated. Assumptions about likely outcomes are made, and the process proceeds. The design process is dynamic and the tactics intricate.

> **Research Issue:** *Can the fast-moving tactics of a multiparty design process be captured? Does doing this affect the tempo of the design process and its concomitant tactics?*

---

4. See Bucciarelli (1994) for several case studies illustrating the social texture of design teams.

5. The inherent social nature of the design process means that the content, as well as the use, of design rationale is social in nature. We will return to this issue in Section 4.

Designers like to work with concrete representations of the design (drawings, notations, models, prototypes), first giving a concrete response to the needs of the design problem and then checking it. The design process consists of cycles of construction and reflection (Schön, 1983). These cycles occur at many levels: minute-by-minute sketching and looking, daily discussions with colleagues, weekly status meetings, tests on prototypes, reviews by clients, formal checkpoints in a development process, and so on.

Research Issue: *Can these natural points of reflection, telling, and accounting in the design process be used to generate explicit design rationale?*

1.3. The Life Cycle of Design

If we consider the life cycle of a designed artifact, we see that the design phase is just one small part. The life cycle is different for different kinds of artifacts and for different domains, but roughly they all involve:

a requirements phase,	where the design problem is initially defined, often in an explicit requirements document;
a design phase,	where the artifact is shaped and which culminates in a detailed specification of the artifact;
a building phase,	where the actual artifact is implemented, constructed, or manufactured;
a deployment phase,	where the artifact is marketed, sold, and put into user settings;
a maintenance phase,	where the artifact is serviced for repairs and enhanced as needed; and possibly
a redesign phase,	where the artifact is used as the basis of a design effort to produce a new artifact.

Each design phase is dependent on previous phases, not only the concrete outputs of those phases, but also on understanding the reasoning of those phases. For example, maintenance, which is the longest and costliest phase, needs to understand the artifact so that it can be fixed and enhanced in ways that are consistent with the assumptions of the original design. The people involved in different phases of the life cycle are quite different, both in technical competence and in temperament, which makes communication between phases even more difficult. Each phase of design could be enhanced by access to an understanding of the rationale in previous phases.

Research Issue: *How can designers be motivated to create design rationale for the future benefit of later players in the design life cycle?*

We might well ask: who is the "designer" anyway? There are people in many roles in the design process that can lay claim to being designers. There are creative or integrative designers, technical designers, and design managers. With respect to the life cycle, there are, for example, maintainers who enhance the artifacts. In participatory design situations, users can play some role as designers.

Research Issue: How can design rationale methods and tools be used to expand the role and voice of various stakeholders in a participatory design process?

1.4. Design Domains

Design is practiced in many different domains, centered on technical disciplines (e.g., mechanical engineering) or classes of artifacts (e.g., aeronautical engineering). A lot of domain-specific knowledge is needed, and the practices of design are different in different domains. Even within domains, designers and organizations tend to specialize. Useful design tools need to be domain-specific, but many of the principles behind the tools are generic. The observations about the design process in this section are intended to be generic.

Research Issue: How deeply can we understand design as a generic activity? How far can we characterize the structure of design abstracted from specific domains?

This book explores many generic issues of design rationale, as is appropriate at this early stage of research on the topic. However, the authors in this book are motivated by particular design domains. Most of the authors here are concerned with the domain of human–computer interaction: they want to improve the design of computer and information systems by helping the design process better deal with human needs and use. Other authors are focused on various engineering domains, such as mechanical and software engineering.

2. DESIGN RATIONALE

Design research has traditionally been concerned with systematizing the design process—its tools, techniques, methods, and management—for artifacts and their specifications. This research has changed dramatically over the past 25 years. The "design methods" of the 1960s sought to provide general representations and formal decompositions and clustering techniques. The design methods movement failed, in the end, to be applicable to real problems (see Jones, 1970). In the 1970s and 1980s, there was a growing recognition that design is not just the solving of difficult problems, but a kind of problem-solving with distinctive properties, as we have described in Section 1 (see also Cross, 1984). This led to what Rittel (in Cross, 1984) has called "second-generation" design methods: methods that assume distributed expertise, the need for discovery, and the centrality of argument and multiple perspectives in all design work. This led to the notion of an explication of the rationale behind design.

The earliest specific proposal for design rationale method—and one that is still influential in current work in design rationale—is the "issue based information system" (IBIS), developed by Rittel (e.g., Rittel & Weber, 1973). IBIS seeks to capture the issues that arise in the course of design deliberation, along with the various positions (or alternatives) that are raised in response to issues, and the arguments for and against the positions. Explicitly articulating such a network of deliberation produces a locally structured record of the reasoning in the design process which supports the recall of decisions and their rationale. This makes the decisions more understandable to designers at a later time and exposes them to reflection and reconsideration. Many of the chapters in this book build on Rittel's IBIS representational scheme.

2.1. What Is Design Rationale?

What does the term design rationale refer to? We intend this book to present design rationale in a broad sense. It is easy to get confused, because the term is used in many different senses in the different chapters of the book; and no one of them is standard in this stage of research on the topic. To summarize what the different sense are, let us consider what a dictionary-style definition might look like:

design rationale—*n.*
1. An expression of the relationships between a designed artifact, its purpose, the designer's conceptualization, and the contextual constraints on realizing the purpose.
2. The logical reasons given to justify a designed artifact.
3. A notation for the logical reasons for a designed artifact.
4. A method of designing an artifact whereby the reasons for it are made explicit.
5. Documentation of (a) the reasons for the design of an artifact, (b) the stages or steps of the design process, (c) the history of the design and its context.
6. An explanation of why a designed artifact (or some feature of an artifact) is the way it is.

The first sense of design rationale simply refers to the reasons, however expressed, for the design of an artifact. Although, as we have seen, there is considerable complexity in design, this sense of the term implies a straightforward attempt to express the actual reasons for the design, as understood by the designer or analyst.

The second sense of design rationale, having to do with justification, implies that the rationale is constructed for some purpose (e.g., to persuade a client, to convince a review committee, etc.).

The third sense of design rationale refers to the notation for the rationale. The notation, which can be a formal, semiformal, or informal, is a representation or format for recording the reasons for designs by articulating the logical relationships between the elements of the design and its context (e.g, requirements, criteria, etc.).

The fourth sense of design rationale refers to a method for designing, in which a design rationale notation is used as a process-facilitation tool for guiding the design process to carefully consider the reasons for certain choices or steps in the design process. A design rationale method is justified if it can help designers break conceptual logjams, make more considered decisions, etc.; thus its benefits are immediate. A design rationale method is used tactically, where it will benefit the process; it is not a "complete" design method (Olson & Moran, 1995).

The fifth sense of design rationale refers to various aspects of design documentation. (a) Most narrowly it refers to just those parts that give reasons for the design (such as the outputs of design rationale methods or design reviews). It is difficult to separate design rationale in this narrow sense from other design documentation, and it is undesirable to do so. (b) Documentation of the steps in the design process, such as would be recorded in a design notebook, is rationale in that it shows how the design was arrived at. (c) Most broadly, a complete historical documentation of a design endeavor is design rationale, because it provides a view of the larger forces (organizational, social, political, cultural) at work that affected the design. A good historical account requires reflection and some detachment from the design process.

The sixth sense of design rationale refers to an explanation that is generated in response to some question or issue about the design. A design rationale explanation must be generated, because the design documentation, even when it attempts to explicitly include rationale, cannot anticipate all possible questions. However, design documentation is a key resource for interpreting the design; and it should be as complete as practically possible.

2.2. The Capture, Representation, and Use of Design Rationale

Design rationale must somehow be captured or constructed if it is to be dealt with explicitly as a resource for design. The techniques for the capture, as well as the use, of design rationale depend heavily on the representation, for example whether it is formal or informal. And the use of design rationale depends on the accessibility and organization of the design documentation.

The space of possible techniques for the capture, representation, and use of design rationale is rich; and the space of issues surrounding them complex. We do not attempt to enumerate these here, but merely to suggest a couple of the possibilities and the issues of practice. The substance of the chapters of this book is to explore these issues.

Formal capture techniques, that is, those that produce formal representations of rationale, constrain expression to a prespecified set of categories and relationships; and thus many aspects of design are not captured. This can inhibit the design process by a subtle tendency to constrain deliberation. However, there are times when a certain amount of constraint (guidance) can facilitate deliberation. The benefit of formally represented design rationale is that it enables the aid of computational systems in organizing, interpreting, accessing, and using the rationale thus encoded.

Informal capture techniques (such as keeping design notebooks with shared documentation technologies) allow a wide range of considerations to be captured, although considerable discipline is require to effectively employ these. Passive capture techniques (such as videotaping) can be used to relieve design participants from having to deal with capture as yet another thing to do. There seems to be a tradeoff between expending resources initially at capture, thus producing nicely organized and indexed documentation, and deferring the effort of searching and organizing until later, when there is a need for the rationale.[6]

In either case, it is useful, indeed crucial, to integrate the tools that deal with design rationale into the day-to-day practices of design. Tools that support communication (such as email or shared document systems) can provide capture as a side-effect. Standard design and analysis tools (such as CAD or simulation) can be integrated with capture tools to allow the designer to leave a process trail that captures a great deal of local rationale. Tools that embody domain-specific knowledge can be used to encode the standard rationale in that domain. Thus rationale capture can be focused on the more interesting nonstandard reasoning. The general strategy for getting design rationale into practice is to embody rationale capture in tools that are of immediate utility to designers.

To use design rationale, design documentation must be accessible. The issue goes beyond capture; the documentation needs to be treated as an evolving and *living design memory*.[7] There needs to be organizational processes in place for maintaining this memory. Various technologies can help (e.g., hypermedia databases or fully indexed document management systems), but more important is the need for participants to become responsible for the creation, maintenance, and continuing relevance of the memory. That is to say, it must become part of the design culture, an issue to which we shall return in Section 4.

---

6. But see Minneman et al. (1995) who show that indexing can be integrated with informal capture.

7. This term is from Terveen, Selfridge, and Long (1995), who present tools and processes for building and maintaining such a memory.

3. ORGANIZATION OF THIS BOOK

The current research on design rationale pushes in several different directions and explores different viewpoints about the nature and use of design rationale. This book contains a representative sample. Most of the current research is concerned with the representation of rationale, both its form and content. Although there is a shared goal of codifying rationale, the approaches being explored are different. Some emphasize the capture of rationale as a by-product of the design process, whereas some stress that the rationale itself must be constructed. Some emphasize the variety of considerations that should be incorporated into rationale, whereas others are concerned with deriving rationale from a scientific base (more rigorous, but narrower in scope). Other research is concerned with how design rationale relates to design practice. This includes explorations of tools for capturing rationale and design tools for using it, as well as empirical studies of the use of design rationale in real design situations.

The chapters in this book are self-contained research contributions to various aspects of design rationale research; together they cover the range of relevant research issues. The sequence of chapters proceeds bottom-up, proceeding from the nitty-gritty of particular notations to their uses in practice to the larger settings for these practices: (a) The first four chapters present specific design rationale notations and methods. (b) Three chapters assess the notations and examine rationalizing behavior in design settings. (c) Three chapters discuss how design rationale can be integrated into design tools and practice. (d) Two chapters show how design rationale can be used for teaching designers. (e) The final three chapters discuss the role of design rationale in organizational settings.

3.1. Different Perspectives on Design Rationale

Chapters 2 to 5 illustrate different perspectives on the nature of design rationale, showing in detail how it can be represented and how it can be used.

Chapter 2 ("What's in Design Rationale?", by Jintae Lee & Kum-Yew Lai) explores in detail the requirements for the expressive adequacy of representations for design rationale by laying out the various aspects of design that other design rationale schemes (e.g., IBIS and its variants) attempt to represent. They present their DRL (Decision Representation Language) notation and show how it represents the various aspects of design rationale. DRL is perhaps the most complete and integrated design rationale notation (and thus the most complex to construct). By carefully representing rationale, they show how we can begin to provide useful computations for designers.

Chapter 3 ("Questions, Options, and Criteria: Elements of Design Space Analysis," by Allan MacLean, Richard M. Young, Victoria M. E. Bellotti, & Thomas P. Moran) describes a simpler notation, QOC (Questions, Options, Criteria), for representing design rationale and shows some of the ways it can be used. The approach behind this notation is called design space analysis, which emphasizes the explication of the space of possible designs (options) and the rationale (criteria) for choosing within this space. In contrast to the IBIS notation, which is geared to capturing deliberation as it happens, the authors stress that QOC representations must be constructed by the designer as an act of reflection on the state of the design process. The chapter presents an empirical study of a design session, and from this further analyzes the kinds of justification found in design.

Chapter 4 ("Deliberated Evolution: Stalking the View Matcher in Design Space," by John M. Carroll & Mary Beth Rosson) takes the perspective of design as an evolution of artifacts in response to new task needs. The authors approach design rationale from a different concern—to embed psychological theory into the evolutionary development of computational artifacts. Their technique is claims analysis. They use the theory to extract implicit claims about the ways various features of artifacts support users' tasks or impact on their experience. By analyzing these claims and their limitations, they are able to iteratively improve the design of the artifact in a more principled way. They illustrate this approach with a case study of the design of integrated browsing facilities in Smalltalk for learning the system and for reusing code.

Chapter 5 ("Problem-Centered Design for Expressiveness and Facility in a Graphical Programming System," by Clayton Lewis, John Rieman, & Brigham Bell) presents a challenge to the notion that design rationale should be conceptualized as abstract issues, criteria, or principles. The authors' perspective is to focus on a representative set of concrete problems that a computational system is intended to address. (Their notion of *problem* is one that the user is trying to solve by using the system, i.e., the design goal for the system is to help the user solve the problem.) These problems are used to guide the exploration of the design space and the evaluation of alternative designs. They illustrate this approach with a case study of designing a graphical programming system.

3.2. Empirical Studies of Design Rationale

The previous chapters presented various notations for design rationale and some initial studies of their use (usually by the authors themselves). Clearly, there are many empirical questions of the usability and usefulness of such notations. Empirical studies are beginning to appear. The next three chapters are some of the first studies.

Chapter 6 ("Analyzing the Usability of a Design Rationale Notation," by Simon Buckingham Shum) presents an empirical study of how different modes of designing affect the use of a design rationale notation. Several software designers were trained to use QOC, and their use of it in a standard design problem was analyzed. They were able to use it at some cost; their use was opportunistic and they were required to cope with several representational tasks. In a second study, where the designer proceeded by evolving a single design (rather than by elaborating options), the notation was less useful, because it could not express the significant relationships among the design elements. Several issues of expressiveness are discussed.

Chapter 7 ("The Structure of Activity During Design Meetings," by Gary M. Olson, Judith S. Olson, Marianne Storrøsten, Mark Carter, James Herbsleb, & Henry Rueter) presents an empirical analysis of ten real software design meetings in corporate settings. The meeting discussions were coded into several descriptive categories. Over one-third of the statements were for clarification, showing the important communicative function of design meetings. Forty percent of the discussion time was spent in the design rationale categories of issues, alternatives, and criteria; and there were many transitions between these kinds of statements. The sequential patterns among the rationale statements, expressed as transitional grammars, were surprisingly similar across all these meetings.

Chapter 8 ("Synthesis by Analysis: Five Modes of Reasoning That Guide Design," by Mark K. Singley & John M. Carroll) presents a self-analysis case study of a design process using various modes of claims analysis. Claims are used as the vehicle through various stages of the design of a computer-based tutoring system for Smalltalk. The design process began with a generic analysis of existing tutoring systems, an analysis of the particular features of Smalltalk, and a study of the behavior of some Smalltalk users; then new system features were envisioned and evaluated.

3.3. Design Rationale Tools in Design Practice

Design practice is specialized in different design domains. Design rationale has no place in standard design practices today. The next three chapters explore the integration of design rationale in the domains of architecture, engineering, and software. One theme that runs through these chapters is that design methods, tools, and rationale are domain specific. Another theme is that it is necessary to integrate rationalizing (and documenting) with creating and constructing, which are the defining acts of design.

Chapter 9 ("Making Argumentation Serve Design," by Gerhard Fischer, Andreas. C. Lemke, Raymond McCall, & Anders. I. Morch) presents a suite of exploratory design tools, including both construction tools and argumentation (rationale) tools, based on an IBIS extension called PHI. A large issue base serves as a knowledge base of generic design knowledge for their experimental domain of kitchen design. The rationale is integrated with a construction kit (called Janus) in which users can quickly lay out and explore kitchen designs; the rationale is brought forth when the user needs to reflect on the state of the design. The authors discuss the role of argumentation in design and describe an architecture for an integrated design environment and the place of design rationale in it.

Chapter 10 ("Supporting Software Design: Integrating Design Methods and Design Rationale," by Colin Potts) explores extensions to the IBIS notation to document different formal software design methods (which tend to specialize for specific domains). New entities are added to IBIS to represent the artifact being designed, fragments ("macros") of rationale are added to represent recurring issues in a specific method, and simulation results are added to the artifacts and rationale. Different views of these representations can be presented, with artifact-centered views being perhaps the most useful to the software designer.

Chapter 11 ("Generative Design Rationale: Beyond the Record and Replay Paradigm," by Thomas R. Gruber & Daniel M. Russell) is an essay on how design rationale might be effectively integrated into engineering design practices. It is also a reflection on some general issues of design rationale. One issue is the uses of design rationale, what kinds of questions does it need to answer. Another issue is how rationale can be generated from typical engineering design representations. A third issue is integration with standard engineering design tools.

3.4. Using Design Rationale for Teaching

Design rationale would seem to be a helpful aid for teaching students or inexperienced designers, because it provides an explanation for why particular design components or features were chosen (or, post facto, why they are appropriate). The next two chapters illustrate ways to use design rationale to teach and facilitate user interface design, a domain where many professionals find themselves struggling with little training.

Chapter 12 ("Rationale in Practice: Templates for Capturing and Applying Design Experience," by George Casaday) presents a method for characterizing generic user interface design knowledge that is between abstract principles and rigid guidelines. The method is deceptively simple, but practical, being based on the experience of facilitating product design teams. The basic vehicle is a template, which represents a pattern of design thinking. Templates can be used in different ways in the design process. Example templates are presented for shaping the design of the user interface, clarifying the goals of the design, and managing the decision process.

Chapter 13 ("HCI Design Rationale as a Learning Resource," by Tom Carey, Diane McKerlie, & James Wilson) reports on explorations in creating a user interface design repositories incorporating design rationale. Such a design repository could be used by both by students and by designers to reuse and adapt existing designs. In one prototype, generic rationale (guidelines) and specific rationale were added to the widget library of a sophisticated user interface toolkit (NeXT Interface Builder). Lessons were learned about the conflict between the widget implementation structure in the toolkit and the need for a use-oriented organization for learning.

3.5. Design Rationale in Organizational Context

How would design rationale techniques fit into the realities of organizational contexts in which most design and development is carried out? Although most of the design rationale techniques presented in the previous chapters are not yet ready for deployment into such contexts, it is appropriate to begin to look at what this might entail. The last three chapters present case studies of designing in industrial settings (including one explicitly using design rationale), describe the nature of design in these settings, and discuss the issues involved in capturing and using design rationale in these settings.

Chapter 14 ("A Process-Oriented Approach to Design Rationale," by E. Jeffrey Conklin & KC Burgess-Yakemovic) is perhaps the earliest case study of the explicit use of design rationale in a real industrial design setting. In order to capture design rationale without disrupting the design process, the authors used a simple indented-text tool, called itIBIS. They used this tool in a field trial in an industrial design project in which a considerable body of rationale was captured over many months. This demonstrated not only that design rationale could be captured, but also that the rationale had considerable value, when examined carefully, in helping them catch and manage several design errors. They discuss how this tool was received and used in the setting.

Chapter 15 ("Organizational Innovation and the Articulation of the Design Space," by Wes Sharrock & Bob Anderson) is an ethnographic case study of an engineering design and development project in a large industrial organization. One author spent 3 months closely observing the project team. The chapter describes the practical day-to-day "design space," that is, the kinds of activities that project members engaged in to advance the design and to keep the project alive. The stringent scheduling, budget, and resource constraints meant that their critical problems were outside the technical realm and required considerable ingenuity: revising the formal development procedures, improvising the normal work practices, and revising the requirements definition. Implications for design rationale are discussed.

Chapter 16 ("Evaluating Opportunities for Design Capture," by Jonathan Grudin) is a cautionary essay on the difficulties in utilizing design rationale in real system development settings, and especially for benefiting in the downstream of the life-cycle of a system, because most developments projects are aborted early in life (and learning from failed projects is difficult at best). Four different system development contexts are considered, and the prospects of capturing design rationale in each is considered, with customized software perhaps having the most promising conditions for success.

4. TOWARD A CULTURE OF DESIGN RATIONALE

Design rationale research is burgeoning across a broad disciplinary front.[8] Indeed, we may be seeing the culmination of Rittel's "second generation." This expansion is encouraged, on the one hand, by recent advances in computer technology that support the capture, storage, distribution, and use of design rationale (networks, hypermedia, document servers, workstation hardware, etc.) and, on the other hand, by a broad consensus about the nature of design (its special status vis-à-vis problem-solving; its collaborative and contextual nature [see Moran, 1994]).

These are exciting indications, but we are clearly still in the early stages of design rationale research. Many of the most fundamental questions have barely been raised. For example:

What types of rationale are there? The chapters in this book talk about "reasons" behind design decisions, but there are many types of reasons. Thus, a "logical" rationale describes the objective issues or arguments for design features, whether a feature satisfies a criterion or whether the criterion is justified for this design problem. But a "social" rationale describes

---

8. This book has its origins in the human–computer interaction community. A workshop on design rationale was held at the ACM CHI'91 Conference on Human Factors in Computing systems and a special issue of the journal *Human–Computer Interaction* was published later that year (Carroll & Moran, 1991). There was a subsequent workshop on design rationale at the CHI'94 Conference, and in the summer of 1995 there was DIS'95, the first ACM Symposium on Designing Interactive Systems: Processes, Practices, Methods, and Techniques. Many other technical communities whose interests and constituencies partially overlap those of human–computer interaction have also begun to focus on design rationale research. In the artificial intelligence community, there was a large workshop at the AAAI'92 Conference (reported by Lee, 1993). The computer-supported cooperative work (CSCW) community has focused increasingly on design meetings as a paradigmatic CSCW domain and on cooperative approaches to designing as a paradigmatic CSCW methodology; the latter now constituting a distinct subcommunity, which has run a series of Participatory Design Conferences. This is just a sampling of the design research. There are many other conferences, as well as journals and newsletters, on design research, methods, and tools, where design rationale issues are occasionally addressed.

the actual design process that culminated in a set of particular features, the collaborative dynamics of the design team, including affective issues and specific incidents (e.g., how a particular feature was abandoned because a particular individual couldn't make it work, how a particular manager selected a feature even though all the logical arguments were against it).

How much rationale is enough? Will every developer have an analyst and a historian noting every hypothetical and every train of thought, videotaping every chance encounter in the hall, collecting every sketch and every note? Rationale is itself designed, and the rationale for *that* design could be explicitly articulated. This is a slippery slope.

Who will create the rationale? Even modest approaches to capturing rationale are now very labor-intensive. Members of a design team may not be inclined to pause and reflect when they might otherwise be sketching and prototyping; management may balk at staffing a designated "rationale analyst." Can a given project afford to have an analyst? Can it afford not to? Many cost-benefit questions need to posed and investigated.

How much design rationale should be made explicit? The initial motivation to create rationale is to support design work by making explicit and preserving reasoning that can be lost. However, it is also true that many human skills and interactions work smoothly in part because everything is not made explicit. Earlier, we appealed to Schön's (1983) notion of reflective practice, but many of his case studies seem to involve verbally inarticulate expert reflection, showing and doing rather than telling and discussing. Is there a gap between reflective practice and rationale?

Who should have access to design rationale? Perhaps it is useful for the individuals directly concerned to share and confront their own social rationales, but should all their colleagues have access to it also? Are people entitled to the reasoning that underlies systems they purchase? Should people be able to review the social processes that created the systems they learn and use?[9]

How domain-specific are various design rationale methods? Most approaches have only been applied in one or a small number of design domains, which confounds domain properties with properties of the approaches.

Will design rationale techniques scale up to and be suitable for real development contexts? There are no studies assessing the use of design rationale methods in real projects, that is, large projects that do not involve the researchers themselves as designers (although Chapter 14 by Conklin and Yakemovic is an important first step in this direction). There are many questions about how work group organizations will accommodate design rationale.

---

9. Carroll, Rosson, Cohill, and Shorger (1995) are studying this by building a history and rationale browser for a community network; members of the community can review, contribute to, and annotate the contributions of others to this database.

Will design rationale change the culture of designing? It may already have done this. Fundamentally, design rationale is an attitude toward designing, not a particular representation or activity. The culture of design rationale emphasizes the reasoning and responsibility of designers, the participation of users, customers, and the public, and the causal consequences of design moves. It transforms design into a systematic inquiry of problem solving, collaboration, and creativity.

NOTES

Acknowledgments. We thank Allan MacLean, Peter Pirolli, and Mary Beth Rosson for comments on earlier versions of this chapter.
Authors' Present Addresses. Thomas P. Moran, Xerox Palo Alto Research Center, 3333 Coyote Hill Road, Palo Alto, CA 94304. Email: moran@parc.xerox.com; John M. Carroll, Department of Computer Science, 562 McBryde Hall, Virginia Tech (VPI&SU), Blacksburg, VA 24061. Email: carroll@cs.vt.edu

---

REFERENCES

Alexander, C. A. (1964). *Notes on the synthesis of form.* Cambridge, MA: Harvard University Press.

Alexander, C. A., Ishikawa, S., Silverstein, M., Jacobson, M., Fiksdahl-King, I., & Angel, S. (1977). *A pattern language: Towns, building, construction.* New York: Oxford University Press.

Brown, J. S., & Duguid, P. (1994). Borderline issues: Social and material aspects of design. *Human–Computer Interaction, 9,* 3–36.

Bucciarelli, L. L. (1994). *Designing engineers.* Cambridge, MA: MIT Press.

Carroll, J. M., & Moran, T. P. (Eds.) (1991). *Design rationale* (Special Issue of *Human–Computer Interaction, 6,* 197–419). Hillsdale, NJ: Lawrence Erlbaum Associates.

Carroll, J. M., Rosson, M. B., Cohill, A. M., & Shorger, J. R. (1995). Building a history of the Blacksburg Electronic Village. *Proceedings of the DIS'95 Symposium on Designing Interactive Systems.* New York: ACM.

Cross, N. (Ed.) (1984). *Developments in design methodology.* New York: John Wiley & Sons.

Goel, V., & Pirolli, P. (1992). The structure of design problem spaces. *Cognitive Science, 16,* 395–429.

Jones, J. C. (1970). *Design methods: Seeds of human futures.* New York: John Wiley & Sons.

Lee, J. (1993). Summary of the AAAI'92 Workshop on Design Rationale Capture and Use. *AI Magazine, 14*(2), 24–26.

Minneman, S., Harrison, S., Janssen, B., Kurtenbach, G., Moran, T. P., Smith, I., van Melle, W. (1995). A confederation of tools for capturing and accessing collaborative activity. *Proceedings of the Multimedia'95 Conference*. New York: ACM.

Moran, T. P. (Ed.) (1994). *Context in design* (Special Issue of *Human–Computer Interaction, 9*, 1–149). Hillsdale, NJ: Lawrence Erlbaum Associates.

Olson, J. S., & Moran, T. P. (1995). Mapping the method muddle: Guidance in using methods for user interface design. In M. Rudisill, C. Lewis, P. G. Polson, & T. D. McKay. (Eds.), *Human computer interface design: Success stories, emerging methods, and real-world context*. San Francisco: Morgan Kaufmann.

Rittel, H., & Weber, M. (1973). Dilemmas in a general theory of planning. *Policy Science, 4*, 155–169.

Schön, D. A. (1983). *The reflective practitioner: How professionals think in action*. New York: Basic Books.

Simon, H. A. (1962). The architecture of complexity. *Proceedings of the American Philosophical Society, 106*, 467–482. Reprinted in Simon (1969).

Simon, H. A. (1969). *The sciences of the artificial*. Cambridge, MA: MIT Press.

Terveen, L. G., Selfridge, P. G., & Long, M. D. (1995). Living design memory: Framework, implementation, lessons learned. *Human–Computer Interaction, 10*, 1–37.

DIFFERENT PERSPECTIVES
OF DESIGN RATIONALE

2

What's in Design Rationale?

Jintae Lee
University of Hawaii

Kum-Yew Lai
McKinsey & Co.

ABSTRACT

A few representations have been used for capturing design rationale. To understand their scope and adequacy, we need to know how to evaluate them. In this article, we develop a framework for evaluating the expressive adequacy of design rationale representations. This framework is built by progressively differentiating the elements of design rationale that, when made explicit, support an increasing number of the design tasks. Using this framework, we present and assess DRL (decision representation language), a language for representing rationales that we believe is the most expressive of the existing representations. We also use the framework to assess the expressiveness of other design rationale representations and compare them to DRL. We conclude by pointing out the need for articulating other dimensions along which to evaluate design rationale representations.

1. INTRODUCTION

As the articles in this issue point out, an explicit representation of design rationales can bring many benefits. Such a representation can lead to a better understanding of the issues involved (Conklin & Burgess-Yakemovic, 1991

Jintae Lee is an Assistant Professor in the Decision Sciences Department at the University of Hawaii. His interest lies in the intersection of cognitive science and information technology. **Kum-Yew Lai** is a consultant at McKinsey & Co.; he is interested in software design.

CONTENTS

[chapter 14 in this book]; Lewis, Rieman, & Bell, 1991 [chapter 5 in this book]), of the design space (MacLean, Young, Bellotti, & Moran, 1991 [chapter 3 in this book]), and of the principles underlying human–computer interaction (Carroll & Rosson, 1991 [chapter 4 in this book]). It can also provide a basis for learning, justification, and computational support in design (Fischer, Lemke, McCall, & Morch, 1991 [chapter 9 in this book]; Lee, 1990a). The extent to which we can actually reap these benefits, however, depends largely on the language we use for representing design rationales. If, for example, design rationales were represented in free text, the benefits we obtained from it would not be different from what we already get from the notes on paper that we take in design meetings. Also, the kinds of computational support that we can provide depends on what a representation makes explicit and how formal the representation is. A few systems have been built and actually used to capture design rationales or arguments (Conklin & Begeman, 1988; Fischer, McCall, & Morch, 1989; Kunz & Rittel, 1970; Lee, 1990a, 1990b; McCall, 1987), and most of them used representations based on the earlier studies of design activity (Kunz & Rittel, 1970) or of argumentation (Toulmin, 1958). However, there is no systematic attempt to justify the choice of these representations or to discuss the rationale for using them.

This chapter is motivated by the following questions: How adequate are the existing representations? Do they allow us to represent easily what we want to represent? In general, how do we evaluate a language for representing design rationales? This chapter attempts to answer these questions by identifying the elements of design rationale that could be made explicit and by exploring the consequences of making them explicit. Laying out these elements provides a framework for placing the existing meanings of design rationale in perspective, as well as providing a framework for evaluating a representation language for design rationales, as we hope to show.

We proceed in the following way. In the next section, we identify the tasks that we might want to support using a design rationale representation.

Throughout the chapter, we use these tasks as a reference against which we evaluate the representations that we discuss. In Section 3, we characterize design rationale by presenting progressively richer models. We start with a simple model of design rationale, where an artifact is associated with a body of reasons for the choice of that artifact. We then elaborate this simple model by incrementally differentiating and making explicit what is implicit in the body of reasons. As we do so, we discuss what each resulting model allows us to do. These models of design rationale provide a framework in which to define the scope of a representation and its adequacy within its scope.[1] In Section 4, we present a language called *decision representation language* (DRL) for representing rationales and use the models to evaluate DRL. In Section 5, we consider other existing languages for representing design and explore some open problems for future research.

2. WHAT DO WE WANT TO DO WITH DESIGN RATIONALE?

To evaluate a representation, we need to know what tasks it is designed to support. The tasks that a design rationale representation can or should support can be described in many ways at different levels of abstractions. For example, Mostow (1985) listed the following tasks: documentation, understanding, debugging, verification, analysis, explanation, modification, and automation. Fischer et al. (1991 [chapter 9 in this book]) point out that documenting design rationales can support maintenance and redesign of an artifact, reuse of the design knowledge, and critical reflection during the design process. MacLean et al. (1991 [chapter 3 in this book]) list two major benefits from design rationale representation: aid to reasoning and aid to communication. The tasks of achieving these benefits are elaborated further in terms of subtasks, such as enabling the designers to envisage the available design alternatives in a more structured way, with the arguments for and against them.

Another way of characterizing the tasks is to list the questions we often need to answer to accomplish the general tasks mentioned before in the design process. To the extent that we want our design rationale representation to help answer these questions, answering these questions becomes the task that

---

1. In this article we use the terms *model* and *representation* in the following way. A model is a conceptual structure, and a representation is a linguistic manifestation of a model. Because the same structure can be described in many ways, a model can have several representations. Therefore, when we want to discuss a structure independent of a particular way of describing it, we use the term *model*. On the other hand, we use the term *representation* to refer to a particular notation for describing the structure.

the representation should support. The following is a set of representative questions that we gathered from our experiments with design rationale (Lee, 1991), from walking through examples (Lewis et al., 1991 [chapter 5 in this book]), and from creating scenarios (Carroll & Rosson, 1990):

- What is the status of the current design?
- What did we discuss last week, and what do we need to do today?
- What are the alternative designs, and what are their pros and cons?
- What are the two most favorable alternatives so far?
- Sun Microsystems just released their X/NeWs server. How would the release change our evaluations?
- What if we do not consider portability?
- Why is portability important anyway?
- What are the issues that depend on this issue?
- What are the unresolved issues? What are we currently doing about them?
- What are the consequences of doing away with this part?
- How did other people deal with this problem? What can we learn from the past design cases?

This list of questions is by no means complete, as there are many possible paths we did not walk through and many scenarios we did not construct (for a taxonomy of questions, see Gruber & Russell, 1995 [chapter 11 in this book]). We also left out those questions that, although important, do not seem to be the job of design rationales to answer (e.g., How can we compute the total cost of this design?). Nevertheless, the questions in the list provide a useful framework for assessing the expressiveness of the different representations. When we discuss the limited or increased expressiveness of a given representation, we refer to those questions that can or cannot be answered as a result. If our task includes answering a question that is not represented in the list, then we can always evaluate the representations by asking whether they would support answering the question and, if not, what additional objects, attributes, or relations would have to be made explicit.

We want to emphasize further that we are assessing only the *expressiveness* of the existing representations of design rationales. That it is desirable to answer the questions in the list does not mean that any representation for design rationales should support answering all of these questions. Each representation must weigh the costs and benefits involved in trade-offs among three general dimensions: expressiveness, human usability, and computational tractability (see Figure 1). These trade-offs should in turn be motivated by the tasks that are intended to be accomplished using the representation. In short, we are not dictating what any existing representation

Figure 1. **Elements in computer-supported activities.**

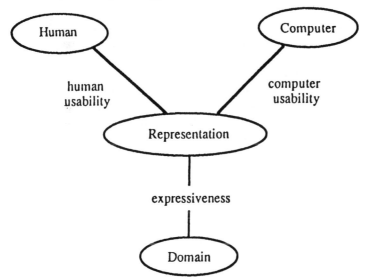

should or should not have. However, we do hope that the analysis presented in this article will help designers of representations for design rationales be more conscious of what their languages can or cannot express and why.

To be sure, we cannot separate our concern with expressiveness entirely from other concerns such as human usability or computational tractability. For example, if a language is meant to be used by people to capture design rationales but is too complex for people to manage, then there is not much point in evaluating its expressiveness. Whether any of the representations we discuss falls into that category is an empirical question. All the languages discussed here actually have been used by people, but that is no guarantee that they will all succeed at their "industrial strength" use (Conklin & Burgess-Yakemovic, 1991 [chapter 14 in this book]). Nevertheless, we believe it would be difficult to evaluate trade-offs among the three dimensions without calibrating individual dimensions such as expressiveness (cf. Levesque & Brachman, 1985 [cited in Brachman & Levesque, 1985], on trade-offs between expressiveness and computational tractability for general knowledge representation).

3. MODELS OF DESIGN RATIONALE

What is design rationale? The term *design rationale* is currently used in at least three different ways: a historical record of reasons for the choice of an artifact (Burgess-Yakemovic & Conklin, 1990), a set of psychological claims embodied by an artifact (Carroll & Rosson, 1990), and a description of the design space (MacLean, Young, & Moran, 1989).[2]

Design rationale often means the historical record of the analysis that led to the choice of the particular artifact or the feature in question. To illustrate, let us take as an example a particular feature of the Macintosh operating system, namely, the placement of all the window commands in the global menu bar at the top of the screen. By a window command, we mean a command specific to a window; for example, SAVE is a window command that saves the contents of the window. A design rationale for this feature in the sense of historical record would be something like:

> The issue of where to put the window commands was raised by Mark on January 20. Kevin proposed the idea of incorporating them into the global menu bar at the top of the screen and pointed out that it saves screen space (e.g., as opposed to putting the commands on each window, as in the Star environment). Julie objected because it requires a long mouse travel from the currently active window in executing a command. But, we decided to have the global menu bar anyway because people generally agreed that the advantage, together with others such as more efficient implementation, outweighs the objection.

We can provide more structure to this historical record, as we discuss in the rest of the chapter. Such structure is usually designed to make explicit the logical structure (e.g., an argument supports a proposal) and/or the historical structure (e.g., a proposal replaces another proposal).

Another meaning of design rationale is the set of psychological claims embodied by an artifact (Carroll & Kellogg, 1989; Carroll & Rosson, 1991 [chapter 4 in this book]), that is, claims that would have to be true if the artifact is to be successful or claims about psychological consequences for the users of the artifact. These claims are different from the historical record; the claims need not be present in the historical record; even if they were, they would have to be extracted from the record and formulated in a testable form. For example, the design rationale in this sense would be something like: "The global menu makes the environment easier to use because it reduces screen clutter," or "Dimming the irrelevant items in the global menu makes it easier to learn about the commands."

The third meaning of design rationale is one used by MacLean et al. (1989), namely, how a given artifact is located in the space of possible design alternatives: What are the other possible alternatives? How are these alternatives related? What are the trade-offs among them? In our Macintosh example,

---

2. The representation used in Burgess-Yakemovic and Conklin (1990) describes logical as well as historical aspects of design rationale, as we discuss later. We associate their work, as well as that of Lee (1990a) and Potts and Bruns (1988), with the historical record only because one of their goals is to capture and document the actual process of design.

the design rationale in this sense would be some description of the logically possible alternatives for placing window commands, how they are related, and what the trade-offs are. It is often difficult to provide such a description in a systematic way, but an example is found in Card, Mackinlay, and Robertson (1990) and in Mackinlay, Card, and Robertson (1990), which provides a vocabulary of the primitives and a set of composition operators for describing the design space of possible input devices. This meaning of design rationale seems different from the first meaning in its emphasis on design rationale not being a record but a construction and from the second in its emphasis not on a particular artifact but on the relation among possible alternatives.

We now develop a series of progressively richer models of design rationale, which provides a framework in which we can place the three different meanings of design rationale. The first two meanings are discussed next. The third meaning of design rationale, as a possibility space, is discussed shortly thereafter.

Design rationale in the most general sense is an explanation of why an artifact is designed the way it is. So, in our first model of design rationale, an artifact is associated with a body of reasons as shown in Figure 2a.

There are different kinds of reasons that we can give for an artifact. The reasons can be historical or logical, roughly corresponding to the meanings of design rationale, respectively, as a historical record and as a set of claims embodied by an artifact.[3] The record of the process that led to the choice of an artifact tells us one kind of reason why that artifact was chosen. The previously shown free text example about the window commands is an example. If we wanted a logical justification, we would have to extract it from the record, but at least such a record tells us the historical circumstances and sequence that led to the design and provides a basis from which to infer the logical reasons. On the other hand, we can represent the logical reasons directly, that is, the reasons justifying the choice of an artifact no matter how or in what order they were articulated. The set of claims embodied by an artifact is an example because these claims would justify the design of the artifact. These claims are logical also in the sense that the context in which they are true has to be made explicit. For example, a claim should not say, "The global menu is better because it leads to smaller implementation" if it really means, "The global menu is better in the context of the Macintosh because it leads to smaller implementation. This is important in a system like the Macintosh, which has a small memory." Extracting these logical reasons

---

3. We believe that the distinction between historical and logical reasons breaks if we push it too far because a purely historical record per se does not really give us a reason. It would give us a reason only to the extent that we can extract some logical structure out of it. Nevertheless, we believe that the distinction is useful for the purpose of evaluating representations because, for a given representation, we would like to know what it makes explicit and what we have to infer from it.

Figure 2. Progressively more differentiated models of design rationale. (a) Model 1: An artifact is associated with a body of all the arguments relevant to the design of the artifact. (b) Model 2: Alternatives and their relations are made explicit, and the arguments about individual alternatives can be differentiated. (c) Model 3: Evaluation measures used and their relations are made explicit, and the arguments about them can be differentiated. (d) Model 4: Criteria used for evaluation and their relations are made explicit, and the arguments about them can be further differentiated in the argument space. (e) Model 5: Individual issues are made explicit, each of which contains the alternatives, evaluations, and criteria used in discussing the issue. A part of the argument space includes the meta-arguments about the issues and their relations.

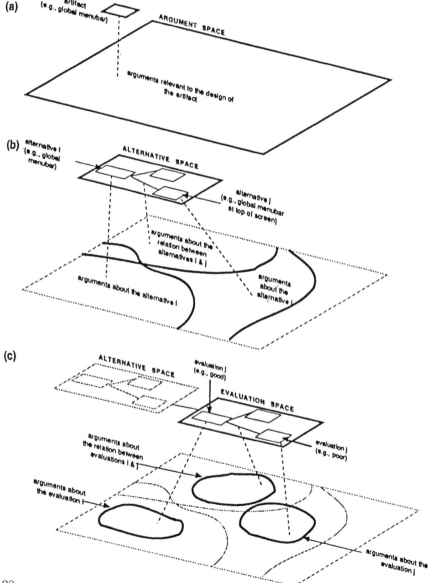

Figure 2. Continued.

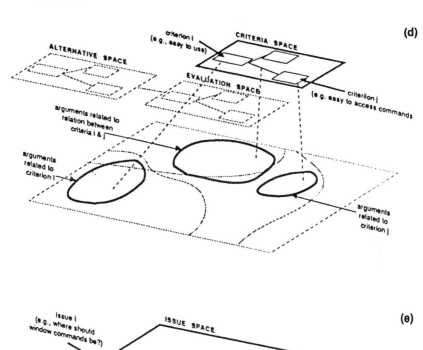

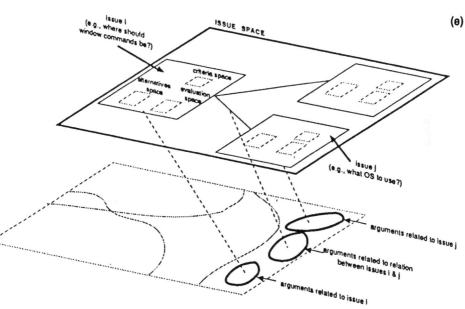

is not an easy task; once identified, however, they provide the advantages of being testable and general.

The internal structure of these reasons can be made explicit to different degrees. At one extreme, the reasons can be completely undifferentiated. An example is the natural language description that we gave earlier as an example of a historical record. If we were to make the historical relations more explicit, we could differentiate further by making explicit these roles and relations such as: Initiator, Second Motioner, Initiates, and Replaces. An example that is not historical is the representation used by Carroll and Rosson (1991 [chapter 4 in this book]) for describing the claims embodied in an artifact. In this representation, the claims themselves are represented in natural language, but the claims are grouped by the questions they answer: What can I do?, How does that work?, and How do I do this? We can also imagine a representation where the logical support relations can be made more explicit by providing such constructs as Logically Implies, Supports, Denies, Qualifies, and Presupposes. We use the term *argument space* to refer what we have called a body of reasons, because the reasons are captured either as a historical record of the various arguments relevant for the design of the artifact or logical arguments underlying the design.

There is much we can do with our first model of design rationale. A representation based on this model can help us answer the question, What did we discuss last week, and what do we need to do today? Such a representation can also help us answer the questions: How did other people deal with this problem? and Can we learn from the past design cases? Carroll and Rosson (1991 [chapter 4 in this book]) provide a good example. They report in detail how their representation of design rationales, mentioned earlier, suggested many issues for redesigning an artifact (the View Matcher in Smalltalk). They discuss how these issues can be couched as a design hypothesis, which can be tested and compiled to form, in the long run, "a contextualized science out of practice" of human–computer interaction.

Our first model, however, does not help very much with the other questions, although we should qualify this statement immediately. Saying that it does not help much is not to say that we cannot answer these questions. Of course, if the user works hard enough, and as long as the representation based on the model has enough information captured, even in the form of natural language free text, we can answer these questions. So the real issue is how much the model itself helps us answer these questions either by helping us see the structure better or by enabling us to define computational services that help us answer the questions. We see later how more differentiated models allow us to answer these questions more easily, although they increase the cost in some other ways (see Conklin & Burgess-Yakemovic, 1991 [chapter 14 in this book]).

Our second model (see Figure 2b) differs from the first by making multiple alternatives and their relations explicit. Design involves formulating several alternatives, comparing them, and merging them, as many of the questions in our list indicate. In our first model, only a single alternative is made explicit at a given time, and the multiple alternatives are present only implicitly in the argument space. Our second model makes these alternatives explicit, including the ones that have been rejected. Once the alternatives become explicit, we can talk about their attributes (e.g., current status such as "rejected" or "waiting for more information"), make the relations among the alternatives explicit (e.g., specialization, historical precedence), define computational operations on them (e.g., comparing alternatives, displaying the alternatives that specialize this alternative), or even argue about whether an alternative is worth considering. The alternatives, other than the one finally chosen, are interesting because many of the issues and the knowledge used in evaluating them are useful in other contexts, for example, when situational constraints change. We use the term *alternative space* to refer to this set of multiple alternatives and their relations.

These relations among the alternatives can also be historical or logical. Historical relations may be not only the linear sequence that we usually describe as versions but also more complex relations such as layers and contexts (Bobrow & Goldstein, 1980). The logical relations may include Specializes, Generalizes, Elaborates, or Simplifies. Or alternatives can be related through a design space (Mackinlay et al., 1990). To the extent that we want a representation to stand for these different alternatives and their relations, we say that the alternative space is within the scope of the representation. gIBIS, for example, seems to include the alternative space within its scope because one of its goals is "to capture alternative resolutions (including those which are later rejected), [and] trade-off analysis among these alternatives" (Conklin & Begeman, 1988, p. 304). The constructs in gIBIS for representing the alternative space consist of: Position, with which we can describe multiple alternatives, and the specialization relation among the Positions.

By now, we have an alternative space connected to the argument space, as shown in Figure 2b. For each of the alternatives, there are arguments describing the reasons for its current evaluation, just as in our first model there are arguments describing the evaluation status of that single alternative, that is, that it was chosen. Some of the arguments can be shared; for example, an argument can support an alternative while denying another, so it is better to think of the arguments about the different alternatives forming a single large argument space, as shown in Figure 2b.

With the representation of the alternative space, we can imagine how we can make a system help us answer some of the questions posed in

Section 2. To answer "What are the alternative designs, and what are their pros and cons?", we can associate an argument space with each of the alternatives through the links such as Supports or Objects To, as in gIBIS. To answer "Why do we even consider this alternative, and how is it related to the one that we discussed last week?", we need to use some historical relations (e.g., Replaces) or structural relations (e.g., Is a part of) among the alternatives.

Once we make explicit multiple alternatives, however, we need to articulate more carefully what the argument space is about (Figure 2b). In our first model, when we had a single artifact (i.e., the chosen one), the argument space contained reasons for the choice of that artifact. Similarly, the arguments for the other alternatives are about why they were not chosen or, to generalize, why they have their particular evaluation status (e.g., "still in consideration," "waiting for more information," "rejected"). These evaluation statuses could be nominal categories (e.g., like the previous examples), ordinal categories (e.g., "very good," "good," and "poor"), or a continuous measure (e.g., the probability that the alternative will achieve a given set of goals).

Therefore, we introduce the *evaluation space* (Figure 2c), where the evaluation statuses are made explicit and interrelated. Usually, we do not and need not specify any elaborate relation among the evaluation measures we use. Often, the implicit ordinal relation among these values (e.g., "very good," "good," "poor," "very poor") is sufficient when we leave it for the human user to assign these values to the alternatives. However, if we want to define any computational service that manages these values, for example, that automatically propagates and merges them to produce a higher level summary, then we need to be very careful about what these values mean. We need to specify the units of measurement, a calculus for combining them, and a model specifying what they mean. Even in the case where these actions are left to humans, for example, if the human user is expected to combine these values to produce a higher level summary measure, then we need to set down what these values mean so that their interpretation does not become arbitrary. To the extent that a representation makes explicit the parts of the evaluation space needed for merging individual evaluations into overall evaluations, we can now answer questions such as: "What are the two most favorable alternatives so far?" and "Sun Microsystems just released their X/NeWs server. How would the release change our evaluations?" We can also explain how an evaluation was made by pointing to the arguments in the argument space behind the artifact in question and by explaining how this particular evaluation measure is derived or computed from them or related to other measures.

Making the evaluation space explicit allows us to differentiate two components of the argument space: (a) arguments about why an alternative has its current evaluation status and (b) arguments about the alternatives themselves (e.g., why we should or should not even consider an object as an alternative or whether this alternative is really a special case of another alternative). That is, as shown in Figure 2c, these different kinds of arguments can be differentiated in the argument space.

Our models so far do not make explicit the criteria used in producing an evaluation. However, the criteria used for the evaluation and their relations are usually quite important to represent explicitly. For example, it is important to know that the argument "We do not need to duplicate menu items" is an argument for the alternative "Global menu at the top of the screen" because of the goal of reducing screen clutter, which is used as a criterion for evaluation. By making this criterion explicit, we can group all the arguments that appeal to this criterion and weigh them against one another. If the criterion changes or becomes less important, then we can do appropriate things to all the arguments that presuppose the goal (e.g., making these arguments less important). Knowing how this criterion is related to others (e.g., "reducing screen clutter" is a way of achieving "easy to use") also allows us to assign proper importance to this criterion or to change its importance when the related criteria change. We use the term *criteria space* to refer to these criteria and their relations. As Figure 2d shows, once we have the criteria space explicit, we can further differentiate the argument space by grouping those arguments that are about the criteria and their relations.

Hence, it is important that a language whose scope includes the criteria space represents the different attributes of the criteria and the relationship among them. For example, it should allow us to represent the importance of these criteria and the synergistic or trade-off relations among them. Some criteria can be subcriteria of another in the sense that satisfying them facilitates the satisfaction of the latter. These subcriteria can be related among themselves in various ways. They can be mutually exclusive in the sense that satisfying one makes it impossible to satisfy others. They can be independent of each other in the sense that satisfying one does not change the likelihood of satisfying others. These subcriteria can be related to their parent criterion in various ways as well. They can be exhaustive in the sense that satisfying all of them is equivalent to satisfying the parent.

With the criteria space represented, we can now see how the system might be able to help us answer questions such as: What if we do not consider portability? or Why is portability important anyway? The answer might be, "If we give up the goal of portability, then the evaluation of the alternative X changes to 'High' because all these claims that argue against X were based on the importance of portability." Or, "Portability is important because it is a subgoal of another important goal, 'Have a wide distribution.' " These answers can be derived from a representation if the representation makes explicit the relation among evaluations, criteria, and arguments. Of course, representation of the criteria is not sufficient for answering these questions. It is not obvious how these questions can be answered, even if some parts of the criteria space are represented explicitly. However, the explicit representation of the criteria space seems a necessary condition if we are to answer these questions. At least, we would have the information necessary to define an operation that will give or suggest the answers to these questions. We briefly describe a few examples of such operations in Section 4.

So far, we have identified and discussed the structure of a single decision underlying an artifact; namely, Which of the alternative designs should we choose? However, with the representation of such local structures alone (viz., its argument space, alternative space, and criteria space), we still cannot ask some of the questions in the list such as: What are the unresolved issues, and what are we currently doing about them? What are the issues that depend on this issue? To answer such questions, we need a more global picture of how individual issues are related. A decision often requires and/or influences many other decisions. For example, a decision can be a subdecision of another if the latter requires making the first decision. A decision can be a specialization of another if the first decision is a more detailed case of the second. It is important to capture how these decisions are related, and we use the term *issue space* to refer to them. A unit in this issue space is, therefore, a single decision that has as its internal structure the other spaces, as shown in Figure 2e. Once we have an issue as an explicit element, we can associate the attributes such as "status" and "actions taken" with issues and answer questions such as "What are the unresolved issues, and what are we currently doing about them?" Representing the dependency relation among the issues will allow us to answer the question "What are the issues that depend on this issue?"

There are still some questions that we have not yet covered, such as: How did other people deal with this problem? Can we learn from the past design cases? We argue, however, that the five spaces identified so far—the spaces of arguments, alternatives, evaluations, criteria, and issue—can contain enough information to answer these questions. We support this argument by showing how these questions can be answered with a language that we have developed for representing design rationales. This language, called DRL, is presented in the next section and is evaluated with respect to the five spaces described in this section.

4. DRL (DECISION REPRESENTATION LANGUAGE)

DRL (see Lee, 1990a) is a language that we have developed for representing and managing the qualitative elements of decision making: for example, the alternatives being considered, their current evaluations, the arguments responsible for producing these evaluations, and the criteria used for the evaluations. We call *decision rationale* the representation of these qualitative elements, and we call a *decision rationale management system* a system that provides an environment for capturing decision rationale and computational services using it. Decision rationale in our sense does not capture some important aspects of design rationale. For example, a design rationale may include the deliberations about how to generate the alternative designs. The scope of DRL, at least for now, does not include the representation of such deliberations.[4] The exact relation between decision rationale and design rationale, however, has yet to be articulated. Nevertheless, we believe that DRL is the most expressive language that has been used for representing design rationales and that it overcomes many of the limitations in the existing languages in a way that is still simple enough for the user. In this section, we evaluate DRL as a design rationale representation language and point out its strengths and limitations as such.

4.1. Description of DRL

Figure 3 shows the object types that form the vocabulary of DRL. Objects of type `Is Related to` and its subtypes can be used to link other objects. For instance, an achieves relation can be used to link an `Alternative` object to a `Goal` object. The legal types that can be linked are shown inside the parentheses following the names of the relations. Figure 4 shows graphically how DRL objects can be linked to each other. Figure 5 shows an example decision rationale represented in DRL. We describe the basic features of DRL briefly and discuss in detail how they allow us to represent the five spaces of design rationale. One should note that the following is not a description of the way the user would use DRL. The actual interface is briefly described later.

A qualification is in order before we proceed further. The vocabulary of DRL presented in this section is a minimal set in the following sense. The initial set included those objects and relations (e.g., alternatives, goals, and claims) needed to represent the important elements of the spaces discussed in the previous section. Then the initial set has been augmented with additional constructs only when they were needed and judged to be

---

4. That is not to say that DRL does not help us generate alternative designs. It does in a couple of ways. DRL allows people to argue about the existing alternative designs, thus helping them to see more clearly their strengths and limitations. It also helps people retrieve the past decisions that contain useful alternative designs that are still useful for the current decision or those that can be so adapted. Furthermore, DRL can represent the relationship between the existing alternative and the new alternative that may have been derived from it. However, being able to generate a new design alternative is still different from representing the rationales for how it was generated.

Figure 3. **The DRL vocabulary.**

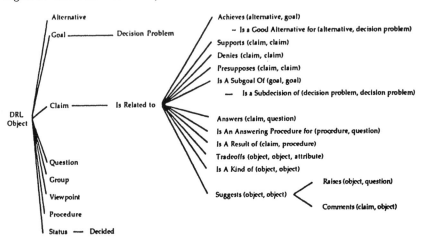

Figure 4. **The structure of a decision graph.**

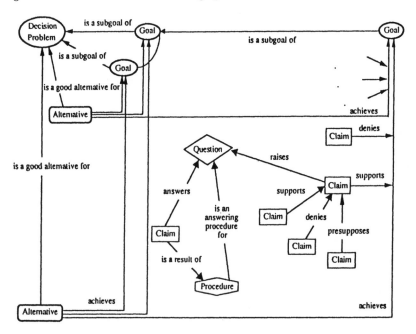

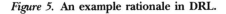

Figure 5. **An example rationale in DRL.**

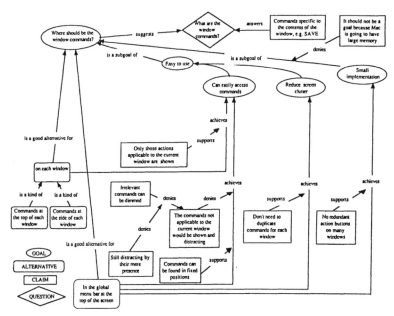

useful for *generic* decision tasks, that is, useful in decisions independent of any particular domain. As such, the DRL constructs presented here are meant to be specialized and augmented further for a specific set of tasks. For example, Lee (1991) described an extension of DRL that supports the task of managing rationales in software engineering.

A `Decision Problem` represented the problem that requires a decision (e.g., where to place the window commands). An `Alternative` represents an option being considered: for example, "In the global menu at the top of the screen." A `Goal` represents a desirable state or property used for comparing the alternatives. A `Goal` such as "Where should the window be?" is elaborated further in terms of its subgoals. For example, "Easy to use" is elaborated into two subgoals, "Can easily access command items" and "Reduce screen clutter."[5] Every relation in DRL is a subclass of `Claim`, as shown in Figure 3. For example, the rightmost `Achieves` link in Figure 5 represents the `Claim` that the `Alternative`, "the global menu at the top of the screen," achieves the `Goal`, "Reduce screen clutter."

We evaluate an `Alternative` with respect to a `Goal` by arguing about the `Achieves` relation between the `Alternative` and the `Goal`, that is, the claim that the `Alternative` achieves the `Goal`. We argue about

---

5. Because goals are described by desired states of the world, the exact meaning of the "Easy to use" subgoal should be "The window commands should be at a place where they are easy to use."

a Claim by producing other Claims that Support or Deny the Claim or by qualifying the Claim by pointing out the Claims that it Presupposes. Each Claim has the following attributes: evaluation, plausibility, and degree. The evaluation of a Claim, represented by the value of its evaluation attribute, is a function of both of its plausibility and degree attribute values. The plausibility of a Claim tells us how probable it is for the claim to be true, and the degree of a Claim tells us to what extent it is true. For example, the degree of the Achieves link between the Alternative and the Goal tells us to what degree the alternative achieves the goal in question. The overall evaluation of an alternative is represented by the degree attribute value of the Is a Good Alternative for link between the Alternative and the Decision Problem, that is, the claim that the alternative is a good alternative resolution for the issue. This degree is a function of the degrees of the Achieves claims that link the Alternative to the different Goals. Not all of the three attributes have to be used for the evaluation. For example, we might require that a Claim be entered only if its plausibility is above a certain threshold and ignore the plausibility once the Claim has been entered. In that case, we can do away with the plausibility attribute, and the evaluation and the degree attributes become synonymous.

There are other auxiliary objects in DRL. A Group object groups a number of objects and has the attribute, "member relations," which tells us how the objects are related. A relation can take a Group of objects rather than a single object. For example, a Goal may be related to a Group of other Goals through a Is a Subgoal of link. The other objects in DRL, such as Question, Procedure, and Viewpoint, represent somewhat auxiliary aspects of decision making such as the questions raised and the procedures used for answering the questions. More details of DRL may be found in Lee (1990a).

DRL has been partially implemented in a system called SIBYL, which runs on top of Object Lens (Lai, Malone, & Yu, 1988). Although the previous description of DRL may seem complex, the actual user interface provided by SIBYL for using DRL is quite simple, and SIBYL has been used for real group decision tasks such as designing a workplace layout. For example, SIBYL makes it easy to create objects like a Decision Problem, Goals, and Alternatives by providing context-sensitive menus and template editors. Once a Decision Problem and some of its Goals and Alternatives are specified, SIBYL displays them in a matrix such as that shown in Figure 6. By mouse clicking on a cell of the matrix, the user gets the menu of all the actions that can be performed on the selected object. For example, by mouse clicking on a Goal, the user can get a menu containing action items such as creating a new subgoal or displaying a goal tree showing how this Goal is related to other Goals.

Figure 6. **User interface for** SIBYL.

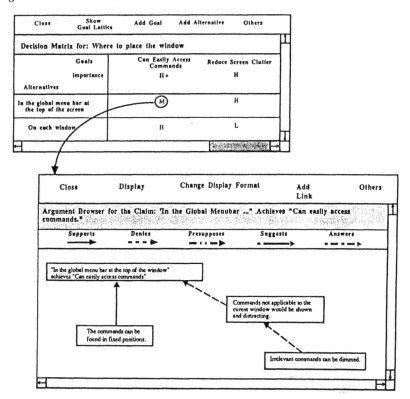

Figure 6 also shows an argument browser displayed when the user chooses the action, "Display Arguments," from a pop-up menu that appears when one of the evaluation cells is selected. The argument browser shows, in a network format, all the Claims that provide reasons for the evaluation, that is, all the claims related to the surrogate claim that the system automatically generates to be argued about, namely, that a given alternative achieves a given goal. By mouse clicking on an object in the argument browser, the user gets a pop-up menu of all the operations that can be performed on the object: such as "Add a Supporting Claim," "Add a Denying Claim," and "Add a Qualifying Claim." When the user chooses one of these actions, the template editor containing a new Claim object appears, and the new object is added to the argument browser with an appropriate link to the object chosen. The user interface that SIBYL provides for using DRL is described in more detail in Lee (1990b). Using the decision rationale represented in DRL, the computer can provide many services, such as managing the dependencies among claims, propagating and merging the plausibilities automatically, providing multiple viewpoints, and retrieving useful knowledge from past decisions (for more details, see Lee, 1990a).

4.2. Evaluation of DRL as a Design Rationale Language

The Argument Space. An argument is represented in DRL as a set of related Claims. A Claim subsumes what other people might call facts, assumptions, statements, or rules. Instead of making these distinctions, which is sometimes arbitrary and difficult to make, a DRL Claim has the attribute, plausibility, which indicates how much confidence we have in the claim. This has the advantage of not imposing a set of predetermined categories on the user and avoiding the ambiguity resulting from the disagreement among people on what facts or assumptions are. When it is desirable to make the distinction, say, between facts and assumptions, we can do so simply or by specializing a claim or by using nominal categories like "fact" and "assumption" as values for the plausibility attribute in different Claims. We can do so after the fact or dynamically by using a numeric measure as the plausibility value and mapping between this measure and the measure based on the nominal categories such as factors or assumptions. Therefore, users do not have to conform to static categories prespecified by the designer of the vocabulary. We discuss different plausibility measures when discussing the evaluation space.

A Claim can be Supported, Denied, or Presupposed by another Claim. These relations among the Claims allow us to respond to a Claim directly without, as in IBIS, having to respond indirectly to the Position that responds to the second Claim. For example, the Claim, "Irrelevant commands can be dimmed," directly denies the Claim, "The commands not applicable to the current window would be shown and distracting," rather than having to be formulated as a Claim for the Alternative in question. These direct relations among the Claims allow us to see the logical and the dynamic structure of the argument more easily. All DRL relations are special types of Claims. For example, when we link Claim 1 to Claim 2 through a Supports relation, we are making the claim that Claim 1 supports Claim 2. Likewise, an Achieves relation from an Alternate object to a Goal object represents the claim that the alternative achieves the goal. Hence, any DRL relation, like Supports, Denies, Achieves, Is a Subgoal of, is a Claim and can be argued about; that is, people can support, deny, or qualify them. For example, "Commands not applicable to the current window would be shown and distracting" is denied by "Irrelevant commands can be dimmed." That the first Claim is denied by the second itself is a relational Claim, which is then denied by "Dimmed commands are still distracting by their mere presence."

The Criteria Space. DRL represents the criteria space fairly well. In DRL, criteria are represented by `Goals`. DRL uses the term *Goal* rather than *Criterion* because, for each criterion, we can always define a corresponding goal (viz., the goal of achieving the criterion) and because we want to convey a richer relationship among these goals than what the term *criteria* usually conveys. For example, a `Goal Is a Subgoal` of another `Goal` if achieving the first `Goal` facilitates the achievement of the second. Subgoals can be related among themselves in various ways; they can be mutually exclusive, independent of each other, or partially overlapping. These relationships are represented by creating a `Group` object and specifying these `Goals` to be its members; the relations among these `Goals` are specified in the "member relations" property of the `Group`.

`Decision Problem` represents the goal of choosing the best alternative. All the other goals for the decision problem are subgoals of the decision problem in the sense that they elaborate what it means to choose the best alternative. For example, the `Goal` "Easy to use" is a subgoal of the `Decision Problem` of our example if we interpret it to mean "Choose the alternative that has the property 'Easy to use.'" In other words, satisfying this goal facilitates the achievement of the goal of choosing the best alternative.[6]

Because the `Is a Subgoal of` relation is a `Claim`, as is any other DRL relation, we can argue about whether a goal is desirable or whether it contributes to achieving another goal by arguing about this relational claim. For example, we can argue about whether small implementation should be a goal at all. In Figure 5, there is an argument, "It should not be a goal because Mac is going to have large memory soon," denying that it is a subgoal of the decision problem; that is, small implementation is not a desirable property that should be used to compare alternatives. The record of these `Claims` and their relations represents the argument space for the goal space. Lee (1990a) discussed how this representation of `Goals` in DRL allows us to create multiple viewpoints and to extract from past decisions knowledge useful to the current decision.

The Alternative Space. DRL represents only parts of the alternative space well. DRL can represent alternatives and the specialization relation among

---

6. The precise semantics of the model underlying DRL are more complicated and are fully explained in Lee (1991). Roughly, for a given decision problem of the form, "What is the best alternative for X?," its underlying interpretation is, "the goal of choosing the best alternative for X." Its subgoal of the form, G, is strictly speaking "the goal of choosing the alternative that satisfies G." An alternative of the form, A (e.g., the global menu bar at top of the screen), is to be interpreted as, "choosing the alternative, A." It is in this sense that a decision problem is the parent of the other goals and that an alternative achieves a goal. This nicety, although important for computational purposes, can be ignored by human users.

them through the Is a Kind of relation. Thus, we can say that "Commands at the top of the window" is a special case of the alternative, "On each window." However, design alternatives may be related in much more complex ways than through the specialization relation. An Alternative can be related to another Alternative via, for example, the following relations: Elaborates, Simplifies, or Is the Next Version of. Alternatives can be related in a more complex way, for example, in the context of a design space. These relations are beyond the current expressive power of DRL.

DRL represents the arguments about the alternative space the same way it represents the arguments about the goal space. We can argue about whether an alternative should be an alternative at all or whether an alternative is really a specialized version of another alternative by creating Claims that deny or support the appropriate relations, such as Is a Good Alternative for or Is a Kind of. The relations also can be qualified by linking them to another Claim via a Presupposes relation. For example, we can say that "commands on each window" is an alternative only if the window system allows the attachment of menu windows to the main window by linking the two claims with a Presupposes relation. One can of course object to this Claim, in turn, by pointing out another way of implementing the window commands at the top of the window.

The Evaluation Space. In DRL, each Claim has the following attributes: evaluation, plausibility, and degree. The evaluation attribute tells us how important the claim is, and its value is a function of both plausibility (how likely the claim is true) and degree (to what extent the claim is true). The overall evaluation of an alternative is represented as the evaluation attribute value of the relational claim, Is a Good Alternative for, between the Alternative and the Decision Problem. This value represents the extent to which the alternative satisfies the overall goal. This value, in turn, is a function of the evaluations of the Achieves relations that link the Alternative to the subgoals of the Decision Problem (i.e., the extent to which the alternative satisfies the subgoals). It is also a function of how the subgoals interact to satisfy the parent goal, such as the extent to which trade-offs and synergies exist among these goals.

DRL does not commit to a particular measure of evaluation. Users of DRL can use nominal categories, numeric measures, or whatever they devise for evaluation. However, such evaluation measures should come with the algorithm for propagating and merging them to produce evaluation measures at a higher level. For example, if we want to use probability as the measure of plausibility, then we should also know how the probability of the two Claims—"The alternative 'In the global menu bar . . .' achieves the goal 'Can easily access commands' " and "The alternative 'In the global menu bar . . .' achieves the goal 'Reduce screen clutter' "—combines over the subgoal relations to produce the probability of the Claim, "The alternative 'In the global menu bar . . .' achieves the goal 'Easy to use,' " given that we also know how these subgoals are related among themselves and to the parent goal. We might try to work out such an algorithm based on Bayes' theorem, for example.

However, as discussed in Section 2, the exact algorithm is important only to the extent that the user can trust the algorithm. That is, if the algorithm is based on many assumptions that the user feels are seriously violated, then the exactness of the algorithm does not contribute much. We might as well concentrate on how to support people for making these judgments. DRL takes this philosophy and tries to help by modularizing and helping to make explicit the relationships that need to be considered for these judgments.

The Issue Space. In DRL, the unit of the issue space is a decision problem. A Decision Problem corresponds to an Issue of gIBIS and a Question in "Questions, Options, and Criteria" (QOC). A decision problem Is a Subdecision of another decision problem if a decision for the first requires a decision for the second. For example, deciding where to place the window commands might require deciding what the window layout algorithm is (e.g., tilting or overlapping). A decision problem Is a kind of another decision problem if the first decision problem is a special case of the second: For example, "Where to place the emacs window commands?" is a special case of "Where to place the window commands?" Of course, we can relate the decision problems through the generic relations such as Is a kind of and Is a part of. We are sure that there are many other possible relations. For example, the Replaces relation in IBIS seems important for describing the dynamic aspect of the issue space. DRL, however, is based on the philosophy that the vocabulary should be extended to tailor the task in hand and that it is better to provide a method for extending the vocabulary as needed rather than to provide constructs that may not be useful in general.

5. RELATION TO OTHER STUDIES

Existing or proposed representations for design rationales range from unstructured (natural language), template-based (Casaday, 1995 [chapter 12 in this book]), to fairly formal (Gruber & Russell, 1995 [chapter 11 in this book]). In Lee and Lai (1991), we assess a number of semiformal representations using the framework developed in this chapter, and discuss the relation between these representations and DRL. Here, we provide a brief summary.

First, there are several representations based on IBIS (Kunz & Rittel, 1970), whose goal is to represent designers' argumentation activities. The most well known among them is gIBIS (Conklin & Begeman, 1988; Conklin & Burgess-Yakemovic, 1991 [chapter 14 in this book]). The units of the Issue Space, the Alternative Space, and the Argument Space are, respectively, Issue, Position, and Argument. gIBIS provides no constructs for representing the Criteria Space.[7] Because criteria are not explicit, we cannot argue about them; we cannot represent the reasons for having these criteria; and we cannot indicate any relationship, such as mutual exclusiveness, among the criteria. Further, when criteria change, there is no easy way to accommodate the changes. It would be more difficult to isolate the real disagreements among people, because the criteria they use in their arguments remain implicit. The explicit representation of goals can also provide modular representations of arguments, multiple viewpoints, and can serve as a basis for relevance matching (cf. Lee, 1990a).

gIBIS' constructs for describing the argument space are also limited in several ways, as we pointed out in Lee and Lai (1991). For example, you cannot qualify an argument. Furthermore, because relations are not claims, as in DRL, there is no way of saying that we agree with A and B but not that A Supports B. The ability to argue about relational claims is important. For example, one may agree that the global menu bar is a bad idea and that seeing irrelevant commands is distracting but not that the second claim supports the first; for instance, the global menu bar does not have to show the irrelevant commands. Also, arguments cannot directly respond to other arguments. Therefore, a query such as "Show me all the arguments that respond to this argument" cannot be computed by the system.

The gIBIS structure has the advantage of being simple at least from the representation standpoint. However, we believe that the foremost criteria for a representation is not whether it is simple, but whether it helps users accomplish their tasks. This capacity to help is in turn determined by the trifactors of human usability, machine usability, and expressiveness. As Figure 1 indicates, there is a certain directionality among these three factors. An expressive language can be made easy to use with an appropriate

---

7. When we say that a representation cannot express some information, we do not mean that people cannot infer that information from the representation. For example, if we keep a detailed enough record in natural language of what happened, or even a video recording of the whole design process, we can always retrieve the information that has been recorded by working hard enough. When we say that a representation cannot express some information, we mean that the representation does not provide constructs that make the information explicit in such a way that helps people easily see the structure or makes it amenable to computational manipulation.

user interface, but it is impossible to make a usable language more expressive. Therefore, it seems that a good starting point is to design an expressive language rather than one that is simple to use. DRL can be viewed as extending gIBIS in several ways: an explicit representation of the criteria space, a richer representation of the argument space, and the provision of an infrastructure for defining evaluation measures.

Procedural hierarchy of issues (PHI; McCall, 1987) overcomes some of the gIBIS limitations by allowing a quasi-hierarchical structure among issues, answers, and arguments. The semantics of the hierarchical relation are different for the different spaces. In the issue space, if Issue A is a child of Issue B, that means A "serves" B—that is, resolving A helps resolving B. In the answer space (i.e., the alternative space), an Answer is a child of another if the first is a more specific version of the second. In the argument space, an Argument is a child of another if the first is a response to the second. Hence, unlike in gIBIS, we can respond to an Argument directly by making it a child node of the Argument. Furthermore, PHI is *quasi*-hierarchical and allows the sharing of nodes (i.e., multiple parents) and cyclic structures. Although the quasi-hierarchy increases the expressiveness of PHI, some important relations cannot be easily expressed in this structure. An Issue cannot specialize another Issue, and an Answer cannot serve another Answer, and so forth. Therefore, the same comments about not explicitly representing relations in gIBIS apply to PHI. DRL can be viewed as pushing further the extensions that PHI made to IBIS by generalizing the hierarchical structure to more complex relations and by making explicit some other elements, especially those in the criteria space.

Potts and Bruns (1988) extended IBIS to represent the derivation history of an artifact design. One starts with an abstract Artifact (e.g., a plan for a formatter), associates with it the Issues that arise in making the plan more concrete, associates with each of the Issues the Alternatives considered (some of which lead to a more concrete plan), and so on until the plan is concrete enough to be implemented. Associated with each Alternative may be a Justification, which is the unit of the argument space. This representation is interesting because it allows us to describe yet another space that we might call the Artifact Space, where the evolving versions of an artifact over time are related. An extension of DRL that incorporates this additional space was presented in Lee (1991).

JANUS (Fischer et al., 1989) is interesting as an attempt to bridge two representations. One of its components, CRACK, uses a rule-based language for representing domain-specific knowledge (e.g., about kitchen design). The other component, ViewPoints, uses PHI to represent the rationale for the decisions made. JANUS integrates the two representations by finding the appropriate rationales represented in PHI for the particular issue that designers face in the construction phase, that is, while using CRACK. Although the current interface is limited to that of locating the relevant parts of the representations, bridging a design rationale representation and a domain representation is a very important topic of research because such a bridge can allow us to represent the relations among the alternatives or the criteria in more domain-specific ways.

QOC is a representation proposed by MacLean et al. (1991 [chapter 3 in this book]) whose constructs map clearly to the framework proposed in this chapter. Question, Option, and Criterion are, respectively, the units of the issue space, the alternative space, and the criteria space. A Criterion (e.g., "reduce screen clutter") is said to be a "bridging criterion" if it is a more specific one that derives its justification from a more general one (e.g., "easy to use"). The units of the evaluation space are links labeled with "+" and "–," corresponding to whether an option does or does not achieve a given criterion. Some constructs for representing the argument space are Data, Theory, and Ad Hoc Theory. One supports the evaluation ("+" or "–") of an Option with respect to a Criterion by appealing to empirical Data (e.g., "The mouse is a Fitts' law device") or to an accepted Theory (e.g., "Fitts' law"). When there is neither relevant data at hand nor existing theory to draw on, the designers may have to construct an Ad Hoc Theory, which is an approximate explanation of part of the domain. MacLean et al. provide an illuminating discussion of the other forms of justifications for design, such as various forms of dependencies and metaphors, although no specific constructs are discussed for representing them.

QOC as we understand it has a number of limitations as a representation language, as we pointed out in Lee and Lai (1991). First, in the argument space, constructs like Data, Theory, or Ad Hoc Theory do not seem to capture many aspects of arguments. For example, it is not clear how an argument such as "Irrelevant commands can be dimmed" should be treated, given that it is neither a piece of empirical data nor a theory. And it is not clear whether and how we can argue about theories or individual claims in theories. In the alternative space, there is a reference to cross-option dependency, but no specific constructs are discussed for representing it. Given the ambiguity about what exactly constitutes the vocabulary of QOC, however, QOC seems to be more of a model than a fully developed representation language.[8] That is, it seems to be an attempt to understand and categorize the elements of design rationale without providing a specific vocabulary for expressing them. This observation is also consistent with the authors' warning against premature commitment to a specific representation (MacLean et al., 1989). Considering the similarity between QOC and DRL in the underlying structure, we hope that DRL

---

8. See footnote 1 about the distinction between *model* and *representation*.

provides a representation language adequate for representing most of the elements that the QOC research has been articulating.

6. CONCLUSIONS

A large body of research in the last two decades or so points to the importance of choosing the right representation for a given task (Amarel, 1968; Bobrow, 1975; Brachman & Levesque, 1985; Lenat & Brown, 1984; Winston, 1984). The task of using and reusing design rationales is no different. The benefits we can get and how easily we can get them depends heavily on the representation we use. The choice of representation is especially important when a human is the user of the representation, as in design rationale capture, because a wrong representation can turn people away from the task altogether, attributing the failure and frustration to the task itself rather than to the inadequacy of the representation used. People might conclude that capturing design rationales is not worth the trouble because it is so hard and because it does not provide enough rewards for the efforts. But the real problem might be that the representation does not allow us to represent easily what we want to represent in a way that can provide much benefit. Thus, it is important that we know how to evaluate a representation for a given task, in our case, for capturing design rationales.

In this chapter, we made a step forward by characterizing the domain of design rationale, that is, by identifying the kinds of elements that form the rationale as well as the relations that hold among them. Characterizing this domain is important because we then know what we can represent, what we have decided not to represent, and what the consequences will be. It also helps us to map the different meanings of design rationale by associating them with the different parts or aggregates of the domain. In other words, it provides a framework for defining the scope and assessing the expressive adequacy of a representation. Using the framework, we defined the existing representations' scope and discussed their adequacy. We have also presented a language, called DRL, which we believe is more expressive than most of the existing languages and overcomes many of their limitations in a way that is still natural to human users. However, that is a testable claim that we plan to investigate empirically by using DRL with many tasks by many users. We also discussed the limitations of DRL, which we hope will be explored by us and others in future research.

The step we made, however, is a small one, and we have a long way to go before we fully understand the important issues in designing an ideal representation for representing design rationales. We provided a framework for evaluating a design rationale representation along one dimension—its expressive power. Even then, expressiveness involves more than being able to represent the elements in the domain explicitly or not. There are many other characteristics, such as the ability to provide abstractions, that are important (see Bobrow, 1975). We need to think about whether these characteristics matter much for the task we have in hand and in what way they matter. Human usability is another Pandora's box that many researchers are opening up. Many chapters in this book address relevant issues: for example, compatibility between empirically identified strategies of capturing rationales and proposed representations (Buckingham Shum, 1995 [chapter 6 in this book]; Singley & Carroll, 1995 [chapter 8 in this book]), patterns of design rationale activities (Olson et al., 1995 [chapter 7 in this book]). Computational tractability, on the other hand, is fairly straightforward to compute. But it requires us to know what services or benefits that a given design rationale representation is supposed to provide, a concern that is emphasized in several chapters of this book (Carey, McKerlie, & Wilson, 1995 [chapter 13 in this book]; Gruber & Russell, 1995 [chapter 11 in this book]; Grudin, 1995 [chapter 16 in this book]). Stepping back even further, a designer of design rationale representation should be aware of the organizational or managerial issues (Sharrock & Anderson, 1995 [chapter 15 in this book]). We need more such studies, more focus on the representation being used, and more systematic categorization of the results. We believe that the benefits from explicit representation of design rationales would more than pay for the efforts that we put into such studies.

NOTES

Background. This chapter first appeared in the *Human–Computer Interaction* Special Issue on Design Rationale in 1991.

Acknowledgments. The project of articulating the elements of design rationale grew out of our frustration with not knowing exactly how to evaluate the existing representations and how to relate them to DRL. Tom Malone suggested that we define the scope of a representation by thinking about which components of a decision matrix it makes explicit. That insight triggered much of the analysis here. Jintae Lee thanks Frank Halasz for providing him with the great environment in which he wrote this article. The comments from the anonymous reviewers, Jack Carroll, Tom Moran, Jeff Conklin, Randy Trigg, and Austin Henderson, were valuable. The chapter was influenced much by the comments from the members of the Argumentation Reading Group at Xerox PARC: Danny Bobrow, Frank Halasz, Bill Janssen, Cathy Marshall, Susan Newman, Dan Russell, Russ Rogers, Mark Stefik, and Norbert Streitz. We also thank the members of the learning group at the MIT AI Lab, especially Patrick Winston, Rick Lathrop, and Gary Borchardt.

Support. This work was supported, in part, by Digital Equipment Corporation, the Natural Science Foundation (Grants IRI-8805798 and IRI-8903034), and DARPA (Contract No. N00014-85-K-0124).

Authors' Present Addresses. Jintae Lee, University of Hawaii, ICS Department, 2565 The Mall, Honolulu, HI 96822. Email: jl@hawaii.edu; Kum-Yew Lai, Southern Garden #1711, Wan Chai, Hong Kong.

REFERENCES

Amarel, S. (1968). On the representation of problems of reasoning about actions. In B. Webber & N. Nilsson (Eds.)., *Readings in artificial intelligence* (pp. 2–22). Palo Alto, CA: Tioga.

Bobrow, D. (1975). Dimensions of representation. In D. Bobrow & A. Collins (Eds.), *Representation and understanding: Studies in cognitive science* (pp. 1–34). San Francisco: Academic Press.

Bobrow, D., & Goldstein, I. (1980). Representing design alternatives. In I. Goldstein & D. Bobrow (Eds.), *An experimental description-based programming environment: Four reports* (Tech. Rep. No. CSL-81-3, pp. 19–29). Palo Alto, CA: Xerox Palo Alto Research Center.

Brachman, R., & Levesque, H. (Eds.). (1985). *Readings in knowledge representation.* Los Altos, CA: Morgan Kaufmann.

Buckingham Shum, S. (1995). Analyzing the usability of a design rationale notation. In T. P. Moran & J. M. Carroll (Eds.), *Design rationale: Concepts, techniques, and use.* Hillsdale, NJ: Lawrence Erlbaum Associates. [Chapter 6 in this book.]

Burgess-Yakemovic, KC, & Conklin, J. (1990). Report on a development project use of an issue-based information system. *Proceedings of the Conference on Computer-Supported Cooperative Work,* 105–118. New York: ACM.

Card, S., Mackinlay, J. D., & Robertson, G. G. (1990). The design space of input devices. *Proceedings of the CHI '90 Conference on Human Factors in Computing Systems,* 117–124. New York: ACM.

Carey, T., McKerlie, D., & Wilson, J. (1995). HCI design rationale as a learning resource. In T. P. Moran & J. M. Carroll (Eds.), *Design rationale: Concepts, techniques, and use.* Hillsdale, NJ: Lawrence Erlbaum Associates. [Chapter 13 in this book.]

Carroll, J. M., & Kellogg, W. A. (1989). Artifact as theory nexus: Hermeneutics meets theory-based design. *Proceedings of the CHI '89 Conference on Human Factors in Computing Systems,* 7–14. New York: ACM.

Carroll, J., & Rosson, M. B. (1990). Human–computer interaction scenarios as a design representation. *Proceedings of the 23rd Annual Hawaii International Conference on System Sciences,* 555–561. Los Alamitos, CA: IEEE Computer Society Press.

Carroll, J. M., & Rosson, M. B. (1991). Deliberated evolution: Stalking the View Matcher in design space. *Human–Computer Interaction, 6,* 281–318. Also in T. P. Moran & J. M. Carroll (Eds.), *Design rationale: Concepts, techniques, and use.* Hillsdale, NJ: Lawrence Erlbaum Associates, 1996. [Chapter 4 in this book.]

Casaday, G. (1995). Rationale in practice: Templates for capturing and applying design experience. In T. P. Moran & J. M. Carroll (Eds.), *Design rationale: Concepts, techniques, and use.* Hillsdale, NJ: Lawrence Erlbaum Associates. [Chapter 12 in this book.]

Conklin, E. J., & Begeman, M. L. (1988). gIBIS: A hypertext tool for exploratory policy discussion. *ACM Transactions on Office Information Systems, 6,* 303–331.

Conklin, E. J., & Burgess-Yakemovic, KC. (1991). A process-oriented approach to design rationale. *Human–Computer Interaction, 6,* 357–391. Also in T. P. Moran & J. M. Carroll (Eds.), *Design rationale: Concepts, techniques, and use.* Hillsdale, NJ: Lawrence Erlbaum Associates, 1996. [Chapter 14 in this book.]

Fischer, G., Lemke, A. C., McCall, R., & Morch, A. I. (1991). Making argumentation serve design. *Human–Computer Interaction, 6,* 393–419. Also in T. P. Moran & J. M. Carroll (Eds.), *Design rationale: Concepts, techniques, and use.* Hillsdale, NJ: Lawrence Erlbaum Associates, 1996. [Chapter 9 in this book.]

Fischer, G., McCall, R., & Morch, A. I. (1989). Design environments for constructive and argumentative design. *Proceedings of the CHI '89 Conference on Human Factors in Computing Systems,* 269–276. New York: ACM.

Gruber, T. R., & Russell, D. M. (1995). Generative design rationale: Beyond the record and replay paradigm. In T. P. Moran & J. M. Carroll (Eds.), *Design rationale: Concepts, techniques, and use.* Hillsdale, NJ: Lawrence Erlbaum Associates. [Chapter 11 in this book.]

Grudin, J. (1995). Evaluating opportunities for design capture. In T. P. Moran & J. M. Carroll (Eds.), *Design rationale: Concepts, techniques, and use.* Hillsdale, NJ: Lawrence Erlbaum Associates. [Chapter 16 in this book.]

Kunz, W., & Rittel, H. (1970). *Issues as elements of information systems* (Working Paper No. 131). Berkeley: University of California, Berkeley, Institute of Urban and Regional Development.

Lai, K.-Y., Malone, T., & Yu, K.-C. (1988). Object lens: A "spreadsheet" for cooperative work. *ACM Transactions of Office Information Systems, 6,* 332–353.

Lee, J. (1990a). SIBYL: A qualitative decision management system. In P. H. Winston & S. Shellard (Eds.), *Artificial intelligence at MIT: Expanding frontiers* (pp. 104–133). Cambridge, MA: MIT Press.

Lee, J. (1990b). SIBYL: A tool for managing group decision rationale. *Proceedings of the Conference on Computer-Supported Cooperative Work,* 77–92. New York: ACM.

Lee, J. (1991). Extending the Potts and Bruns model for recording design rationale. *Proceedings of the 13th International Conference on Software Engineering,* 114–125. New York: ACM.

Lee, J., & Lai, K.-Y. (1991). *A comparative analysis of design rationale representations* (CCS Tech. Rep. No. 121). Cambridge, MA: MIT, Center for Coordination Science.

Lenat, D. B., & Brown, J. S. (1984). Why AM and Eurisko appear to work. *Artificial Intelligence, 23,* 269–294.

Lewis, C., Reiman, J., & Bell, B. (1991). Problem-centered design for expressiveness and facility in a graphical programming system. *Human–Computer Interaction, 6,* 319–355. Also in T. P. Moran & J. M. Carroll (Eds.), *Design rationale: Concepts, techniques, and use.* Hillsdale, NJ: Lawrence Erlbaum Associates, 1996. [Chapter 5 in this book.]

Mackinlay, J., Card, S. K., & Robertson, G. G. (1990). A semantic analysis of the design space of input devices. *Human–Computer Interaction, 5,* 145–190.

MacLean, A., Young, R. M., Bellotti, V. M. E., & Moran, T. P. (1991). Questions, options and criteria: Elements of design space analysis. *Human–Computer Interaction, 6,* 201–250. Also in T. P. Moran & J. M. Carroll (Eds.), *Design rationale:*

Concepts, techniques, and use. Hillsdale, NJ: Lawrence Erlbaum Associates, 1996. [Chapter 3 in this book.]

MacLean, A., Young, R. M., & Moran, T. P. (1989). Design rationale: The argument behind the artifact. *Proceedings of the CHI '89 Conference on Human Factors in Computing Systems,* 247–252. New York: ACM.

McCall, R. (1987). PHIBIS: Procedurally hierarchical issue-based information systems. *Proceedings of the Conference on Planning and Design in Architecture,* 17–22. Boston: American Society of Mechanical Engineers.

Mostow, J. (1985). Toward better models of the design process. *AI Magazine, 6,* 44–57.

Olson, G. M., Olson, J. S., Storrosten, M., Carter, M., Herbsleb, J., & Reuter, H. (1995). The structure of activity during design meetings. In T. P. Moran & J. M. Carroll (Eds.), *Design rationale: Concepts, techniques, and use.* Hillsdale, NJ: Lawrence Erlbaum Associates. [Chapter 7 in this book.]

Potts, C., & Bruns, G. (1988). Recording the reasons for design decisions. *Proceedings of the 10th International Conference on Software Engineering,* 418–427. Washington, DC: IEEE Computer Society Press.

Sharrock, W., & Anderson, R. (1995). Organizational innovation and the articulation of the design space. In T. P. Moran & J. M. Carroll (Eds.), *Design rationale: Concepts, techniques, and use.* Hillsdale, NJ: Lawrence Erlbaum Associates. [Chapter 15 in this book.]

Singley, M. K., & Carroll, J. M. (1995). Synthesis by analysis: Five modes of reasoning that guide design. In T. P. Moran & J. M. Carroll (Eds.), *Design rationale: Concepts, techniques, and use.* Hillsdale, NJ: Lawrence Erlbaum Associates. [Chapter 8 in this book.]

Toulmin, S. (1958). *The uses of argument.* Cambridge, England: Cambridge University Press.

Winston, P. (1984). *Artificial intelligence.* Reading, MA: Prentice-Hall.

3

Questions, Options, and Criteria: Elements of Design Space Analysis

Allan MacLean
Rank Xerox EuroPARC

Richard M. Young
MRC Applied Psychology Unit

Victoria M. E. Bellotti
Rank Xerox EuroPARC

Thomas P. Moran
Xerox Palo Alto Research Center

ABSTRACT

Design Space Analysis is an approach to representing design rationale. It uses a semiformal notation, called QOC (Questions, Options, and Criteria), to represent the design space around an artifact. The main constituents of QOC are *Questions*

Allan MacLean is a senior scientist at Rank Xerox Research Centre in Cambridge, England; he has a background in psychology and HCI, and his current research focuses on the support of work processes in distributed environments. **Richard Young** is a cognitive scientist with interests in human–computer interaction and the modeling of human cognition; he is a Research Scientist at the UK Medical Research Council's Applied Psychology Unit, Cambridge. **Victoria Bellotti** is a cognitive ergonomist studying techniques and applications for supporting collaborative and remote work and design; she is a senior scientist in the Advanced Technologies Group at Apple Computer. **Thomas Moran** is a cognitive/computer scientist interested in human–computer interaction and collaborative system design, with an emphasis on supporting informal intellectual work practices; he is Principal Scientist and Manager of the Collaborative Systems Area at the Xerox Palo Alto Research Center.

CONTENTS

identifying key design issues, *Options* providing possible answers to the Questions, and *Criteria* for assessing and comparing the Options. Design Space Analysis also takes account of justifications for the design (and possible alternative designs) that reflect considerations such as consistency, models and analogies, and relevant data and theory. A Design Space Analysis does not produce a record of the design process but is instead a coproduct of design and has to be constructed alongside the artifact itself. Our work is motivated by the notion that a Design Space Analysis will repay the investment in its creation by supporting both the original process of design and subsequent work on redesign and reuse by (a) providing an explicit representation to aid reasoning about the design and about the consequences of changes to it and (b) serving as a vehicle for communication, for example, among members of the design team or among the original designers and later maintainers of a system. This article describes the elements of Design Space Analysis and illustrates them by analyses of existing designs and by empirical studies of the concepts and arguments used by designers during design discussions. The aim here is to articulate the nature of the QOC representation. More extensive explorations of processes for creating and using it can be found elsewhere (e.g., Buckingham Shum, 1995 [chapter 6 in this book]; MacLean, Bellotti, & Shum, 1993; McKerlie & MacLean, 1993, 1994).

---

1. MOTIVATION FOR DESIGN RATIONALE

The end product of systems design is a concrete artifact, that is, a software and/or hardware system or product. The output of what is normally considered the design process is a description of the artifact, such as a specification for implementing it or blueprint for constructing it. Such descriptions represent the designer's decisions of what the artifact is to be like, but they do not contain any of the designer's thinking and reasoning behind these decisions, that is, the arguments for why the artifact is the way it is. A *design rationale* is a representation for explicitly documenting the reasoning and argumentation that make sense of a specific artifact.

Design rationale is important because an artifact needs to be understood by a wide variety of people who have to deal with it. This variety of people ranges from those who design and build it (e.g., systems analysts, user-interface designers, and software implementers) to those who sell and service it (e.g., trainers and software maintainers) to those who actually use it. What is important for many of these people is not just the specific artifact itself but its other possibilities. For example, a designer decides between different possible ways to shape the artifact; a maintainer wants to change the artifact to respond to a new need without disturbing the integrity of the artifact; a user wonders why this artifact is different from some other familiar artifact. We hypothesize that an important way to understand an artifact is to compare it to how it might otherwise be.

In this chapter, we propose a style of analysis, which we call *Design Space Analysis*, that places an artifact in a space of possibilities and seeks to explain why the particular artifact was chosen from these possibilities. A Design Space Analysis creates an explicit representation of a structured space of design alternatives and the considerations for choosing among them—different choices in the design space resulting in different possible artifacts. Thus, a particular artifact is understood in terms of its relationship to plausible alternative artifacts.

A Design Space Analysis can be arbitrarily elaborate. In this chapter, we take a pragmatic approach in which we use the simplest possible analyses to solve particular problems. Therefore, we propose a very simple notation, called QOC, that focuses on representing the most basic concepts of Design Space Analysis: *Questions*, which pose key issues for structuring the space of alternatives; *Options*, which are possible alternative answers to the Questions; and *Criteria*, which are the bases for evaluating and choosing among the Options.

We argue in this chapter that design rationale analyses of this kind are *coproducts of design*, along with the target artifact being designed. That is to say, documented analyses are themselves artifacts—they are explicit representations that must be designed and created by designers (by whom we mean any of a variety of players who influence the shape of the artifact in the design process, not just people who wear an official badge of "designer"). We argue that designers are capable of Design Space Analysis and that it is a fairly natural style of reasoning for them. It takes discipline and effort, however, to create such representations, and this effort must be targeted carefully at those aspects of the design process where the effort pays off. We further argue that this kind of reasoning cannot be "captured," for it is not simply a historical record of a design process or a structured representation of the dialogue among designers.

We argue that an explicit design rationale can be a useful tool in the design process in a variety of ways: from reasoning and reviewing to managing, documenting, and communicating. However, demonstrating the utility of a design methodology based on analytic rationale representations is beyond the present state of the art in design rationale research. We are at an early stage of this research—developing representations for design rationale theoretically (to understand their ability to encode the appropriate information), empirically (to understand whether designers can create them), and methodologically (to formulate effective procedures for creating them). Our objective in this article is to demonstrate progress on all of these fronts.

The chapter is organized as follows: First, in Section 2, we illustrate the basic elements of Design Space Analysis and the QOC notation by working through two examples and by reflecting on the important characteristics of our approach and how it relates to other work on design rationale. Then we proceed to several advanced topics of Design Space Analysis. In Section 3, we present empirical studies of designers at work and observations of how the structure of thinking in their deliberations relates to Design Space Analysis. In Section 4, we examine issues in representing various modes of justification and the coherence of design in Design Space Analysis. The majority of the chapter illustrates and discusses the properties of QOC-based representation of a design space. It is more concerned with the representation as a product of design than it is with the process of creating that product. We finish, however, by presenting an appendix with some preliminary advice—based on our current experiences, observations, and theory—on some heuristics for carrying out Design Space Analysis.

2. BASICS OF DESIGN SPACE ANALYSIS

In this section, we present the basic concepts of Design Space Analysis—Questions, Options, and Criteria—and describe the QOC notation. We illustrate these concepts and their interrelationships by analyzing two design examples. In the first example (Section 2.1), we examine a part of the user interface of a window system and try to understand the design of a scroll bar mechanism. The second example (in Section 2.2) analyzes two alternative designs for a bank automated teller machine (ATM) to help understand the essential differences between them. We then look (in Section 2.3) at some characteristics of our Design Space Analysis approach and how it compares to other approaches. Finally, we discuss (in Section 2.4) how these kinds of analyses fit into the context of the design process.

2.1. Analyzing an Artifact—A Scroll Bar Example

In this first example, we consider the scroll bar mechanism in the Xerox Common Lisp (XCL) environment, and we note that the scroll bar design is different from many of the more recent window environments. We want to understand why. We do this by "reverse engineering," a rationale for some attributes of the XCL scroll bar.

The windows in XCL provide views onto objects (such as documents) that are too large to be viewed in their entirety, and thus XCL provides scroll bars to control the views of the objects in the windows (see Figure 1). The first step in an analysis is to abstract from the artifact a set of characteristic *features*, which represent the design decisions that were made. We note several features of XCL's scroll bar design, for example:

F1. The scroll bar is normally invisible and only appears when it is needed for scrolling.
F2. The scroll bar is fairly wide.
F3. The scroll bar indicates the position of the view in the window.
F4. The scroll bar indicates the relative size of the view in the window.

There are, of course, many more features that characterize this particular design, and we could analyze them all. But we want to focus here on the first feature—that the scroll bar appears only when needed. How this works is that the scroll bar is normally invisible, but it appears when the cursor is moved from inside the window over the left edge (to just the location

Figure 1. A window in the Xerox Common Lisp (XCL) system. The scroll bar
appears when the cursor is moved out of the left edge of the window.

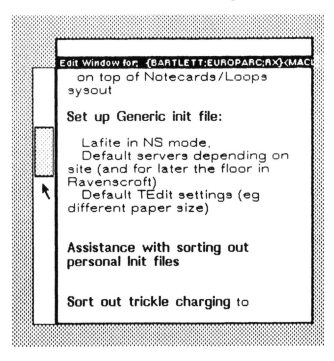

where the scroll bar appears). This is an interesting design decision, and
it is different from most of the more recent window systems that have their
scroll bars permanently fixed to the windows (e.g., the Macintosh). Our
problem is to understand why this might have been done in XCL.

Questions and Options

We address the problem by analyzing the design space to see how a
decision to have an appearing scroll bar could have been made, to understand whether it was a good decision, and to determine what the tradeoffs
might have been.[1] Actually, the width of the scroll bar is a related feature,
and so we analyze these two features together. The method for a Design
Space Analysis is to view each feature as only one Option available among a
set of other Options, to pose Questions for structuring the Options, and to

---

1. The methodology used in this example is the "rational actor model" (Allison,
1971), in which we assume that the designer was acting rationally by making the
most rational choice available to meet his or her criteria. Thus, we reason backwards
from our knowledge of the designer's choice to what the criteria must have been.

enumerate the Criteria that determine the choice of particular Options. We use the QOC notation to represent this analysis.

Any specific Design Space Analysis must be placed in context. We begin with the assumption that there has been a decision to use a scroll bar as the user-interface technique to control viewing. From this decision, there follow several Questions in order to elaborate the decision into more detail. One obvious Question is how wide is the scroll bar? For now we only need to consider two qualitatively different Options: that the scroll bar could be relatively wide or relatively narrow. This is all we need to analyze the issues. We already know that there are Options to make the scroll bar permanently fixed to the window or to make it appearing, and we need to formulate a Question to relate these Options. Let us simply pose the Question as how should the scroll bar be displayed? (This is an interesting Question, because it would be easy just to assume that the scroll bar is permanent.) If we choose the Option of an appearing scroll bar, then as a consequence we have the further question of how to make the scroll bar appear? There are a number of possible Options for this. Some kind of a scroll button, either on the keyboard or on the screen, could be used to make the scroll bar appear. Alternatively, we could use some kind of a "natural" cursor movement, such as a movement of the cursor over the edge of the window to where the scroll bar will appear. In this way, we have begun to articulate the design space.

Figure 2 shows the QOC notation for the design space thus far. The notation uses a node-and-link diagram to portray the Questions, the Options, and their relationship. The links show the Options that respond to the Questions and the consequent Questions that follow from the Options. The Options chosen (i.e., design decisions made) for the XCL environment are shown boxed. Figure 2 begins to provide a context for understanding these decisions. The alternative Options represented give an understanding of how the design could have been different (e.g., a permanent, narrow scroll bar). The Questions organize the Options by highlighting the critical dimensions along which the Options differ. The role of the Questions is to delineate local contexts within the design space to help ensure that like Options are compared with each other. The set of Questions provides a structured space of Options. Good, incisive Questions help to open up novel Options. That is to say, the role of the Questions is generative and structural, not evaluative. We must next consider how to evaluate the Options and to rationalize the decisions.

Criteria and Assessments

Choosing among the various Options requires a range of considerations to be brought to bear and reasoning over those considerations. The most important elements for organizing this reasoning are *Criteria*, and these must

Figure 2. A fragment of the design space for the XCL scroll bar using the QOC notation, in which Questions and Options are enumerated and related. The boxed Options are the decisions made in the design of the XCL environment.

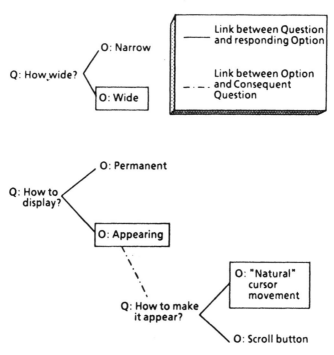

be added to the design space. Criteria represent the desirable properties of the artifact and requirements that it must satisfy. It is important to articulate clearly the Criteria, because they make clear the objectives of the design. They form the basis against which to evaluate the Options.

The QOC notation provides a way to represent an *Assessment* of whether an Option does or does not satisfy a Criterion. For example, one may wish to claim that a particular Option is good because it does not take up much "screen real estate" but that, on the other hand, it is not as easy to hit with the mouse. In this case, there are two Criteria: screen compactness and ease of hitting with the mouse, and there is a positive Assessment of that Option against the first Criterion and a negative Assessment against the second Criterion. The array of individual Assessments provides a context for making an overall judgment of the suitability of the Options. We give a more extensive discussion of the nature of Criteria in Section 4, but note for now that the Criteria are worded so that positive Assessments (i.e., satisfying the Criteria) are always what is desired. For example, we would not state the just-mentioned Criterion as difficulty of hitting with the mouse.

Figure 3. **A QOC representation of the design space for the XCL, elaborated from Figure 2 to include Criteria and Assessments. The boxed Options are the decisions made in the design of the XCL environment.**

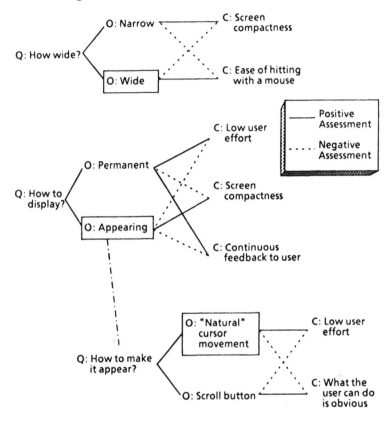

Figure 3 shows the design space of Figure 2 expanded to include some relevant Criteria. In addition to the Criteria just mentioned, there are the Criteria of low user effort (i.e., that the required user action is quick and easy), continuous feedback to user (i.e., that the user has a continuous visual indication of the state of the window view), and what user can do is obvious (i.e., that it is made clear to the user what actions are allowed). Figure 3 shows the questions and Options to which the Criteria apply. Each Criterion is applied to all the Options for a given Question, but different sets of Criteria are applicable to different Questions. Note, however, that many Criteria are applicable across multiple Questions. This is important, as is discussed in Section 4. In the QOC diagrams presented in this article, a Criterion used for multiple Questions is represented multiple times, under each relevant Question, to simplify the spatial layout of nodes and links.

The Assessments are shown as labeled (positive or negative) links between the Options and the Criteria. For example, the Option of a per- manent scroll bar is linked positively to (i.e., assessed positively against) the Criterion of continuous feedback to user (because the visual feedback from such a scroll bar is always available to the user), but it is linked negatively to screen compactness (because a permanently visible scroll bar always take up screen space).

Now that we have laid out all the elements of the rationale, let us return to the problem we posed at the beginning of this example: Why was an appearing scroll bar chosen for the XCL environment? It may seem like a dubious decision, because the appearing Option satisfies only one of the three Criteria, whereas the permanent Option satisfies two of the three. This view is too crude, but it gives us an opportunity to discuss some of the properties of Design Space Analysis. The QOC notation is not intended to yield obvious choices by simply counting Assessments; rather, it provides the basis for a more complex discussion of the tradeoffs. One possible reason for choosing the appearing scroll bar is that screen compactness was considered to be the most important Criterion. If this were the case, however, it does not seem to be consistent with the choice of a wide scroll bar.

Another issue is the strength of the Assessments. If the positive Assess- ments favoring the choice of the appearing scroll bar were stronger (in some sense) than those supporting the permanent scroll bar, then this could explain the choice. In this example, we only distinguish positive from negative Assessments. We could very well use 3, 5, or 7 levels of assessment values (e.g., strongly positive, positive, mildly positive, neutral, mildly nega- tive, negative, strongly negative). We are not against this in principle, but we feel that the complexity in making such Assessments usually outweighs the gains in understanding the overall picture, which is the real objective of the analysis. In this case, it is hard to differentiate the Assessments beyond what we have done. One reason is that other design decisions may influence the Assessment. For example, the negative Assessment of the appearing scroll bar against low user effort is difficult to evaluate more precisely without knowing more about the appearing Option. What we need to do in this case is to explore the appearing Option in more detail by Assessing the Options for how to make it appear? We note that the "natural" cursor movement Option was chosen, which is positive against low user effort. This particular version of the appearing Option in fact mitigates the negative Assessment of it against low user effort; that is, it makes the Assessment less negative, if not neutral.

This points to another property of Assessments—that the Assessments of different Options against a given Criterion are usually *relative*. For example, a wide scroll bar is relatively easier to hit with the mouse than a narrow one, and we represent this by positive and negative Assessments, respectively, against the ease of hitting with the mouse Criterion. Relatively speaking, wide is better than narrow, and this is all we need to say at this level of specification of the Options.

Returning to the appearing scroll bar issue, we note that the Assessments against the screen compactness and continuous feedback to user Criteria are fairly clear between the two Options. But the relative Assessment against low user effort is minimized. One might even argue that there is no difference—that the effort of scrolling by bringing up the appearing scroll bar with the "natural" cursor movement is no different from the effort of moving the cursor to a permanent scroll bar. The result is that low user effort should not be considered a very crucial Criterion for this Question.

We can see from this discussion that there is an *interaction* between the Assessments across different Questions. We have already seen that the Assessment of the appearing scroll bar depended strongly on the Option chosen for how to make it appear? There is also an interaction between the importance of Criteria across Questions. For example, the importance of the screen compactness Criterion depends on the choice of how to display? This Criterion is much less important if the appearing scroll bar is chosen because it does not clutter the screen all the time, as a permanent scroll bar would. This observation explains the apparent contradiction noted earlier as to whether the screen compactness Criterion was important or not to the XCL designer. The conclusion from this analysis is that screen compactness did seem to be the determining Criterion for choosing the appearing scroll bar.

This discussion illustrates the argument-based nature of Design Space Analysis. (Indeed, that has been a major emphasis of earlier presentations of our approach; see MacLean, Young, & Moran, 1989.) The previous analysis, kept simple for expository purposes, can be elaborated or challenged (as perhaps most readers at this point would like to do). The objective of the QOC notation is to lay open for argument the elements of the rationale (see Section 2.3 for how the argument can be represented). The QOC notation allows us to pinpoint where the debatable issues are. The components of the QOC should therefore be treated not as fixed, structural relations but, rather, should be regarded as ongoing concerns, or even provocations. These issues are discussed further in Section 4.

2.2. Comparing Alternative Designs—An ATM Example

Let us now consider another analysis in which our problem is to compare two alternative designs of a bank ATM. ATMs are relatively simple devices with which most people are familiar, yet there is considerable variety in the

Figure 4. **The steps required to get cash from the SATM and the FATM.**

The SATM
1. Push card into slot
2. Type PIN number when prompted
3. Select "Cash Withdrawal" (from the several services offered)
4. Select "Another Amount" (you could have selected one of five preset amounts)
5. Type in amount required and press the Enter key
6. Select "No" (when asked if you would like to request another service)
7. Remove card from slot
8. Take cash from drawer and receipt from slot

The FATM
1. Select cash amount (must be one of six preset amounts)
2. Insert card
3. Remove card
4. Type in PIN number
5. Take cash and receipt from drawer

---

design of ATMs, both in what facilities they provide and in the ways people interact with them. One interesting contrast is between a standard ATM (SATM) and a new fast-cash ATM (FATM) recently introduced by a British bank. The SATM offers a range of services, such as balance enquiries, new checkbooks or statements, and cash withdrawals. The FATM provides only for cash withdrawal. However, more than just restricting services, the procedure for using the FATM is different from the procedure for the SATM. Figure 4 shows the steps required to get cash from the two ATMs.

Our task is to compare these two designs to understand what the advantages of the new FATM design are. We do this with a Design Space Analysis. Our challenge is to produce a design space that captures both designs.[2] We do this by identifying parallel features that characterize the differences between the two designs. These features are represented as alternative Options. We then formulate Questions that characterize the dimensions of these differences and Criteria to evaluate them. This analysis is summarized in a QOC diagram in Figure 5.

The most obvious dimension of difference has to do with the Question what range of services offered?, where the Options are full range for the SATM and cash only for the FATM. It is interesting to

---

2. Note that scroll bar analysis also embodies two contrasting designs (although that was not its primary motivation)—the wide, appearing scroll bar of the XCL system and a narrow, permanent scroll bar found in many other systems. In fact, it was interesting to note, in a more extensive analysis of the design space around the XCL scroll bar than presented here, that the resulting design space included key Options for no less than four different existing scroll bar designs.

Figure 5. **A QOC representation summarizing the distinctions between the SATM and the FATM. The Options representing the SATM decisions are indicated by solid boxes, and the FATM decisions are indicated with the dashed boxes.**

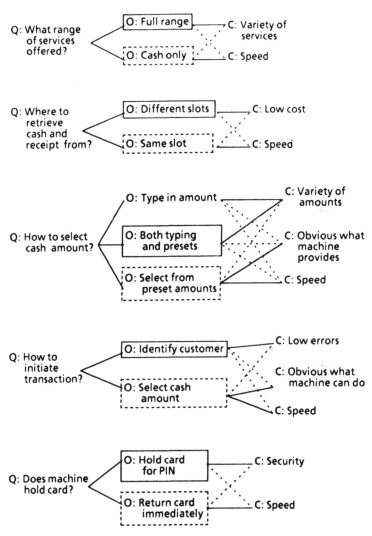

view the FATM design as challenging the standard Options of the SATM. To the Question where to retrieve cash and receipt from?, the FATM says why not take them from the same slot. Note that many features of the FATM design challenge not only the physical design of the SATM but also the procedure for using it. For example, the FATM starts off by having the customer select cash amount (rather than the SATM's having to first identify customer), which raises the question of how to initiate the transaction? Also, the FATM challenges whether the machine needs to hold onto the customer's bank card while it gets the customer's PIN and validates it.

Next we need to identify Criteria to help us understand the pros and cons of each design. Clearly, speed is the major Criterion motivating the FATM design and is relevant for all the Questions. However, speed is traded off against a variety of different Criteria in different cases, such as low cost, low errors, variety of services. For example, utilizing the same slot to get the receipt is costlier (because it involves hardware changes), whereas initiating the transaction with select cash amount could cause errors (because someone could select an amount and leave the machine in a nonneutral state for the next customer to come along). It appears that there may be some benefits in addition to speed—or example, making it more obvious what the machine can do. The FATM has several new and independent features, but some of the Assessments of their being better for speed are arguable. For example, the argument that initiating by select cash amount is faster is based on the assumption that preparing the cash takes time for the machine to do and this time can be overlapped with the customer's identification. The argument that the select from preset amounts Option offers a speed advantage over the SATM's both typing and presets relies on disallowing slower typing selections and facing the customer with less choice. This may of course mean that some customers are unable to get their preferred amount of cash and could spend more time trying to work out how to get the amount they want.

This representation of the design space raises the more basic issue of what problem the FATM is attempting to solve: Is it the time per transaction that has to be reduced or is it important to restrict the kinds of transactions allowed? The analysis suggests both are relevant but does not really tease them apart. It is also clear that there are other possible Options for many of the Questions; therefore, other solutions may be better. Rather than suggesting definitive conclusions, this first cut at the design space sets the frame for further exploration. In fact, we carry out such an exploration in our empirical studies presented in Section 3, where we observe designers working on the same problem and expand the design space on the basis of their deliberations.

2.3. Characteristics of QOC Representation

Now that we have seen some concrete examples of QOC notation, we can step back and consider the important characteristics of our approach to design rationale. In pointing out these characteristics, we can compare and contrast our approach with other approaches to design rationale. These characteristics highlight many different aspects of our approach—

representational properties of QOC, structural features of QOC that suggest the kinds of tools in which to embed the notation, and properties implicating the design process. We consider representational and structural characteristics here and consider the design process in Section 2.4.

Design Space Focus. The QOC representation emphasizes the systematic development of a space of design Options structured by Questions. The rationale in Design Space Analysis is built on the comparison of alternative Options. We can constrast this with the "claims analysis" approach of Carroll and Rosson (1991 [chapter 4 in this book]). Their approach is to evolve a design by exploiting the positive aspects of claims of its effectiveness while addressing the negative aspects. It is difficult at this point to compare the elements of their analysis with ours, but one obvious difference is that they continuously refine a single design, whereas we explicitly advocate the development of a space of alternatives.

Focus on Criteria. The QOC representation brings the objectives for the design, in the form of the Criteria, into explicit focus. Design rationale schemes derived from issue-based information systems (IBIS; Kunz & Rittel, 1970), such as gIBIS (Conklin & Begeman, 1989), do not explicitly bring forth criteria. IBIS has Arguments for and against Positions, but IBIS Arguments only implicitly refer to what we would call Criteria. Criteria per se are not proper objects of the IBIS notation. There are other proposals for characterizing the objectives of design. Lewis, Rieman, and Bell (1991 [chapter 5 in this book]) proposed that concrete problems are an effective way to represent design objectives. Others (e.g., Carroll & Rosson, 1990, 1991 [chapter 4 in this book]) have proposed the representation of design objectives by scenarios.

Coproduct of Design. A Design Space Analysis is not a record of the design process, but rather it is a coproduct. As an artifact in its own right, the QOC itself has to be designed. The rationale representation (a QOC) is created along with the descriptive representation (e.g., a specification) or the artifact itself (e.g., a prototype). Designers are clearly capable of producing such analyses: for example, Botterill's (1982) rationale for the IBM System/38 and Johnson and Beach's (1988) rationale for the style sheets in the ViewPoint office system. These rationales emphasize a logical rather than a chronological account. The argumentation that makes such accounts coherent has itself to be carefully crafted; it does not simply emerge from a historical record of the design process.

This approach can be contrasted with the IBIS-derived systems, such as gIBIS (Conklin & Burgess-Yakemovic, 1991 [chapter 14 in this book]) and Procedural Hierarchy of Issues (PHI; McCall, 1986), whose purpose is to capture the history of design deliberations. Although the Issues, Positions, and Arguments of IBIS appear to be like our Questions, Options, and Criteria, they are quite different. Their Issues are general purpose in that they can be about any topic that comes up in design discussions, and their Positions similarly are general purpose. QOC Options, on the other hand, are specifically *design* options, and Questions are specifically to structure the design space. Also, as noted before, their Arguments are quite different from our Criteria. Because they are capturing discussion on the fly, the resulting argumentation structure is less coherent and succinct. Clearly, there is a tradeoff here between the effort to construct and the resulting coherence of the rationale. That being said, having an IBIS-encoded history would be a valuable resource in building a Design Space Analysis. In fact, we could see a QOC representation as a condensation of an IBIS history that brings out the most important elements of the history for logical argumentation. (This is similar in concept to Parnas & Clements', 1986, notion of faking the design history.) Such a QOC analysis could suggest logical deficiencies in the discussions and could be used as a basis for directing the discussion into dealing with them. Thus, we see IBIS and QOC as having complementary roles.

Embedded in Design Activity. A corollary of being a coproduct of design is the symbiosis between the descriptive and rationale representations. The two kinds of representations are linked in that the chosen QOC Options represent selected features (i.e., those selected to be analyzed) of the artifact in the descriptive representation. Design is viewed as going back and forth between the two representations, intermingling the processes of construction and reflection in what Schön (1983, 1987) called "reflection in action." We envisage QOC representations being used in this kind of design context. In the same spirit, Fischer, Lemke, McCall, and Morch (1991 [chapter 9 in this book]) provided demonstrations of how rationale-based tools can be integrated into "design environments," which are construction kit facilities, along with active "critics" that give advice when conditions in the proposed design raise particular design issues. This advice from the critics is then backed up with a domain-specific knowledge base of rationale issues that are brought forth to explain the advice to the designer. The designer can also augment the rationale to include specifics of the current design situation. In our terms, their rationale representation is coupled with the descriptive representation of the artifact being designed.

Semiformal. The QOC notation is best regarded as semiformal. The basic QOC concepts (Questions, Options, Criteria) and their relations provide a formal structure for representing the design space, giving the representation a strongly diagrammatic style; however, the actual statements within any of the nodes of the diagrams are informal and unrestricted. QOC diagrams can quickly become messy and difficult to manage, even at 50 nodes, and computer-based tools are necessary. Hypertext systems (see Conklin, 1987) promise support for this kind of notation. The advantage of general purpose hypertext tools is that they allow informal annotations, as well as explicit structural relationships, to be easily represented. A disadvantage is that structures are tedious to create, and little support is provided for fully utilizing the structure. Nevertheless, useful results have been obtained with hypertext both in helping with the generation and organization of ideas (e.g., VanLehn, 1985) and in producing coherent structures for later examination (e.g., Marshall & Irish, 1989). Some of the examples in this article were created in the NoteCards system (Halasz, Moran, & Trigg, 1987), and the figures are based on NoteCards network browsers.

Argument Based. As we have seen in the earlier sections, rationale is based on argumentation, not proof. In principle, any elements of a QOC representation can be queried or challenged. Rather than being fixed, they are intended to be justifiable, based on further arguments that can be opened up to inspection, thereby making it possible for flaws in the original thinking to be identified and thus the representation improved. Behind all of the QOC elements (Questions, Options, Criteria, Assessments), there may be arguments supporting or objecting to their presence or their characterization. The arguments behind the Assessments are perhaps the most important. Figure 6 illustrates how arguments relate to Assessments. In Section 4, we elaborate several different kinds of arguments for justifying Assessments. In this article, we seldom show arguments explicitly in our diagrams (but see Figures 8 and 11). However, arguments are often discussed informally in the text, as may be seen in Sections 2.1 and 2.2.

The closest representation to our QOC notation is Lee and Lai's (1991 [chapter 2 in this book]) Decision Representation Language (DRL). DRL has a carefully defined semantic representation of decision elements and their relations. QOC's elements map very closely to DRL's (Questions are Decision Problems, Options are Alternatives, Criteria are Goals, Assessments are plausibilities of "Achieve" relations, Arguments are Claims). We expect any further elaborations of QOC to follow DRL's representation pretty closely (however, we are wary of the usability implications of proliferating distinctions in the representation). DRL is implemented in a system called SIBYL and is being used to explore various kinds of computational services over DRL structures, such as dependency management (Lee, 1990), where the consequences of, say, a revised assessment can be traced back to the decisions that were based on it. This seems to be a productive

Figure 6. The relationship between arguments and QOC Assessments. This diagram says: Argument 1 gives the reason for the relative Assessment that Option 1 is better than Option 2. Argument 2 challenges the Assessment of Option 2 (although it does not dispute the relative Assessment of Argument 1). Argument 3 supports Argument 1, whereas Argument 4 objects to it.

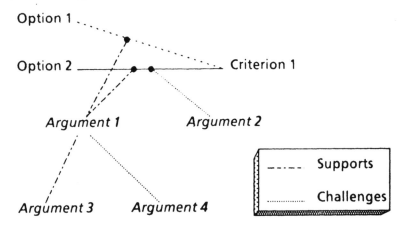

way to enhance the value of rationale representations by providing tools to aid design.

Expandable Detail.   The argument basis is an example of an area where the QOC representation can be expanded to an arbitrary level of elaboration. Although such details can be important, they can also get in the way of seeing the overall picture. The expandable-detail characteristic refers to the ability to work with manageable and comprehensible QOC diagrams that include only those parts of the rationale relevant to the particular issues at hand. Any part of a QOC diagram should be able to be regarded as a summary or shorthand for a more extensive story that may be probed for more details. This implies that adequate tools for this kind of representation must have good browsing, filtering, and display capabilities.

Purposeful.   This is the other side of the coin from expandable detail. It is not practical to represent every possible detail, and it is not useful to do so.[3] A QOC representation need not be a complete specification of a design. It may not be worth it to include well-understood or noncontentious

---

3. We seem to be misunderstood on this point by Gruber and Russell (1995 [chapter 11 in this book], Section 4.1). We would, further, agree with their point that a representation such as QOC is better viewed as a resource for constructing explanations rather than as embodying complete explanation in itself.

parts of the design space. It may be sufficient to provide a Design Space Analysis for only parts of the space—where difficult issues were encountered during design, where nonobvious solutions were adopted, where it is felt that maintainers may need a clear understanding, where it is known that critical parameters of the design need careful monitoring, and so forth. In many situations, it may be appropriate to produce a rationale after the event, such as a maintainer creating a QOC to "reverse engineer" a part of a system and preserving it for future maintenance. Understanding the various purposes of Design Space Analysis requires us to consider the roles that it can play in the design process.

2.4. Design Space Analysis in the Design Process

Two issues come to mind in considering how an explicit Design Space Analysis can be useful and practical in the design process. One is the cost of creating analyses. The other is scaling analyses up to large design projects. It is too early in this research to have answers to these issues, but our current views on how such analyses may be used in design take them into account. One approach, being investigated by Fischer et al. (1991 [chapter 9 in this book]), is to create a large knowledge base of design rationale, which requires a considerable battery of system tools and a well-defined and well-maintained corpus of preencoded, domain-specific design rationale. We are wary of trying to build a single grand rationale structure for a design. Rather, we see a series of smaller Design Space Analyses being created during the course of a design project. Each analysis would be a focused effort serving a local purpose (and thus having a local payoff) in the design project while at the same time contributing to an overall documentation of the project's design rationale. Templates of the type proposed by Casaday (1995 [chapter 12 in this book]) are one promising way of supporting design from this perspective.

Explicit Design Space Analyses can be useful in design in a variety of different ways. First we consider the analyses as communication vehicles in the design process. Then we look at other classes of design activities.

Communication. Design is a social process that usually involves a variety. of individuals, each with different skills and objectives. Communicating a shared understanding of the design is a crucial aspect of managing the design process. Explicit documentation of the rationale involved in the design should be a useful aid for communication between members of the design team, between designers and their users, and between the current design team and future design teams that want to build on or reuse parts of the current design. QOC representations should be effective communication vehicles, because they are simple enough to be understood by a variety of people, they are flexible enough to represent a variety of issues from a variety of viewpoints, and they are explicit enough to expose assumptions that can be challenged by others. However, the social subtleties in communication, such as what information people are willing to make explicit, should not be underestimated.

Perhaps the best way to see the importance of communicating an understanding is to see it as spanning the life cycle of a software design project. An understanding has to be shared among a wide variety of players and functions: marketing, requirements analysis, system design, user-interface design, implementation, documentation, sales, training, customer support, user customization, software maintenance, system administration, new releases, contracted enhancements, and system redesign. Design Space Analysis should help different teams within a project clarify their concerns to each other. Even within a design team, members should be better able to understand decisions with which they were not personally involved or be reminded of decisions in which they were involved earlier. Maintenance, including the adaptation to new user requirements after release as well as bug fixes, is a particularly important phase of design. Some estimate that this can occupy as much as 90% of the effort in the software life cycle (e.g., Balzer, Cheatham, & Green, 1983; Martin, 1977). Conklin (1989) claimed that as much as half of the effort in maintenance is understanding the system in order to make effective fixes and enhancements. Documented Design Space Analyses can help maintainers foresee the consequences of proposed alterations by making clear what decisions, tradeoffs, and evaluations will be affected by the change. Communication between designers and end users is also important. As we move toward more customizable systems (e.g., Mackay, 1991; MacLean, Carter, Lövstrand, & Moran, 1990), it is important to make clear not only what features can be customized but what the consequences of any change (customization) might be.

Again, Design Space Analyses require effort to create them. A major problem, as Grudin (1988, 1995 [chapter 16 in this book]) pointed out, in using them solely to enhance communication is that the people who create them are not the ones who benefit directly from them (a problem that also applies to software documentation). Therefore, we need to consider other aspects of the design process and utilize more immediate motivations for getting the analyses created. We consider the roles of Design Space Analysis in the creation, evaluation, reflection, and management of design.

Creation. We expect Design Space Analysis to facilitate innovation and reasoning in the design process by helping designers generate, represent, and think through, in a disciplined yet flexible way, their decisions—alternatives to them, the arguments for and against them, their implications, and the interrelations among them. The function of Design Space Analysis, however, is not just to facilitate making decisions. The process of developing QOC analyses exposes assumptions, raises new Questions, challenges Criteria, and points to ways in which new Options can capitalize on the strengths and overcome the weaknesses of current Options.

Reflection. Although it is difficult to be very analytical in the "heat" of the creative phases of designing, design projects are punctuated by reviews, reports, and presentations; these are natural times for standing back and reflecting on the state of the design. Design Space Analysis is an appropriate framework to help structure such reflection—for example, justifying design decisions and considering other opportunities for exploration. The QOC representation provides a uniform format with which to produce these reviews, reports, and presentations, thereby making them easier to store, cross-reference, and index, thus documenting the rationale behind the design project.

Management. Design Space Analysis should be useful in different aspects of project management. Breakdowns in design often occur because of designers' cognitive limitations (Guindon, Krasner, & Curtis, 1987). Effectively managing the complexity of design could affect both the quality and the efficiency of the design process. For example, Design Space Analysis provides a representation in which to incorporate design requirements (e.g., as Criteria) and constraints (e.g., as selected Options that impinge on the resolution of other Questions), both initially and as they change, and it can provide a map of the explorations over the design space. Embedded in appropriate tools, it could help track how well the explorations are satisfying the requirements. Because Design Space Analysis explicitly represents a design space, it is well suited to tracking changes.

It is an empirical question whether any of these proposed uses for Design Space Analysis will work in real design projects.[4] Our first modest empirical step, presented in the next section, is to understand how the concepts and structure of Design Space Analysis fit the pattern of reasoning that designers naturally exhibit. In doing this, we are using Design Space Analysis in yet another way—as a research tool to understand the structure of design reasoning.

---

4. Since this chapter was originally written, Shum (1991b; Buckingham Shum, 1995 [chapter 6 in this book]) has reported empirical work with designers which explores the limits of achieving some of the claims made here. In addition, McKerlie and MacLean (1993, 1994) have reported work from a practical design setting which suggests that some of these benefits can indeed be achieved.

3. EMPIRICAL STUDIES OF DESIGN REASONING

Design Space Analysis and the QOC representation give us a useful way to organize the information about the context of reasoning surrounding a design. However, it is not clear how compatible the QOC representation is with the ways designers naturally talk about design. Because our goal is to use Design Space Analysis in the design process, it is clearly important that we understand how it might fit into the ways designers actually work. In this section, we present empirical studies of designers at work, and we use Design Space Analysis to structure the content of their discussion.

We examine two observational studies of professional software designers considering the design of the bank ATMs that we presented in Section 2.2. The first study (called ATM1) has been analyzed in considerable detail, and it is the primary focus of this section. The second study (called ATM2), carried out at the University of Michigan, has had less detailed analysis—we only draw on it for a brief example.

In these studies, we used pairs of professional software designers, so their natural activity of discussion exhibited their reasoning. The pairs had worked together in the past and, thus, did not have to spend any time adjusting to each other. The studies were carried out "in the zoo" (halfway between an artificial laboratory task and uncontrolled free behavior "in the wild"). The designers worked on a fairly natural design problem, but a problem of our choosing, carried out in a meeting room that was set up for video recording.

The problem given to the designers in the first study (Jaimie and Donald) is shown in Figure 7. Their task was to analyze the FATM (which had been proposed in response to queues building up at the SATMs), to critique it relative to the SATM, and to suggest alternative designs if appropriate. Debriefing after the session confirmed that neither designer had ever seen an actual FATM. This task was methodologically attractive, because it enabled a small but complete problem to be tackled in a relatively short time and because it naturally involved rationalization activity.

We recorded the session on video tape. The two designers sat in a room by themselves alongside an electronic whiteboard, which they used heavily. They spent about 45 min on the problem on their own, then about 10 min summarizing their conclusions to us, and then about 15 min on a debriefing during which they told us their backgrounds and experience (both as software designers and as users of ATMs). The debriefing confirmed that they felt the problem and the setting were natural for them.

We transcribed the video tape and categorized the behavior into Design Space Analysis elements. These elements were then structured into a design space using the QOC notation. We used this exercise to help us understand the extent to which the discussion can be represented using QOC and to give some insights into phenomena that do not naturally fit into Design Space Analysis terms.

Figure 7. **The problem presented to the designers in the ATM studies.**

Standard ATM

The National Barklands Bank (NB) Automated Teller Machine (ATM) is a fairly typical ATM. If you want to get cash from it, you would go through the following steps

- Push card into slot.
- Type PIN number when prompted.
- Select *Cash Withdrawal.* (from the several Services offered).
- Type in amount required and press *Enter* key.
- Select *No* (when asked if you would like to request another service).
- Remove card from slot.
- Take cash from drawer, and receipt from slot.

But

The NB bank noticed that long queues sometimes built up at these standard ATMs. They asked their design staff to see if they could speed the process up. Their proposed design (FATM) presents the customer with the following procedure:

The Fast ATM (FATM)

- Select cash amount. (Must be one of six preset amounts).
- Insert card.
- Remove card.
- Type in PIN number.
- Take cash and receipt from drawer.

Your task . . .

You are brought in as design consultants by NB, who would like to know whether you think they have produced a successful design for the FATM. We would like you to analyse the new design and

(1) summarise for us what you feel are the main advantages and disadvantages of the FATM;
(2) suggest any further improvements to the design, or better design alternatives.

3.1. Protocol Analysis

Encoding the Protocol

The video record was transcribed into a verbal protocol with annotations of the nonverbal activity so that the protocol could be understood without referring back to the video tape. Our focus was on analyzing the content of the session and building a coherent representation of the main ideas in the discussion.[5] Asides and redundant remarks were filtered out, and the protocol was segmented into 358 assertions, each of which captured a substantive point in the discussion. The assertions were numbered sequentially to give a convenient index into the session.

Before going into detail, it is worth briefly characterizing the flow of the session as a whole: The designers' discussions ranged over a number of topics, and they tended to go back and forth between them in an unstructured fashion. They jumped straight into the task and started off by trying to understand the differences between the two ATMs described. During this period, they did not identify a good reason for the FATM to use a different order of steps and maintained that the SATM order was better (Assertions 1 to 103). They then moved toward a design solution that attempted to reduce the number of steps and time required for each step. In attempting to resolve the conflict between maintaining a range of services and reducing customer transaction times, they devised a proposal for a switchable ATM that the bank staff could set into a fast-cash mode, thus restricting services during busy periods (Assertions 104 to 214). This was followed by a review of the design (Assertions 215 to 260). They realized that they did not fully understand why queues were building up and discussed possible reasons (Assertions 261 to 303). Finally, they went over the details of their proposed design (Assertions 304 to 358).

Categorizing the Assertions

One way to get a feel for the designers' style of reasoning and its relation to Design Space Analysis is to categorize the assertions into something close to the QOC elements. The vast majority of the session consisted of assertions about the substance of the design problem; only 4% of the assertions were nonsubstantive (e.g., discussion of how to make use of the whiteboard). We classified the substantive assertions into three broad categories—options, issues, and justifications—which are related to the QOC elements in ways we discuss shortly. Inevitably with data of this sort, reliable classification is difficult. Categorization relies crucially on maintaining and interpreting the context in which the assertions were raised. The grain of analysis is such that the discussion of one point often spreads across many assertions, and conversely a few assertions (about 4%) were so ambiguous that they had to be classified into two categories (see also, MacLean, Bellotti, & Young, 1990).

---

5. Olson and Olson (1991) reported an analysis of the ATM1 study that focused on the meeting dynamics of the session. Their goal was to analyze present practice to look for opportunities for technology intervention. Their analyses focused on the coordination of group activity (e.g., stating goals, agenda setting, activity tracking) and on the development of issues (e.g., stating the goal and generating, structuring, evaluating, and selecting ideas).

We observed that the designers talked frequently about specific Options. In all, 38% of the assertions were categorized as Options in the QOC sense. It was relatively easy to categorize these, for example:

> Well my favorite would be just to have a "fast cash" button at the point where you said "select cash withdraw." (Assertion 107)

The Options were usually discussed in isolation, with relatively little structure linking them; that is, there were no Questions to structure the Options. Questions in the QOC sense were hardly ever stated; the nearest example was:

> Is there any way we can improve on preset amounts? (Assertion 190)

There were some questions of a yes-or-no variety, such as:

> Do you want the receipt? (Assertion 44)

Such a question does not provide much help in exploring a space of possible Options. However, we categorized 8% of the assertions as "issues," which were the nearest thing to Questions. In addition to examples like those just shown, these were more akin to issues in the IBIS sense, for example:

> What if you asked for fifty quid, and you've only got thirty quid in your bank account? (Assertion 339)

Finally, 54% of the assertions were categorized as "justifications." Justification is used here broadly, covering Criteria, Assessments, and arguments, as well as other forms of justifying. For example, the following captures an Assessment that relates to an explicit Criterion:

> Well, they've speeded it up by taking away the other services. (Assertion 142)

Some explicit tradeoffs between different Criteria were considered:

> It is basically because otherwise you trade off speed against security. (Assertion 333)

On the other hand, a range of more general forms of argument was used extensively. Some consisted of questionable rationalizations, such as:

> If you're going to spend three quarters of the money, you might as well spend a hundred percent and have two full blown machines. (Assertions 138 to 140)

Some arguments appealed to analogy with other situations:

You could say five items only like in supermarkets or whatever. (Assertion 187)

Some arguments utilized small usage scenarios:

You've got to remove the card before you take the cash. (Assertion 161)

Finally, some arguments attempted to build an ad hoc theory of why queues might be forming:

Is it because there's loads of people, or is it because people are fumbling with their cards? (Assertions 266 to 267)

In fact, 18% of the assertions were related to this ad hoc theory, and we return to it shortly. We discuss the more general forms of justification in Section 4.

Representing the Reasoning Structure

The categorization of assertions shows that a great deal of the discussion in the session looks as if it has the flavor of Design Space Analysis. However, it gives us no sense of the structure of their reasoning. To impose some structure, we begin with an a priori analysis of the design space comparing the SATM and the FATM—the one presented in Section 2.2 (see Figure 5)—and we see how the discussion fits into that framework. This is presented in a QOC diagram in Figure 8.

Assertion numbers (in italics) are placed in the diagram in Figure 8 to show where the discussion talked about the various points represented. The position of the assertion numbers on the diagram gives some idea of how each assertion was categorized. Assertion numbers placed directly adjacent to Options or Criteria indicate explicit reference to them. Those placed in the areas to the left of Options indicate discussion of more general issues, and those placed on the Assessment links between Options and Criteria indicate discussions involving justification.

More than three quarters of all the assertions in the protocol are represented in Figure 8. At times, there is not a one-to-one mapping between assertions and points in the diagram, so some assertions appear twice. Of the assertions not represented here, most cover details that are not captured in this diagram (e.g., the possibility of having an interface that did not use a CRT screen).

Several points can be noted from this representation of the discussion:

1. Some discussion takes place around nearly every point we had in the earliest analysis.
2. New items (indicated in bold) that were not in the a priori analysis emerged in the discussion. The new items mostly involve more detailed exploration of the consequences of Options at the top level.

Figure 8. QOC diagram representing the structure of the discussion of the ATM1 design session. This diagram is based on the a priori analysis diagrammed in Figure 5. The bold items are new. The italic numbers represent the assertion numbers in the protocol.

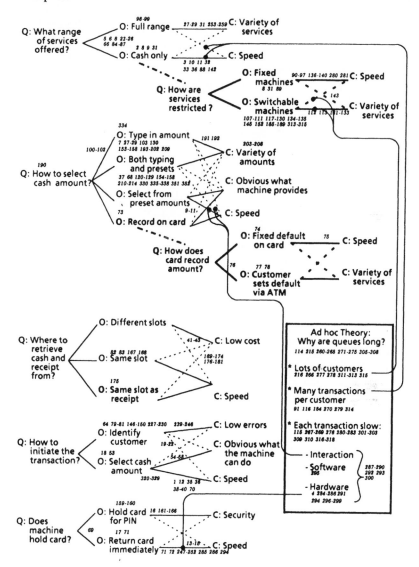

3. A considerable amount of discussion revolved around the ad hoc theory of why queues might be forming.

4. Figure 8 does not emphasize the order in which items were discussed, although it can be inferred from the assertion numbers. It is clear that the logical structure of the design space represented here does not match the chronological structure of the design discussion. Items are often revisited several times throughout the discussion (see Olson & Olson, 1991).

3.2. Some Phenomena in Design Reasoning

We have established the relationship between the content of the design discussion and the elements of Design Space Analysis. There is clearly not complete overlap, but there is sufficient correlation to suggest that it would not be unreasonable to expect designers to work with the QOC concepts. However, there are also a number of areas where the overlap is relatively weak. Some suggest that Design Space Analysis could perhaps improve the effectiveness of design practice. Others suggest ways in which a process to support the use of Design Space Analysis in design practice could be developed. We now discuss some of these.

Ad Hoc Theories

As the session goes on, the designers realize more and more that they do not know why the queues are building up at the SATMs, and they eventually agree that this is essential information to be able to tackle the design problem adequately:

> What's causing the long queue. Is it people just going through these steps, or is it people adding options to other services, and then using the other options? (Assertions 114 to 116)

The information is not given in the problem statement, and they eventually ask the experimenters for more information but are told that no more is available. They then spend more than 5 min in a very analytic phase building an ad hoc theory of why queues might be long, based on their own knowledge of the world. This is summarized in the box in the QOC diagram in Figure 8. Three classes of reason for queues being long are shown: lots of customers, many transactions per customer, and each transaction slow. (A fourth class of reason to do with user errors and lack of knowledge is not represented.) Each transaction slow further subdivides into three subsidiary reasons: hardware reasons (e.g., reading the card and counting money), software reasons, and reasons to do with the interaction between the customer and the machine. As they are developing it, the designers use the ad hoc theory to help them understand reasons behind the design of the original FATM and to revisit some of their own design.

> People doing lots of transactions. Transactions take a long time. And that's what they've worked on, that's what they've solved. . . . That was that FATM. (Assertions 279 to 281b)

> People not knowing what to do, is to make a simpler user-interface so that's that. Too many people is to provide more machines, with a potential for a "fast light" machine. (Assertions 309 to 313)

Figure 8 shows some of the ways in which the details of the theory are used to argue about the Assessments represented within the core QOC notation. Given the nature of this ad hoc theory, it is not surprising that all of the reasons given relate to Assessments of the speed Criterion. The status and use of this kind of reasoning are similar to the idiosyncratic views designers have of users reported by Hammond, Jørgensen, MacLean, Barnard, and Long (1983). It is based on their own experience and insights rather than on more objective information—of course it is all too common in design that all relevant information is simply not available, as was the case on this occasion. This kind of reasoning has the status of an argument, which backs up the QOC representation of the design space, as we discussed earlier. Its role is also very similar to the role of scientific theories that we discuss in Section 4.

Design Biases

There was a strong tendency for the designers to look for evidence to confirm their initial biases—a well-known phenomenon in the psychology of thinking (e.g., see Wason, 1968). This can have the effect of dismissing possible good Options, because no positive Criteria are considered, or of proceeding with poor Options, because no negative Criteria are considered. It is interesting to note that in keeping with Wason's observations, and in spite of the dangers, there is no evidence of the designers changing their minds about any of the possibilities they considered during the session. We saw in the previous section that, when they questioned why queues might be building up, they used their analysis to confirm that their novel design was good. Even when this line of argument helped them recognize a problem that the FATM had addressed ("too many transactions"), they did not reevaluate it as a solution. We do not intend to suggest that there were flaws in the argument they presented—we simply want to remark on the fact that they never directly asked what the down side of their proposed solution might be.

On the other side of the coin, there were aspects of the proposed design solution given to them (i.e., the FATM) that they seemed resistant to accept from the outset—for example, the initiation of the transaction by selecting the cash amount:

> Yes, and why select the cash amount first? The problem with that is what happens if someone selects the cash amount and then goes off somewhere else. You're going to have to have some sort of cancel button or something, or I suppose you could just press another amount. I would have thought they would, um, worry people. (Assertions 18 to 22)

They identify a possible problem with the interface and immediately think about how it might be circumvented. They come up with plausible solutions but, nevertheless, do not appear to reevaluate the problem in light of them. Later, when discussing the order of steps more generally, they suggest another problem:

> I think there's a disadvantage. In fact you could write that down. If you have a different order on different machines, this is going to, er, you know, people are going to have to learn two different ways of doing things; they're not going to like that. (Assertions 39 to 40a)

Later, it is clear that they have formed a firm opinion of the relative merits of the two machines when they refer to the SATM as the "good one":

> So having one good one, and one only fast cash doesn't really seem to be a good idea does it? (Assertion 137)

The main point of these examples is that the designers make no explicit attempt to explore reasons why the FATM might be a good design, in the same way as they did not attempt explicitly to question why the switchable ATM might be a poor design. It is important for an objective design discussion to consider both pros and cons. However, we must also recognize that it can be difficult to look for arguments, or indeed evaluate available information, to counter current beliefs.

Emergence of New Designs

The most innovative result of the ATM1 design session was the proposal for a "switchable" ATM, that is, an ATM that can be switched from full to restricted service. Let us trace the development of this novel design to see how it emerges out of the reasoning process. Considering the implicit Question what services offered?, the Option of a full range of services versus some sort of restriction is discussed at various times throughout the session, both in general and in moving toward a new switchable

ATM. The FATM's cash only Option (see Figure 6) is rejected fairly early on:

> So those are the restrictions; you've got to go for cash, and you've got to go for a "select amount", and that's supposed to save heaps of time. Don't believe it somehow. (Assertions 9 to 10)

It is gradually replaced by a notion of a restricted range of services. The first step is to incorporate the fast-cash notion into the SATM interface:

> Well, there are a lot of different ways of doing it. Well my sort of favorite would be just to have a sort of a fast cash button at the point where you said "select cash withdraw" . . . would be this idea where you had sort of a preliminary screen come up which would just have five different amounts of fast cash, or do something else. (Assertions 106 to 109)

The designers are then very satisfied at being able to give access to "all the other things" while not having more buttons to worry about:

> And then you'd have exactly the same number of buttons as on this one, or actions, and yet you'd have access to all the other things, and then you could do them on the same machine. (Assertions 110 to 113)

However, they realize that if they give access to all other possible services at the time, queues may build up with customers carrying out multiple transactions. They come up with a notion of having some fixed machines that offer only cash alongside the full range machines. Note the use of the analogy with a fast check-out in a supermarket for customers who only want a few items. (We discuss such roles of analogy more fully in MacLean, Bellotti, Young, & Moran, 1991.)

> What you could have is just like you've got in the supermarkets. If you've got say three of these machines all the way along, you could actually have a little light above it which said "Cash withdrawal only" and then you could sort of . . . we remove the other options, yes? (Assertions 117 to 120)

Eventually, they come up with the notion of a single switchable machine:

> New Design "A", One; same machine for ATM and FATM. (Assertion 145)

They then combine these ideas and flesh out a design that uses a light to signify which mode the machine is in:

> So we could use (waves his hands around in the air, and they both laugh) a "Fast light" (he writes " 'fast' light-operated . . ." presumably by the bank staff. Yes, and you could say, like five-items-only in supermarkets or whatever.

. . . there'd be a light above it and also other options wouldn't appear on the screen. (Assertions 185 to 189)

Toward the end of the session, they check their design out against some of the reasons for queues building up:

[If the problem is] Too many people [the solution] is to provide more machines, with a potential for a "fast light" machine. [If the problem is] Lots of transactions per person [the solution] is again this fast light thing. (Assertions 311 to 315)

Note how the new design gradually emerges throughout the entire session. There is a continual interplay between analysis and innovation, and the various phases are interspersed with much discussion of other aspects of the design. It is interesting to note the conciseness of the representation of the key points of this discussion in Figure 8. All of the substantive points, apart from the details of the fast-light design are summarized under the what range of services offered Question.

A second example comes from the ATM2 study, where the problem was left more open ended by describing only the SATM and the problem of queues building up and asking for a new design. The designers in the ATM2 session quickly got into exploring ways of restricting services. Figure 9 is a QOC representation of some of their exploration. They quickly move beyond the kind of restrictions suggested by the ATM1 designers and discuss restricting services either in terms of the number of services customers are allowed (e.g., a customer can only use one service during a single transaction) and/or by restricting the services available to a subset (e.g., cash only).

The ATM2 designers provide a third example of innovation toward the end of the session. One of them suddenly sees a connection with the way a local bagel store handles its lengthy queues while customers are waiting in line. It has an employee work along the queue, explaining the choices available and helping customers still in their orders on a form. The customers hand over their forms when they reach the service counter, enabling their requests to be processed more speedily. Familiarity with this arrangement prompts a similar suggestion for the queues outside the bank. (Note again the powerful role analogy plays in suggesting the solution—see also MacLean et al., 1991.) Although the notion turns out not to be sensible in detail for the ATM problem, it plays a crucial role in leading the designers to generate their most innovative proposal—the idea of having "active" bank cards that customers program directly while they are waiting in line. This Option is, in fact, very similar to the defaults on a card that the ATM1 designers considered briefly and then abandoned (cf. Figure 9 with Figure 8). An interesting feature of this Option is that

Figure 9. **Part of the QOC analysis from the ATM2 protocol (new items are indicated in bold).**

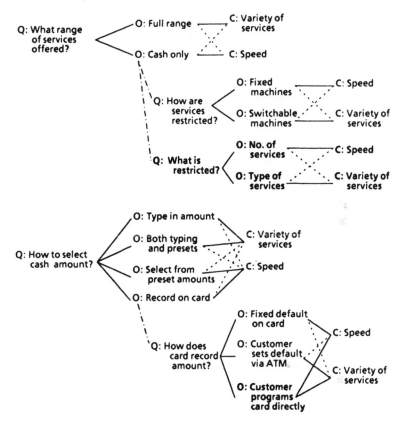

it resolves the tradeoff between the Criteria of speed and variety of services as it satisfies both of them. It is intriguing to note how when the output from the two design sessions is combined in this way, yet more powerful solutions emerge. The notion of defaults could have prevented the ATM2 customers from having to program the card each time they used an ATM, and the notion of a programmable card might have made the defaults more attractive to the ATM1 designers.

3.3. Summary of Findings

We can learn a lot by looking at design activity from a QOC perspective. When we relate the contents of the protocols in both studies to a single QOC representation, there is considerable overlap between what the designers talk about and the concepts we use in Design Space Analysis. The data suggest areas where the use of Design Space Analysis could be beneficial. We, not the designers, provided the logical structure around which the material was organized as there was little relationship between the chronological structure of the session and the logical structure of its content. If designers used QOC to structure their deliberations, we believe that they could improve their reasoning by working with a structure more logically related to the design problem. The success of itIBIS for structuring design discussions (see Conklin & Burgess-Yakemovic, 1991 [chapter 14 in this book]) suggests that this claim is not unreasonable. Moreover, the remarkably systematic reasoning structures which appear to underlie design meetings (Olson et al., 1995 [chapter 7 in this book]) suggest that the overheads could be less than might be expected, if only designers could be helped to perceive the structure that is already present. In the data of ATM1 and ATM2, searching Questions that might have helped with structuring the design space were barely represented. At best, the discussion seemed to be structured around individual Options rather than groups of related Options. Exploration of the design space was haphazard—some points were revisited many times during the session, whereas others that looked promising were dropped. We saw that pros and cons were not thoroughly explored, leading to the potential for biased judgments to be made.

If designers could provide a logically structured output, such as we used to summarize the session, it could make the reasoning behind the design clearer to other people and it could help the designers to reason about the design while they were creating it—for example, by keeping track of what they discussed and by helping them to see missing or inconsistent parts of the design representation. We point toward how Design Space Analysis might be used to encourage such reasoning in Section 5.

Much of the talk in the ATM1 session was shown to be devoted to assertions that justified possible design options. However, many of the justifications (including the ad hoc theory) do not fit comfortably into the QOC concept of Criteria. Section 4 expands on the concept of Criteria and considers how a range of different kinds of justification relate to Design Space Analysis.

4. JUSTIFICATION IN DESIGN SPACE ANALYSIS

We have seen a variety of instances and issues of justification—the argumentation used to evaluate design alternatives. In this section, we discuss a variety of topics under this important notion. Section 2 introduced Criteria as the basis for evaluating Options, beginning with the Assessment of individual Options against individual Criteria. In Section 3, we observed that designers use a variety of forms of justification during design, not all of them appealing to Criteria. We also saw in the examples of Section 2 that there are dependencies across design decisions and that these dependencies provide another basis for justifying decisions. In this section, we take a closer look at these various kinds of justification and explore their place in Design Space Analysis. We begin with a closer look at Criteria.

4.1. Understanding Criteria

The examples we have used so far might suggest that Criteria are fairly simple well-defined entities. In fact, it can take considerable effort to find suitable Criteria and phrase them in such a way that they achieve the desired impact. Several issues have to be taken into account. It is important to characterize a Criterion at a suitable level of abstraction relative to the Options being considered. The concept of a Bridging Criterion is used to talk about how the Criterion can be worded so as to make its relationship to the local design space clear while also being clear about the more general Criteria to which it relates. If the Criterion becomes too heavily entangled with details of the local Options, however, its status can become unclear, so it is important to understand some basic properties of Criteria that help make them useful. Finally, Criteria play a crucial role in understanding tradeoffs in the design space. If appropriate ones are to be found, key issues can be characterized as simply as possible.

Bridging Criteria

Just as Options can be justified by being Assessed against Criteria, one Criterion can be justified by reference to another. Criteria differ in their degree of generality, and the more specific Criteria can inherit their justification from the more general ones. This is portrayed in Figure 10 as an extended QOC diagram. It is more convenient to assess the Options against Criteria that are specific to the design being analyzed instead of having to spell out the justification all the way back to a set of very general Criteria. One example in the case of the scroll bar is the Criterion of ease of hitting with the mouse. Such Criteria are called *Bridging Criteria* because they bridge between specific aspects of the design (e.g., the use of a mouse) and broad, general Criteria such as speed and accuracy. Unlike General Criteria, Bridging Criteria are typically of narrow applicability and are invented for their relevance to a particular class of designs. The Criterion of ease of hitting with the mouse arises not from considerations of usability in the abstract but, rather, assumes that decisions have already been made to use a mouse and a scroll bar of a certain kind. In return for their lack of breadth, Bridging Criteria offer the advantages of being easy to work with and of encapsulating a possibly complex set of arguments and interdependencies into a single entity within the analysis.

Figure 10. Moving from Bridging Criteria to General Criteria. The Criteria in the dotted boxes are Bridging Criteria. The Criteria become more General toward the right.

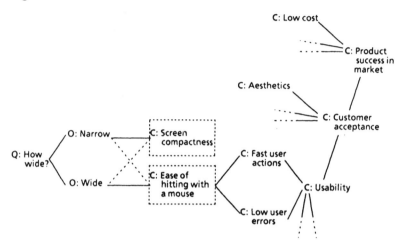

Bridging Criteria are justified by exhibiting their impact on more general Criteria. Thus we see, in Figure 10, ease of hitting with the mouse as being relevant to the width of the scroll bar by arguing that it contributes in turn to the more General Criteria of fast user actions and low user errors. As with several other aspects of Design Space Analysis, there is no fixed stopping point to this process; that is, there are not necessarily any fundamental Criteria for which further justification cannot be asked. Rather, what happens as we ask for justification repeatedly is that we get driven to broader and broader Criteria, which have less and less direct relevance to interface design. We see that fast user actions and low user errors contribute to usability (which is surely true, although not very helpful), which in turn contributes to customer acceptance, which in turn contributes to product success. Thus, as the Criteria become more general, they tend to become the province of the financial analyst and marketing strategist more than of the interface designer.

Some Properties of Criteria

It is hardly practical to try composing a formal definition of what is meant by a Criterion, but it is worth listing some properties that help identify appropriate Criteria. We illustrate example properties by exposing a stage we went through when we were first analyzing the scroll bar. For a while, we tried to work with a notion of wide for mousing, which was a sort of composite of what we call in Figure 3 the Option wide and the Criterion ease of hitting with a mouse. The notion was evidently responding to the intuitive idea that it is a good thing to make the scroll bar wide, so that it can be selected easily with a mouse. But attempting to use it as a "pure" Criterion led to a variety of problems that were resolved only when it was broken into the two separate components.

Observations that suggest relevant properties of Criteria include:

1. A Criterion measures a property of the artifact that the designer *controls only indirectly* by exercising choices over Options. (Options are under the direct control of the designer, because they represent possibilities that the designer can select.) The would-be Criterion of wide for mousing fails because one and the same notion appears to offer both direct control (i.e., over the width) and a measure of an aspect of evaluation (i.e., the "mousability"). The Criterion ease of hitting with a mouse is acceptable because there is already a commitment to use a mouse in the design space being considered. (However, if a purpose of the design space was to explore different pointing devices, a mouse would be an Option under the control of the designer and so ease of hitting with a mouse would not be an acceptable Criterion. In the spirit of the discussion in the previous section, a more General Criterion such as ease of hitting would be required.)

2. A Criterion must be *unconditional* in the sense that, other things being equal, the greater the extent to which the Criterion is met, the better the design. Wide for mousing fails this test. It suggests "the wider the better," but it is clear that there is a negative impact of a wide bar on the use of screen space. On the other hand, ease of hitting with a mouse is acceptable because it is clearly a desirable characteristic and does not prejudge a class of solution that is likely to have negative consequences. For example, a small target that attracts the cursor when it gets near it could be another plausible solution supported by the Criterion.

3. A Criterion must be *evaluative*; that is, it must be a measure of some property of the artifact, with a definite sense of higher Assessment values being better. Notice that the notion of wide for mousing is unclear in this regard: The width in itself is neutral, and only the mousability is evaluative.

4. As an extension of Observation 3, it is convenient to think of a Criterion as *potentially yielding a quantitative value*, even if only on an ordinal scale. One could even consider assigning actual numbers to Assessments, combining them as if they were Expected Utilities, and applying the apparatus of mathematical decision analysis. We do not judge that course worth pursuing in detail, but taking the idea of numerical values as a metaphor proves useful to help sharpen the concept of Criterion. The

notion of wide for mousing certainly fails this test. What would its putative value be measuring? Would it be a width, for example, expressed in millimeters, or some measure of mousability?

Tradeoffs Between Criteria

Design is always a matter of working through conflicting constraints, and one of the most important uses of Design Space Analysis is to understand the tradeoffs between different requirements. We are now very familiar with the characteristic visual pattern of parallel solid links and crossing dashed links in the QOC diagrams, which expresses the simplest tradeoff structure: two Options against two conflicting Criteria. Shum (1991a) pointed out the importance of a design notation clearly expressing such functional roles as this.

To make the tradeoff structure clear, it is important to choose the Options carefully. A characteristic example is the case where a Question is about a continuous-valued parameter, such as how wide should the scroll bar be? We noted in Section 2.1 that we chose the qualitative values wide and narrow (rather than, say, 2 mm, 4 mm, 8 mm, 16 mm) to make clear the tradeoffs. Once these are clear, other Questions can probe other aspects of the width (e.g., should the width be a multiple of the grid size?). Eventually a small set of specific candidate values can be evaluated.

Another interesting property of Options is that they are often regarded as characteristic representatives from the set of possible Options rather than as specific candidates. An example is found in the ATM1 study, where the designers were trying to decide how many values to offer for the preset amounts of money that can be withdrawn from the ATM:

> They probably are twenty and fifty—they are the most common ones. In fact that would probably do, almost. Just a big red button and a big blue button. (Assertions 198 to 200)

Our focus on understanding tradeoff structures implies that this statement is not to be taken literally as suggesting that the ATM should have one big red button and another big blue one but, rather, as representing an Option of "very few." It is treated by the designers almost as a caricature of a possible Option and serves effectively to make a clear distinction between the possible amounts. Similarly, the "twenty" and "fifty" cash amounts should not be read as a concrete proposal of those particular values but, rather, as a place holder for "two appropriate amounts." The concrete wordings of the Options help the designers retrieve relevant considerations but are clearly intended as representatives rather than as specific commitments.

4.2. Further Justification

An important feature of the Assessments of Options against Criteria is that they can be challenged and that justifications can be given for and against them. We saw these represented in Section 2.3 as arguments that support or challenge the Assessments. The relationships between specific and general Criteria that we saw in Section 4.1 are also subject to justification in the form of arguments. The justifications embodied in arguments appeal to evidence of various kinds: logical, theoretical, empirical, and so forth. In this section, we explore some of these different types of justification and how they relate to Design Space Analysis.

Theory and Data

One way to justify is to appeal to empirical data, accepted theory, or both. This is indeed the main route by which theory and data are incorporated into Design Space Analysis. A simple example, diagrammed in Figure 11, is provided by the Assessments in the scroll bar analysis between the width Options and the Criterion ease of hitting with a mouse. The Assessment is that a wider scroll bar is more easily hittable. (Note that this is a relative assertion that is represented by a pair of Assessments, one positive and one negative, from the Criterion to each of the width Options.) How do we justify this Assessment? We can appeal to some experimental work showing that a mouse used to hit an on-screen target can validly be treated as a device to which Fitts's Law applies (Card, English, & Burr, 1978), and we can argue that the scroll bar can be considered such a target. Further, we can appeal to Fitts's Law itself (see Fitts, 1954), which is an empirical law giving an indication of the difficulty of hitting a target relative to the amplitude of the movement and the size of the target. As it turns out, in this case, the evidence does more than merely support the claim. It also fleshes out the relationship by providing, should we want it, a means for quantifying and calculating the effect of a scroll bar's width on the time it takes to hit it.

A broader range of examples illustrating the incorporation of theoretical input into a QOC design space can be found in Bellotti (1993). In most cases, however, no existing theory or relevant data will be available. The designers will have to construct an approximate explanation by formulating an ad hoc theory or collecting some "quick and dirty" data in order to produce a convincing Design Space Analysis. We saw an example of this kind of theorizing in the ATM1 design session in Section 3, where the designers proposed an ad hoc theory that the long queues at ATM machines could be due to customers carrying out too many transactions at the machine; they proposed a new machine that could be set to allow a restricted range of transactions (see Figure 8), which they assessed

Figure 11. **Data and theory used as arguments to support Assessments against a Criterion involving the ease of hitting an on-screen target.**

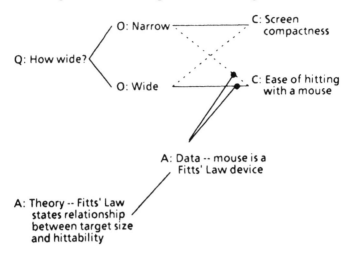

to be better on speed. They used the ad hoc theory as a form of argument to justify this Assessment (this is shown in Figure 8 by a link from the ad hoc theory argument to the Assessments against speed). It should be noted that the correctness of the restricted range Option depends crucially on the correctness of the ad hoc theory. If most customers are not making multiple transactions, then restricting the number of allowed transactions will have little effect on the speed.

Models, Analogies, and Metaphors

Another form of justification is conformance with a model, analogy, or metaphor based on something outside of the design itself. In some cases, the analogy may apply broadly across much of the interface, as with the notorious "desktop metaphor." More frequently, it applies to just certain aspects of the interface. Appeals to analogy as part of the design discussion have considerable effect on the process of design and the generation of ideas, as we saw in Section 3. For the present focus on justification, the important point is that appeals to such conformity are used by designers to justify particular Options. We discuss the roles analogy can play in design more fully in MacLean et al. (1991), so we only give two brief examples here. First, analogies can have the same role as theories and data in a QOC representation. For example, if we are asked to justify why the programmable ATM card will speed up use of the ATM, we can appeal to the bagel store analogy (described in Section 3) to show that a similar separation of placing requests and receiving services works in another context. For the purposes of using such a justification, it of course does not matter whether the original inspiration for the design came from that analogy or whether the analogy was purely a post hoc justification. Second, an analogy can be the source of more detailed justification. Consider developing a QOC analysis to support the source of the analogy (e.g., the bagel story). This could be transferred in suitably adapted form to the target of the analogy (the ATM). The technique deployed in the bagel store works because it allows all the decision making and interaction to have been completed by the time the customer reaches the counter. This fragment of explicit rationale could be transferred intact to the context of the ATM. We thus represent why the analogical situation works—not just claim that it works.

Scenarios

Typically, a scenario involves envisaging what it would be like to use the artifact being designed. An important property of scenarios is that they generate a context of use that emphasizes variables not apparent from a static description of the artifact. As with analogies, scenarios can justify a design in two ways. At the holistic level, a scenario can justify a particular solution by demonstrating that an envisaged mode of use will work. Goel and Pirolli (1989, p. 28) observed their subjects making extensive use of what they called "scenario immersion" for the evaluation of design possibilities. We noted several occurrences of scenarios being used in this way in the ATM1 and ATM2 studies. The ATM2 designers, for instance, undertook a fairly comprehensive usage scenario where they imagined the operations a user would have to execute to select a service, enter a PIN, select an amount of cash to withdraw, and so on. At a more detailed level, a scenario can evoke new Criteria that the design should meet. This is most likely to happen when a scenario shows up flaws in a proposed solution. For example, in visualizing the steps required to use a new ATM, the designers might reach the point where the customer wants to select another service, only to realize that the proposed interface neither displays the services available nor provides a means of reaching them. Such a scenario highlights two things that are expressible in QOC: It suggests additional Options to resolve the problem and a new Criterion of ensuring that relevant facilities are accessible. These uses of scenarios are consistent with the walkthrough methodologies proposed by Lewis, Polson, Rieman,

and Wharton (1990) and by Lewis et al. (1991 [chapter 5 in this book]).
The main difference is one of emphasis—we are focusing on the representation of the design space, and Lewis et al. are focusing on a scenario-
based design process.

4.3. Relations Across the Design Space

We turn now to the issue of the interrelationships between different
design Options. In Section 2, we saw that Options are structured into a
tree of Questions and Options, showing which Options are relevant to
each Question and what Questions are relevant to choosing certain Options. But we also saw in the examples in Section 2 that no coherent or
satisfactory design can result from considering each Option in isolation.
Instead, each Question has to be answered in the light of the answers
already given (perhaps tentatively) to other Questions. That is to say, the
evaluation of Options is a function not only of the local decisions but also
of the nonlocal interdependencies between Options. This section describes
how Design Space Analysis deals with these interdependencies.

Internal Consistency—Generic Questions

One kind of justification frequently offered for Options appeals to a
notion of "internal consistency." In Design Space Analysis terms, this means
that a Question in one part of the design space should be answered in
the same way as similar Questions elsewhere. Consistency can be imposed
on user interfaces by formal rules, such as the grammar rules of the
Command Language Grammar (Moran, 1981) or the Task Action Grammar (Payne & Green, 1986). A simple example of a consistency issue in
the scroll bar domain concerns the assignment of functions to mouse
buttons. For left and right movement, there is an obvious compatible
mapping onto the left and right buttons, but for vertical movement the
choice is less clear. Suppose it is decided (on some grounds, perhaps with
supporting arguments) to use the left button for scroll up and the right
to scroll down. If up and down movement is to be carried out with the
mouse in another part of the design, then clearly the same mapping should
be used. This is represented in Design Space Analysis by considering some
Questions as being generic, and the Options considered and the Criteria
used reflect the generic nature of the Question. The decisions made for
these Questions are then considered as generic decisions. A specific design
Question can be recognized as being an instance of a *Generic Question*, and
it would "inherit" the generic decision. (Of course, it is always possible to
have an exception to the rule, but this would involve finding good local
reasons for not taking the generic decision.) Therefore, having a class

structure of generic Questions is the way this kind of consistency is represented in Design Space Analysis.

Cross-Question Constraints

Another form of interdependency is where an Option chosen for one Question directly affects the choice of Options for another Question. It can easily happen that the relevance of a Criterion or the existence of an Option depends on a decision made elsewhere in the space. We saw this first in the scroll bar analysis in Section 2.1, where the decision of whether to choose an appearing scroll bar strongly depended on just how it was decided to make it appear. Another example is from the discussion of Bridging Criteria in Section 4.1, where we saw that the Criterion ease of hitting with the mouse applied to the Question how wide should the scroll bar be? assumes that it has already been decided to use a mouse to access a scroll bar. Bridging Criteria keep this dependence implicit—it being precisely the job of a Bridging Criterion to summarize the impact of a number of considerations in a single entity—but there are also many cases where the dependence is best made explicit.

A dramatic example of cross-Question constraint is seen in the basic decisions for a simple text editor. Two Questions concern what names should be given to the editor functions and how should the functions be invoked by the user (e.g., typing commands, invoking menus, clicking on icons, etc.). These two Questions appear to be independent. But what if the Option chosen for invocation is to type single-letter abbreviations? A constraint suddenly appears that the names chosen must have different initial letters! This kind of constraint acts in many ways like a Criterion, except that it *must* be satisfied for the design to be coherent. We call something that has this kind of impact on another part of the design space an *Export*. In the part of the design space where its impact is felt, we refer to it as an *Import*. This gives a convenient way of representing each part of the design space separately without having to keep track of explicit links between them. The main role of Exports and Imports is to help simplify the representation by allowing the analysis to be broken into modular pieces, with the Exports and Imports representing interdependent assumptions between modules.

Global Impact of Criteria

Criteria themselves serve as one way of representing relationships across the design space, because many Options in different places can be influenced by the same Criterion. In our own Design Space Analyses, we often find it helpful to draw up a single list of the Criteria appealed to. When elaborating a new part of the design space, such a list acts as a menu suggesting Criteria that may be relevant and encouraging the use of existing Criteria where appropriate (rather than creating new ones). In cases of a tightly determined design, we typically find that just a small number of Criteria has a pervasive influence on the design, being appealed to from many different places. In our analysis of the FATM rationale, for instance, because the purpose of the redesign was to deal with each customer more quickly, the Criterion speed is much in evidence, participating in tradeoffs against a variety of other Criteria in different places.

Criteria can also have a more global impact on the design space. The relative emphasis given to different general Criteria is a major determinant of the overall style and orientation of a design. In this respect, Criteria have a role similar to that of requirements in Newman's (1988) analysis of interface style. For example, if provide feedback is made more important than response speed, it can have a big effect on the kind of user interface that results. Similarly, one can see how giving a lot of weight to Criteria concerned with usability and ease of learning can lead to a general-purpose, easy-to-use interface such as that of the Apple Macintosh, whereas giving greater prominence to Criteria having to do with the close match of the design to the requirements of a particular task, and an emphasis on the efficiency of usage by trained users, can lead to specialized interfaces such as those found in airline booking systems.

5. CONCLUSIONS

For the most part, this chapter has concentrated on a QOC representation as a product (or at least a coproduct) of design. We illustrated the basic QOC elements in Section 2. We showed how they relate to the kinds of discussion that take place in an unstructured design meeting in Section 3, and in Section 4 we discussed several design-related issues of varying degrees of formality and showed how they relate to Design Space Analysis. We believe that it is critical at this stage of design rationale research not to confuse product and process. It is essential that we be clear about the kind of product we are trying to produce before we suggest processes for creating it. We hope that this chapter has served to articulate the nature of a QOC representation clearly and has shown how it relates to a variety of the more general design-related concepts with which we are familiar.

It is worth noting that a major contribution of this chapter has been in using Design Space Analysis as a technique for understanding designs through analysis (in Section 2) and for characterizing some aspects of design activity (in Section 3). The contributions at this stage are research techniques as much as they are techniques for use in design. Clearly, however, our goal is to develop techniques that can be used in the design process. It is no doubt obvious that an impediment to other people using our approach is that we have placed little emphasis on *how* to go about carrying out a Design Space Analysis.

Our primary focus in this chapter has been on the properties of the notation and the resulting design representations. However, we are beginning to consider the *processes* involved in creating a Design Space Analysis. There is no strict methodology for creating a design space, such as a top-down sequence. To try to follow a strict procedure does not work. An analysis is developed in all places at once by a mixture of inspiration and reflection, as ideas pop up, get understood, and fit into place. On the other hand, the process of developing a rationale is not random. There are systematic steps that can aid the process enormously. One approach (derived from our own analysis work and our observations of designers at work) involves focusing on parts of the QOC representation in principled ways to see how to augment it. We can formulate such principles as *heuristics* for would-be analysts to guide them in building QOC representations. The Appendix gives a brief overview of this approach to creating Design Space Analyses and gives examples of such heuristics. MacLean, Bellotti, and Shum (1993) describe their use in more detail within a model of the design process structured around gathering, organizing, and reasoning with design information.

We hope that the ideas we have put forward are useful in their present state insofar as they express a representation for design that can be picked up fairly easily by others. As we have seen, the semiformality of the QOC representation means that Design Space Analysis can be adopted, even without computer-based tools, simply by using pencil and paper. We hope that both researchers and designers will find it useful as a way of thinking about design, doing design, reasoning about design, and recording and communicating the arguments behind design. What is needed now is an examination of the practical implications of Design Space Analysis by groups other than ourselves, and indeed we know of several investigators who are exploring aspects of Design Space Analysis in teaching, research, and design. We look forward to seeing how the ideas presented here, inevitably under-specified, are interpreted by others and applied in practice.[6] This kind of experience will produce the input we need to move toward process-oriented techniques for using Design Space Analysis in design practice.

---

6. Some examples of the interpretation and application of QOC appear elsewhere in this book. Casaday (1995 [chapter 12 in this book]) presents a template which encapsulates and makes concrete aspects of the QOC representation, and Carey et al. (1995 [chapter 13 in this book]) use it to structure a toolkit to support inexperienced designers.

NOTES

Background. This chapter first appeared in the *Human–Computer Interaction* journal, Vol. 6, 1991.

Acknowledgments. Thanks to Judy Olson and Gary Olson for access to the data from the ATM2 study carried out at the University of Michigan with the assistance of Robin Lampert and Mark Carter. Thanks also to the designers who took part in the studies. We are grateful to the following people for many helpful comments on earlier drafts of this chapter: Liam Bannon, Tom Carey, Jack Carroll, Steve Draper, Jonathan Grudin, Mik Lamming, Jintae Lee, Judy Olson, Simon Shum, and two anonymous reviewers. We are also grateful to our colleagues in the AMODEUS project for many helpful discussions.

Support. The work reported here was partly funded by the European Commission as part of the AMODEUS project, Esprit Basic Research Action 3066.

Authors' Present Addresses. Allan MacLean, Rank Xerox Research Centre, Cambridge Laboratory, 61 Regent Street, Cambridge CB2 1AB, England. Email: maclean.cambridge@rxrc.xerox.com; Richard M. Young, MRC Applied Psychology Unit, 15 Chaucer Road, Cambridge CB2 2EF, England. Email: rmy@mrc-apu.cam.ac.uk; Victoria M. E. Bellotti, Apple Computer, Advanced Technology Group, 1 Infinite Loop, Cupertino, CA 95014. Email: victoria@atg.apple.com; Thomas P. Moran, Xerox Palo Alto Research Center, 3333 Coyote Hill Road, Palo Alto, CA 94304. Email: moran@parc.xerox.com.

REFERENCES

Adams, J. L. (1974). *Conceptual blockbusting.* San Francisco: W. H. Freeman.

Alexander, C. (1964). *Notes on the synthesis of form.* Cambridge, MA: Harvard University Press.

Allison, G. T. (1971). *Essence of decision: Explaining the Cuban missile crisis.* Boston: Little, Brown.

Apple user interface guidelines. (1987). Reading, MA: Addison-Wesley.

Balzer, R., Cheatham, T. E., & Green, C. (1983). Software technology in the 1990's: Using a new paradigm. *IEEE Computer, 16*(11), 39–45.

Bellotti, V. M. E. (1993). Integrating theoreticians' and practitioners' perspectives with design rationale. *Proceedings of InterCHI'93,* 101–106. New York: ACM.

Botterill, J. H. (1982). The design rationale of the System/38 user interface. *IBM Systems Journal, 21*(4), 384–423.

Buckingham Shum, S. (1995). Analyzing the usability of a design rationale notation and its interaction with modes of designing. In T. P. Moran & J. M. Carroll (Eds.), *Design rationale: Concepts, techniques, and use.* Hillsdale, NJ: Lawrence Erlbaum Associates. [Chapter 6 in this book.]

Card, S. K., English, W. K., & Burr, B. J. (1978). Evaluation of mouse, rate-controlled isometric joystick, step keys and text keys for text selection on a CRT. *Ergonomics, 21,* 601–613.

Carey, T., McKerlie, D., & Wilson, J. (1995). HCI design rationale as a learning resource. In T. P. Moran & J. M. Carroll (Eds.), *Design rationale: Concepts, techniques, and use.* Hillsdale, NJ: Lawrence Erlbaum Associates. [Chapter 13 in this book.]

Carroll, J. M., & Rosson, M. B. (1990). Human–computer interaction scenarios as a design representation. *Proceedings of HICSS-23: 23rd Hawaii International Conference on System Science,* 555–561. Los Alamitos, CA: IEEE Computer Society Press.

Carroll, J. M., & Rosson, M. B. (1991). Deliberated evolution: Stalking the View Matcher in design space. *Human–Computer Interaction, 6,* 281–318. Also in T. P. Moran & J. M. Carroll (Eds.), *Design rationale: Concepts, techniques, and use.* Hillsdale, NJ: Lawrence Erlbaum Associates, 1996. [Chapter 4 in this book.]

Casaday, G. (1995). Rationale in practice: Templates for capturing and applying design experience. In T. P. Moran & J. M. Carroll (Eds.), *Design rationale: Concepts, techniques, and use.* Hillsdale, NJ: Lawrence Erlbaum Associates. [Chapter 12 in this book.]

Conklin, J. (1987). Hypertext: An introduction and survey. *IEEE Computer, 20*(9), 17–41.

Conklin, J. (1989). Design rationale and maintainability. *Proceedings of the 22nd International Conference on System Sciences,* 533–539. Los Alamitos, CA: IEEE Computer Society Press.

Conklin, J., & Begeman, M. L. (1989). gIBIS: A tool for all reasons. *Journal of the American Society for Information Science, 40,* 200–213.

Conklin, E. J., & Burgess-Yakemovic, KC. (1991). A process-oriented approach to design rationale. *Human–Computer Interaction, 6,* 357–391. Also in T. P. Moran & J. M. Carroll (Eds.), *Design rationale: Concepts, techniques, and use.* Hillsdale, NJ: Lawrence Erlbaum Associates, 1996. [Chapter 14 in this book.]

Fischer, G., Lemke, A. C., McCall, R., & Morch, A. I. (1991). Making argumentation serve design. *Human–Computer Interaction, 6,* 393–419. Also in T. P. Moran & J. M. Carroll (Eds.), *Design rationale: Concepts, techniques, and use.* Hillsdale, NJ: Lawrence Erlbaum Associates, 1996. [Chapter 9 in this book.]

Fitts, P. M. (1954). The information capacity of the human motor system in controlling amplitude of movement. *Journal of Experimental Psychology, 47,* 381–391.

Goel, V., & Pirolli, P. (1989, Spring). Design within information processing theory: The design problem space. *AI Magazine,* pp. 18–36.

Gruber, T. R., & Russell, D. M. (1995). Generative design rationale: Beyond the record and replay paradigm. In T. P. Moran & J. M. Carroll (Eds.), *Design rationale: Concepts, techniques, and use.* Hillsdale, NJ: Lawrence Erlbaum Associates. [Chapter 11 in this book.]

Grudin, J. (1988). Why CSCW applications fail: Problems in the design and evaluation of organizational interfaces. *Proceedings of the CSCW '88 Conference on Computer-Supported Cooperative Work,* 85–93. New York: ACM.

Grudin, J. (1995). Evaluating opportunities for design capture. In T. P. Moran & J. M. Carroll (Eds.), *Design rationale: Concepts, techniques, and use.* Hillsdale, NJ: Lawrence Erlbaum Associates. [Chapter 16 in this book.]

Guindon, R., Krasner, H., & Curtis, B. (1987). Breakdowns and processes during the early acquisition of software design by professionals. In G. M. Olson, S. Sheppard, & E. Soloway (Eds.), *Proceedings of the Second Workshop on Empirical Studies of Programmers* (pp. 65–82). Norwood, NJ: Ablex.

Halasz, F., Moran, T., & Trigg, R. (1987). NoteCards in a nutshell. *Proceedings of CHI + GI '87: Human Factors in Computing Systems,* 45–52. New York: ACM.

Hammond, N., Jørgensen, A. H., MacLean, A., Barnard, P., & Long, J. (1983). Design practice and interface usability: Evidence from interviews with designers. *Proceedings of the CHI '83 Conference on Human Factors in Computing Systems*, 40–44. New York: ACM.

Hayes, J. R. (1981). *The complete problem solver.* Philadelphia: Franklin Institute Press.

Johnson, J., & Beach, R. J. (1988). Styles in document editing systems. *IEEE Computer, 21*(1), 32–43.

Kunz, W., & Rittel, H. (1970). *Issues as elements of information systems* (Tech. Rep. No. S–78–2). Stuttgart: University of Stuttgart.

Lee, J. (1990). SIBYL: A qualitative decision management system. In P. H. Winston & S. Shellard (Eds.), *Artificial intelligence at MIT: Expanding frontiers* (Vol. 1, pp. 104–133). Cambridge, MA: MIT Press.

Lee, J., & Lai, K.-Y. (1991). What's in design rationale? *Human–Computer Interaction, 6*, 251–280. Also in T. P. Moran & J. M. Carroll (Eds.), *Design rationale: Concepts, techniques, and use.* Hillsdale, NJ: Lawrence Erlbaum Associates, 1996. [Chapter 2 in this book.]

Lewis, C., Polson, P., Rieman, J., & Wharton, C. (1990). Testing a walkthrough methodology for theory-based design of walk-up-and-use interfaces. *Proceedings of the CHI '90 Conference on Human Factors in Computer Systems*, 235–242. New York: ACM.

Lewis, C., Rieman, J., & Bell, B. (1991). Problem-centered design for expressiveness and facility in a graphical programming system. *Human–Computer Interaction, 6*, 319–355. Also in T. P. Moran & J. M. Carroll (Eds.), *Design rationale: Concepts, techniques, and use.* Hillsdale, NJ: Lawrence Erlbaum Associates, 1996. [Chapter 5 in this book.]

Mackay, W. (1991). Triggers and barriers to customization. *Proceedings of the CHI '91 Conference on Human Factors in Computing Systems*, 153–160. New York: ACM.

MacLean, A., Bellotti, V., & Young, R. M. (1990). What rationale is there in design? *Proceedings of the INTERACT '90 Conference on Human–Computer Interaction*, 207–212. Amsterdam: North-Holland.

MacLean, A., Bellotti, V., & Shum, S. (1993). Developing the design space with Design Space Analysis. In P. Byerley, P. Barnard, & J. May (Eds.), *Computers, communication and usability: Design issues, research, and methods for integrated services* (pp. 197–220). Amsterdam: Elsevier.

MacLean, A., Bellotti, V., Young, R., & Moran, T. (1991). Reaching through analogy: A design rationale perspective on roles of analogy. *Proceedings of the CHI '91 Conference on Human Factors in Computing Systems*, 167–172. New York: ACM.

MacLean, A., Carter, K., Lövstrand, L., & Moran, T. P. (1990). User-tailorable systems: Pressing the issues with Buttons. *Proceedings of the CHI '90 Conference on Human Factors in Computing Systems*, 175–182. New York: ACM.

MacLean, A., Young, R. M., & Moran, T. P. (1989). Design rationale: The argument behind the artifact. *Proceedings of the CHI '89 Conference on Human Factors in Computing Systems*, 247–252. New York: ACM.

Marshall, C. C., & Irish, P. M. (1989). Guided tours and on-line presentations: How authors can make existing hypertext intelligible for readers. *Proceedings of Hypertext '89*, 15–26. New York: ACM.

Martin, J. (1977). What to plan for to manage the future of your data center. *Canadian Datasystems, 9*(3), 28–32.

McCall, R. (1986). Issue-serve systems: A descriptive theory of design. *Design Methods and Theories, 20*(3), 443–458.

McKerlie, D., & MacLean, A. (1993). QOC in action: Using design rationale to support design. *SIGGRAPH Video Review*, Issue 88, 1993. Abstract in *Proceedings of InterCHI'93*, 519. New York: ACM.

McKerlie, D., & MacLean, A. (1994). Reasoning with design rationale: Practical experience with Design Space Analysis. *Design Studies, 15*, 214–226.

Moran, T. P. (1981). The command language grammar: A representation for the user interface of interactive computer systems. *International Journal of Man-Machine Studies, 15*, 3–50.

Newman, W. M. (1988). The representation of user interface style. In D. M. Jones & R. Winder (Eds.), *People and computers IV: Designing for usability* (pp. 123–144). Cambridge, England: Cambridge University Press.

Olson, G., & Olson, J. (1991). User centered design of collaboration technology. *Journal of Organizational Computing, 1*, 61–83.

Olson, G. M., Olson, J. S., Storrosten, M., Carter, M., Herbsleb, J., & Rueter, H. (1995). The structure of activity during design meetings. In T. P. Moran & J. M. Carroll (Eds.), *Design rationale: Concepts, techniques, and use*. Hillsdale, NJ: Lawrence Erlbaum Associates. [Chapter 7 in this book.]

Parnas, D. L., & Clements, P. C. (1986). A rational design process: How and why to fake it. *IEEE Transactions on Software Engineering, 12*(2), 251–257.

Payne, S. J., & Green, T. R. G. (1986). Task action grammar: A model of the mental representation of task languages. *Human–Computer Interaction, 2*, 93–133.

Polya, G. (1957). *How to solve it* (2nd ed.). Garden City, NY: Doubleday.

Schön, D. A. (1983). *The reflective practitioner: How professionals think in action*. New York: Basic Books.

Schön, D. A. (1987). *Educating the reflective practitioner*. San Francisco: Jossey-Bass.

Shum, S. (1991a). Cognitive dimensions of design rationale. In D. Diaper & N. V. Hammond (Eds.), *People and computers VI* (pp. 331–344). Cambridge, England: Cambridge University Press.

Shum, S. (1991b). *A cognitive analysis of design rationale representation*. PhD thesis, Department of Psychology, University of York.

Simon, H. (1981). *The sciences of the artificial* (2nd ed.). Cambridge, MA: MIT Press.

Smith, D. C., Irby, C., Kimball, R., & Verplank, W. (1982, April). Designing the Star user interface. *Byte*, pp. 242–282.

VanLehn, K. A. (1985). *Theory reform caused by an argumentation tool* (Tech. Rep. No. ISL–11). Palo Alto, CA: Xerox Palo Alto Research Center.

Wason, P. C. (1968). Reasoning about a rule. *Quarterly Journal of Experimental Psychology, 20*, 273–281.

APPENDIX. CREATING A DESIGN SPACE ANALYSIS

The chapter discussed several aspects of Design Space Analysis. We introduced the basic elements of the analysis and presented the QOC notation for representing it, and we presented several ways of extending the analysis to cover a variety of modes of justification. We saw that this kind of analysis is not unlike the naturally occurring discussion that takes place among designers, but we also observed that naturally occurring discussion seems to fall short of a logically coherent rationale when compared to a Design Space Analysis. Although there are times during design when trying to conform to a logical structure would be inhibiting, there are also many times when it would be helpful to improve the quality of design reasoning. We believe that the explicit use of Design Space Analysis can help give structure and discipline to design reasoning. There are several kinds of tasks in design where such discipline could be helpful: preparing a presentation or review of a design project, providing a map to keep track of the territories explored in a project, helping generate and evaluate new ideas during design meetings, and so on.

There is no strict methodology for creating a design space, such as a top-down sequence. To try to follow a strict procedure would not work. An analysis is developed in all places at once by a mixture of inspiration and reflection, as ideas pop up, get understood, and fit into place. On the other hand, the process of developing a rationale is not random. There are systematic steps that can aid the process enormously. These steps (derived from our own analysis work and our observations of designers at work) involve focusing on parts of the QOC representation in principled ways to see how to augment it. We formulate the steps as a set of heuristics for would-be analysts to guide them in building QOC representations.

We call this advice *heuristic* because creating the design space, as well as creating the artifact, is a discovery process; in fact, it is one and the same discovery process. The heuristics reflect general considerations about problem solving (e.g., see Hayes, 1981; Polya, 1957) and creative processes (e.g., see Adams, 1974) applied to the specific task of creating a design space in QOC notation. We start by presenting some "local" heuristics, which are aimed at helping us locally expand the notation. We then present some more "global" heuristics, which are aimed at dealing with larger patterns in the notation.

A1. Local Heuristics for Design Space Analysis

The purpose of local heuristics is to give advice for how to reason in the area of a single Question to enhance understanding of the design or to try to find a better solution. They provide for expansion of the QOC notation in a link-by-link and node-by-node manner.

The generation of possible design Options can be aided by having a clear understanding of important issues. The right Question highlights relevant issues and encourages the generation of appropriate Options, so:

Heuristic 1: Use Questions to generate Options.

However, it is usually difficult to formulate directly incisive Questions. We have noted that possible Options seem to spring to mind, apparently in isolation. Asking oneself "to what Question is the Option an answer" and reflecting on its important or novel features can lead to good Questions:

Heuristic 2: Use Options to generate Questions.

Going back and forth between Questions and Options (using Heuristics 1 and 2) is consistent with Schön's (1987) discussion of generating possible solutions, reflecting on their characteristics, and generating better solutions.

Insights into the structure of the design space can be gained if we can identify appropriate Options, by which we mean Options that bring out distinctive features in the set of possibilities:

Heuristic 3: Consider distinctive Options.

In some situations, considering extreme solutions is a way to find distinctive Options. Such Options can be a useful way to "shake up" our view of the design space, both to aid understanding it (e.g., by making clear the tradeoffs between the Criteria) and to help generate new solutions within it.

The heuristics so far have emphasized the exploration of alternative Options. The exploration of a range of Criteria is also necessary to provide a balanced view of the pros and cons of proposed Options:

Heuristic 4: Represent both positive and negative Criteria.

As a minimum for exploring the design space, we recommend that each Option have at least one Criterion against which it is assessed positively and one against which it is assessed negatively. This allows us to understand the tradeoffs. It is important to recognize such tradeoffs when they exist, but there is clearly more to design than simply evaluating tradeoffs, so we could express this heuristic more generally as, "Look for the downside as well as the upside."

Tradeoffs do not have to lead to compromise. Design should be creative—avoid compromises by trying to find Options that bypass them. Sometimes the alternative Options can be combined into a single Option with the advantages of each. The heuristic for doing this is:

Heuristic 5: Overcome negative, but maintain positive, Criteria.

A2. Global Heuristics for Design Space Analysis

The purpose of the global heuristics is to help us look beyond the local region of the representation. These heuristics are aimed at dealing with broader design issues: modularizing the design, looking for emergent design possibilities, and addressing the coherence of the design as a whole.

There are usually too many degrees of freedom in moving toward a design solution, and the decisions are highly interrelated. Extensive problem structuring is required to determine a fruitful way of framing the problem (e.g., Schön, 1983). The most common strategy for tackling such problems is to subdivide the problem into smaller more manageable components or modules (e.g., Alexander, 1964; Simon, 1981). In software projects, this decomposition is most obvious when different teams are given different parts of the project to work on, but the same basic strategy applies even when individuals tackle design problems. In practice, complete isolation between modules is impossible to achieve, so we try to define modules that maximize the interaction within them and minimize the interaction between them. In Design Space Analysis, cross-Question constraints (imposed by dependencies between Options) are perhaps the most critical feature in defining modules:

Heuristic 6: Identify Options that generate dependencies.

One of the main advantages of a Design Space Analysis is that several possible designs are captured within the same representation. This can provide insights into combinations of Options that have not been previously considered together but that might lead to an improved design:

Heuristic 7: Look for novel combinations of Options.

Section 4 argued that Criteria play an important role in shaping and maintaining the overall coherence of a design. It is therefore important to:

Heuristic 8: Design to a set of Criteria.

A strategy we have found useful for creating a Design Space Analysis is to identify a list of General Criteria that are most important for the design and from these to formulate a set of more specific Bridging Criteria to use in evaluating local Options. We have observed that systematic application of Criteria is rare in design practice, and attention to this heuristic would help designers clarify their motives and objectives.

A final issue is to note that design is as much about inventing themes as making specific design decisions. Locally optimized decisions do not add up to good overall design. Many core decisions must be made to establish a consistent overall policy of the design. In Design Space Analysis terms, it is vital to:

Heuristic 9: Search for generic Questions.

Generic Questions serve to select aspects of the design to frame their analysis in terms of the overall impact on the design rather than on local considerations. The responses to such Questions will determine the overall coherence and consistency of the design. For example, several commercial desktop systems are built on a set of style rules or specifications that define their "look and feel," the most prominent being the original design of the Xerox Star (Smith, Irby, Kimball, & Verplank, 1982) and the Apple Macintosh (Apple User Interface Guidelines, 1987).

4

Deliberated Evolution: Stalking the View Matcher in Design Space

John M. Carroll
Mary Beth Rosson
Virginia Polytechnic Institute and State University

ABSTRACT

Technology development in human–computer interaction (HCI) can be interpreted as a coevolution of tasks and artifacts. The tasks people actually engage in (successfully or problematically) and those they wish to engage in (or perhaps merely to imagine) define requirements for future technology and, specifically, for new HCI artifacts. These artifacts, in turn, open up new possibilities for human tasks, new ways to do familiar things, and entirely new kinds of things to do. In this chapter, we describe psychological design rationale as an approach to augmenting HCI technology development and to clarifying the sense in which HCI artifacts embody psychological theory. A psychological design rationale is an enumeration of the psychological claims embodied by an artifact for the situations in which it is used. As an example, we present our design work with the View Matcher, a Smalltalk programming environment for coordinating multiple views of an example application. In particular, we show

John Carroll is a cognitive psychologist interested in the analysis of human learning and problem solving in human–computer interaction contexts and in the design of methods and tools for instruction and design; he is Professor of Computer Science and Psychology and Head of the Computer Science Department at Virginia Polytechnic Institute and State University. **Mary Beth Rosson** is a cognitive psychologist interested in the mental activities associated with complex design tasks and in the analysis and development of methods and tools to support such tasks; she is Associate Professor of Computer Science at Virginia Polytechnic Institute and State University.

CONTENTS

how psychological design rationale was used to develop a view matcher for code reuse from prior design rationales for related programming tasks and environments.

1. TASKS AND ARTIFACTS

In 1605, Sir Francis Bacon (1605/1970) called for a "natural history of trades" (p. 10). He urged that technical tools, techniques, and processes be made more public and explicit. This was one element in his broader project of developing practical science, which hinged on the assumption that if such knowledge could be more systematically considered and integrated, human progress would necessarily result. Thus, Bacon suggested that new concepts and inventions would result "by a connexion and transferring of the observations of one Arte, to the use of another, when the experiences of several misteries shall fall under the consideration of one man's minde" (p. 10).

Half a century later, when the Royal Society of London and the French Académie des Sciences were established, Bacon's proposal was high on their agendas. However, the Royal Society quietly but promptly abandoned the project, and the Académie des Sciences pursued it rather haltingly over the next century, culminating in the publication of Diderot's *Encyclopédie ou Dictionaire Raissoné des Science, des Artes, et des Métiers*. Although Diderot's objectives were explicitly those Bacon had originally articulated, the final product of the *Encyclopédie* substantially lagged behind the leading edge of technology, and its directive impacts on technological evolution were relatively minor (Ferguson, 1977).

In this chapter, we are disciples of Bacon in an approach to human–computer interaction (HCI) that we call "deliberated evolution": We want to integrate understanding systems and building systems within a single framework of research/development. Our focus is on the psychological import of HCI artifacts (workstation hardware, operating systems, application programs, user interface displays and devices, and so forth). We are exploring the directive role that psychological design rationales of HCI artifacts can play in design work. In Bacon's terms, our interest is in the connection and transferring of one design to a use in another design by means of explicit design rationale and systematic design methods.

There are many reasons why Bacon's original proposal failed to guide effectively 17th-century technology evolution (Ferguson, 1977). One factor is that it just took too long to develop the natural history of trades. Our case study stresses the utility of concurrently designing new technology and analyzing its scientific rationale. We discuss the View Matcher, a software tool originally designed to address learning problems inherent in the Smalltalk language environment, which was subsequently redesigned as a tool to facilitate code reuse by skilled Smalltalk programmers. The several cycles of View Matcher design exemplify the deliberated evolution of technology that we see as Baconian.

Most technical activity in HCI can be framed as transaction between tasks and artifacts. The tasks people actually engage in (successfully or problematically) and those they wish to engage in (or perhaps merely to imagine) define requirements for future technology and, specifically, for new HCI artifacts. These artifacts, in turn, open up new possibilities for human tasks, new ways to do familiar things, and entirely new kinds of things to do. They also create new complexities of learning and performance; new interactions among tasks; and, of course, new errors and other difficulties for users. The new task eventually devolve into requirements for further technology evolution, provoking further transaction (see Figure 1).[1]

Examples of this are pervasive in HCI, but particularly good ones inhere in the particularly momentous technological developments in the field. Consider the spreadsheet. The first electronic spreadsheet, VisiCalc, was brought to market in 1979. It clearly embodied a simple yet powerful response to a set of extant tasks: table-based calculations. Placing the spreadsheet in an electronic medium permitted accurate calculation and convenient handling, but it did much more. It opened up important new possibilities for table-based calculation tasks. Electronic spreadsheets facilitated projective analyses (e.g., what-if reasoning about altered conditions and abductive reasoning from a desired result to conditions that could produce it). Users could easily alter values and recalculate. Indeed, spread-

---

1. Though most technology development is evolutionary (Basalla, 1988), some clearly is not. Following a suggestion by Randy Smith, we identify our concern here as "normal" technology development, by analogy with Kuhn's (1970) notion of "normal science."

Figure 1. The transaction between tasks and artifacts.

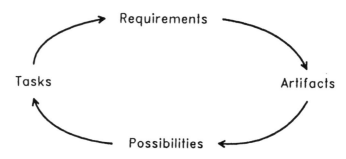

sheets even afforded a kind of ad hoc work integration: Users could type a memo describing an analysis right into a spreadsheet cell (Mack & Nielsen, 1987).

This evolution of spreadsheet tasks can be viewed as successfully altering requirements for spreadsheet systems. Thus, in the early 1980s, Context MBA provided integrated support for windows, graphing, word processing, and file management, for example, displaying the spreadsheet in one window and a graph of selected cells or a report in another. Lotus 1-2-3 introduced natural order recalculation (in which cell dependencies determine the order of recalculation), easing the overhead of what-if explorations. These advances, in turn, can be seen as encouraging further task evolution. For many users, the spreadsheet environment became a fulcrum for work: planning, communicating, accessing information, reporting, and presenting. It is now typical for spreadsheet systems to support multiple windows, to integrate support for text and graphics, and to share data with other programs. Many spreadsheets offer a range of recalculation options to facilitate projective analysis, and some offer a "solver" function that takes a specification of a desired result and suggests how to obtain it.

If we take the transaction between tasks and artifacts seriously as the evident framework of technology evolution in HCI, we can ask how this structure might be more deliberately managed and directed. In the development of the spreadsheet, successive task–artifact excursions were not, as far as we know, guided by a public and explicit "natural history" of the spreadsheet, to borrow Bacon's term. Thus, even the summary understanding encapsulated earlier was developed after the main points were already embodied in new spreadsheet designs (Licklider, 1989; Mack & Nielsen, 1987). Nevertheless, the prior tacit—or at least private—understanding of spreadsheet tasks, and attendant technological limitations and possibilities, worked pretty well in guiding the evolution of spreadsheet applications.

Figure 2. A task–artifact framework for design in HCI.

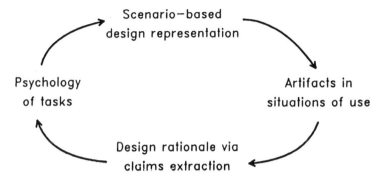

Could a better Baconian tool produce even better results? Our approach to this has been to augment directly the transaction manifest in current practice with explicitly managed tools and representational techniques. We want to enhance the natural ecology of HCI technology evolution but not too radically so as to distort or undermine what is in essence a fairly successful framework for design work and practical science (and to which designers are now committed through their practice). We are seeking streamlined techniques to supplement this framework with additional deliberate analysis and record keeping.

Figure 2 schematizes our augmented task–artifact framework for HCI research/development. Briefly, we propose guiding the discovery and integration of design requirements by means of a scenario-based methodology (e.g., Carroll & Rosson, 1990, 1992a; see also Carroll, Thomas, & Malhotra, 1979; Guindon, 1990; Wexelblat, 1987). We propose capturing and projecting possibilities of use in a psychological design rationale, developed by claims analysis (e.g., Bellamy & Carroll, 1992; Carroll & Kellogg, 1989). We further suggest that, by collating and abstracting such design rationales across collections of user interaction scenarios and artifacts, a pertinent psychology of user tasks can be constructed (Carroll, 1990a; Carroll, Kellogg, & Rosson, 1991).

In essence, we are attempting to enhance design practice by prescribing an incrementally more systematic version of the very practice we now can observe. For example, user interaction scenarios are already widely used for envisioning systems before they are built (they supplement and sometimes supplant textual specifications in many development groups). Scenarios are required for the design of task-oriented training materials and documentation (Carroll, 1990b) and for the design of user-testing instruments (Roberts & Moran, 1983). Similarly, developers need to place their work within a context of ideas and techniques to understand what they have accomplished and what they want to accomplish subsequently. Such

Figure 3. A View Matcher window for the blackjack game application. The Stack View is in the upper left; the Application View is in the upper right; the Class Hierarchy View is in the lower right; the Commentary View is in the lower left; the Inspector View is above the Commentary View. Note that, in the actual Smalltalk/VPM implementation, views are distinguished by color and grey level.

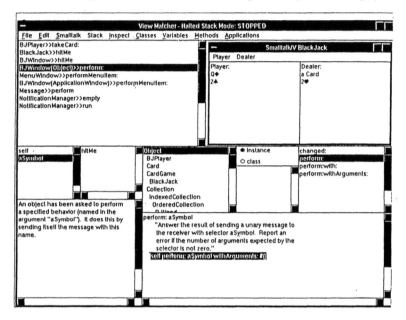

efforts are already common and important in HCI, although often informal (e.g., Smith, Irby, Kimball, Verplank, & Harslem, 1982).

In the balance of this chapter, we develop an example to illustrate how our notions of scenario-based design; task psychology; and, especially, in the context of this edited volume, psychological design rationale can work together within the task–artifact framework for research and practice in HCI.

2. A VIEW MATCHER FOR LEARNING SMALLTALK

The View Matcher, shown in Figure 3, is a structured browser for Smalltalk/V (Carroll, Singer, Bellamy, & Alpert, 1990). It presents multiple views of a running Smalltalk application, for example, a blackjack game. The various views are derived from the major system tools in Smalltalk (the debugger, the class hierarchy browser, and the inspector), but they are simultaneously opened and displayed with the application graphics and are jointly updated whenever the user interacts either with the application or with one of the tools. The View Matcher is intended to coordinate a programmer's developing understanding of Smalltalk and its environment. It has been incorporated into a minimalist curriculum for Smalltalk used at IBM (Rosson, Carroll, & Bellamy, 1990).

The design of the View Matcher was based on our understanding of the possibilities for learning inherent in the Smalltalk environment. This understanding rests on our analysis of new users as active learners, opportunistically pursuing personally meaningful goals, and trying to make sense of their experiences (Carroll, 1990b). A learning process of this sort entails three characteristic user concerns: one in which the learner has no idea what goals to pursue and wonders "What can I do?"; a second in which the learner finds and investigates a system object, wondering "How does this work?"; and a third in which the learner has some concrete goal in mind and wonders "How do I do this?" in the system.

Each of these user concerns can be instantiated for the Smalltalk environment in one or more specific user interaction scenarios (see Figure 4). Such scenarios reify typical or critical usage episodes and can be generated both through empirical observation and from task analysis. Because scenarios provide a description of the artifact embedded in a context of user background knowledge, goals, and reactions, they are an appropriate framework for analysis of the artifact's psychological consequences.

2.1. Extracting Learning Claims From Smalltalk

For illustration, consider the "Starting up Smalltalk" scenario in Figure 4. The system menu of Smalltalk/V offers "demos," the selection of which allows the learner to run a half dozen animated graphic demonstration programs. Including any function in a system, and indeed offering it on the main menu, embodies claims about what is useful or desirable to users. Including demos on the Smalltalk system menu embodies the claim that the demos will be helpful in some way (illuminating, motivating, task orienting, etc.) to learners pursuing the "What can I do?" concern (see Claim 5a in Figure 5).

Trade-offs pervade all design discourse, and typically the claims we can infer from a design include associated "downsides." Thus, the potential benefits of the demos are balanced by considerations that could limit these benefits: As the scenario conveys, the system may not support adequate exploration of the demos (and accordingly they could fail to be illuminating, motivating, etc.), or the demos themselves might be unrepresentative Smalltalk applications (and accordingly they could provide poor learning models).

Several aspects of this claim are important to note. First, it is not a claim about the intentions of the designers of Smalltalk; it is a claim about psychological consequences for users of Smalltalk. Our approach to design

Figure 4. User interaction scenarios for learning Smalltalk. The italicized labels indicate the high-level goal of the scenario; the text describes the goals, understanding, and reactions of a user pursuing this goal in the Smalltalk environment.

What Can I Do?

Starting up Smalltalk: A new user sees "demos" on the system menu and reasons that a demo will be a good way to see what Smalltalk programs are all about. It turns out that the demos have some flashy graphics, but you just have to sit back and watch. The learner wonders where the demo programs are but is not sure how to find them. The learner gives up on the demos but remembers someone saying that classes are an important concept in Smalltalk and that there is a "browse classes" choice on the main menu. When this is selected, the class hierarchy browser appears—after making a few random selections here, the learner sees what appears to be Smalltalk code and hunkers down to read through it.

How Does It Work?

Analyzing card-hand display: A colleague has given the learner an example application, a blackjack game, to learn from. The learner plays with the game and wonders how it does its display updates. The user is unable to find anything that looks like a "display" method in either the BlackJack class or its superclass CardGame. Eventually, the learner discovers an *open* method that mentions a BJWindow class and, after looking there, finds a *display* method in its superclass ApplicationWindow. The learner cannot make sense of what it does and has no idea how to tell whether it is actually used by the blackjack application.

How Do I Do This?

Changing card-hand format: While playing with a blackjack game, a learner realizes that the cards are not being displayed effectively; they are listed horizontally, and, for a large hand, this means that not all are visible. A vertical listing would solve this problem. The learner remembers that the class hierarchy browser uses such a format and decides to use it as an example. After reading a lot of code, the learner eventually discovers that the browser uses something called a ListPane, a class he or she remembers seeing once before, somewhere under Window. The learner locates this class in the browser and begins reading about it, trying to understand how it might fit into blackjack.

---

rationale does not attempt to externalize the designer's reasoning process (cf. Conklin & Burgess-Yakemovic, 1991 [chapter 14 in this book]), but, rather, attempts to externalize psychology embodied in the designed artifact and its use. We are pursuing analytical interests in the tradition of psychologists like Gibson (1979) and Simon (1981) who stress the causal role of the external environment in shaping behavior and experience. Second, the claim contains rich and specific information about a major design feature of the Smalltalk environment. Our objective is that the information be detailed enough to support further design work (in deliberate contrast to

Figure 5. **Psychological claims embodied in the design of Smalltalk/V. The claims, all of which pertain to learning the Smalltalk language environment, are structured into a principal upside (in Roman type) and one or more associated downsides (in italic type) and are organized under the learner concern in the context of which they typically arise.**

What Can I Do?

5a. Exploring demos helps new users learn by doing
(but scripted demos offer little for the user to do)
(but the demos may not be paradigmatic applications)
(but learners may have difficulty finding the corresponding code)

5b. Exploring system tools helps new users learn by doing
(but the tools may be too complex for learners to understand)
(but learners may have difficulty finding the corresponding code)

5c. Browsing and editing in the class hierarchy browser establishes core Smalltalk programming skills
(but may reduce time spent on instantiating and analyzing objects)

How Does It Work?

5d. The primacy of the class hierarchy directs attention to inheritance relationships among objects
(but learners may not be familiar with a class's superclasses)
(but it may reduce attention to other important relationships, e.g., the user interface framework)

5e. The object communication summarized in the message execution stack allows learners to decompose application functionality
(but learners may not know how or when to explore the execution stack)
(but learners may be unable to interpret the communication patterns)

5f. The debugger's message execution stack supports learning transfer from procedural programming experience
(but may encourage learners to rely on a procedural model of computation)

5g. Inspecting an object's instance variables across time supports mapping between its state and behavior
(but learners must first find or create a useful object to inspect)
(but learners may have difficulty managing and integrating many inspections of an object)

5h. Analyzing an application with multiple system tools (e.g., as in the debugger) supports convergent reasoning
(but integrating information across tools may be difficult)
(but jointly managing the application and tool windows may be difficult)

How Do I Do This?

5i. Interesting features of existing applications evoke subgoals
(but a particular subgoal may be difficult or impossible to pursue)

5j. Existing applications and methods provide templates for the design of new functionality
(but learners may have difficulty finding or understanding examples)
(but users may be less likely to pursue other more appropriate possibilities)

5k. Class and method names evoke analogical subgoal mappings
(but some spurious mappings may be pursued)

5l. Navigating the class hierarchy supports unintentional learning
(but searching for specific classes can be frustrating)
(but the size of the hierarchy may intimidate learners)

---

traditional guidelines, which are often too general to guide directly design problem solving). Third, the claim is qualitative, informal, and incomplete. For example, the psychological consequence "helps new users learn by doing" leaves it open as to whether the demos instruct the learner (i.e., present information), motivate the learner to seek information, or merely suggest a task orientation toward learning Smalltalk (indeed, there are other possibilities). In this case, we did not feel warranted in hypothesizing a more specific attribution of consequence.

Finally, in three distinct senses, the claim is an empirical hypothesis. First, it is part of our analysis of the psychological design rationale of the Smalltalk/V language and environment. As in any analysis, we could be wrong: The psychological science that warrants this claim could turn out to be wrong about the role of action and exploration in learning; conversely, the science might be fine, but our appeal to it in this case could turn out to be erroneous; or—quite likely—both the science and our analysis are fine as far as they go, but they may overlook other, perhaps more important, psychological consequences of the demos (e.g., consequences that might only emerge from direct user studies of Smalltalk learning).

Second, to the extent that our analysis is correct, the demos claim is an empirical hypothesis made by Smalltalk as a theoretical entity. Elsewhere, we have argued that the designed artifacts of contemporary HCI are the most successful and appropriate theoretical entities in the developing science of HCI (Carroll, 1989a, 1989b; Carroll & Campbell, 1989). The demos claim codifies a part of the empirical theory of programmers and programming activity that is embodied in the design of the Smalltalk language and environment.

Third, the demos claim is part of our higher level hypothesis that the level of design rationale depicted in Figure 5 is coherent. It is clear that an artifact as complex as Smalltalk embodies a virtually infinite number of psychological claims about its learners and learning situations. As in any other analytical project, our analysis must abstract and generalize in order to be codified at all. Our objective in constructing an analysis like that in Figure 5 is to capture a rich and coherent working representation of a design situation that is at the same time abstracted and abbreviated enough to be heuristic.

Note that the claims are not intended to be taken as privileged in any methodological sense. Psychological design rationale does not incorporate formal discovery procedures or any means of guaranteeing a priori that a given analysis is correct (no empirical science meets these positivist requirements; see Feyerabend, 1988). Indeed, it is likely that the possibilities for user experience and action codified in a psychological design rationale cannot ultimately be grounded in any objective epistemology (i.e., any view of knowledge dissociated from meaningful situations). Our expectation is that the appropriate epistemology here (and elsewhere in psychology) must be relativized to the commitments people make toward contexts of experience and action (Polanyi, 1958; Schön, 1982; Suchman, 1987).

More broadly, we see neither the domain of our analysis (essentially cognitive psychology) nor its vocabulary as privileged. As discussed later, many other analyses would be possible and may indeed be crucial. The limitations of our analysis stem merely from our interests, the limits of our competence, and the incomplete state of our analytical project. We make this caveat to discourage positivist distractions for the reader and to lay our philosophical cards on the table. In this chapter, however, our use of psychological design rationale is modestly empirical and is focused on its heuristic application in design argumentation.

This framework may be similar to others described in this volume. Our notion of *user concern*, which we take to be an abstraction across more specific user scenarios, could also be seen as an abstraction of what Lewis, Rieman, and Bell (1991 [chapter 5 in this book]) call *target problems*; their notion of *doctrine*, the knowledge one must have in order to use an artifact, seems to correspond to a subset of what we call *claims* (in purporting to be usable, the artifact embodies claims that its users have or can generate its doctrine). One might also draw correspondence between our schema of "user concern, artifact feature, and psychological consequence" and MacLean, Young, Bellotti, and Moran's (1991 [chapter 3 in this book]) schema of "questions, options, and criteria." The key differences seem to be that our analysis is at a higher level (with respect to both artifact structure and psychology) and is more bound to context of use: User concerns pertain to whole tasks, whereas questions pertain to constituent methods; artifact features and psychological consequences are analyzed as a set under the scope of a given user concern, not independently as are options and criteria (see also Buckingham Shum, 1995 [chapter 6 in this book]). At some point, it will be important to pursue and resolve these connections, but we do not presume to be able to do that now.

To a great extent, psychological design rationale, as we understand it and have defined it earlier, merely systematizes extant practice. Many of the classic case studies in HCI (e.g., Smith et al., 1982) are brimming with psychological claims, arguments, and attributions (e.g., the key claim that graphical presentation of an interface metaphor can facilitate learning and using by supporting analogical reasoning). Perhaps even more impressive is the extent to which descriptions of computational artifacts are unwittingly psychological—many discussions of object-oriented systems, for example, include promises or claims about the naturalness of the paradigm for humans without any psychological account of what is involved in such an attribution (see Rosson & Alpert, 1990).

The approach we are pursuing seeks to improve this practice in two ways. First, case studies often innocently conflate the designer's intentions with psychological consequences of the design as experienced by users. This is understandable in work that is both analytical and memoirlike in nature, and surely designers' intentions can provide key insights toward analyzing their work. But often there is a distinction: Misconceptions and frustrations of users confound the best of intentions. Our goal is to help recognize and respect this distinction so that each can play an appropriate role in understanding and guiding design. Second, the various claims, features, and attributed psychological consequences described in case studies often overlap and interact so intricately in the exposition that it can be difficult to see just what is being asserted and to extract general and applicable principles. Our hope is that claims analyses like that in Figure 5 can impose a discipline on the construction of psychological design rationales in order to clarify what is being asserted, thereby making it more feasible to apply what has been learned.

For example, Claims 5c and 5d in Figure 5 capture a Smalltalk issue that has been discussed informally for years: The primacy of the class hierarchy browser (the system browser in Smalltalk–80: a tool that provides access to and editing of Smalltalk code) emphasizes (a) the importance of browsing and editing code and (b) directs programmers' attention to the inheritance hierarchy within which this code exists. Both of these are important consequences for users of Smalltalk: Much of programming in Smalltalk consists of browsing and editing code, and understanding and taking advantage of inheritance relationships are key to the development of modular and elegant designs. Making these tasks salient to novices, however, tends to obscure other activities, for example, analyzing the functional relationships realized through an object's instance variables (e.g., the relationship of the underlying application to its user interface).

The remaining claims in Figure 5 have a structure analogous to those we have commented on. We refer to them only as necessary in developing our design case study.[2]

2.2. Designing the View Matcher

Our psychological design rationale for Smalltalk with respect to the three learning concerns suggests many issues for redesign. Our consideration of these issues can be couched as a design hypothesis: a learning presentation of Smalltalk that preserves or enhances the constructive psy-

---

2. Our development of psychological design rationale, our understanding of the psychological design rationale for Smalltalk, and our design of the View Matcher for learning have developed in concert since 1988; the presentation here reflects the current state of all this work. A brief description of our Smalltalk analysis circa 1988 can be found in Carroll and Bellamy (1989); a description of the View Matcher circa 1988 can be found in Carroll (1990b, pp. 267–272).

chological consequences of the claims in the rationale but that addresses the associated downsides of these claims.

For example, providing dynamic examples (e.g., the demos) is a good idea, but these examples are undermined by lack of support for exploring their design and implementation and by the fact that they are not representative Smalltalk applications (users cannot interact with the demos; they are simply graphic scripts that "play out" and are implemented as individual methods rather than as independent applications). Thus, we envisioned a presentation of Smalltalk that incorporated more paradigmatic examples and that better supported exploration of its design and implementation.

Offering a variety of interactive tools for analysis of running applications supports the "How does it work?" concern (see Claims 5e–5h in Figure 5). Learners can open an inspector on an object to examine its internal state; in an error situation, they can analyze application functionality in the debugger's message execution stack. They can open multiple tools to compare and contrast different information sources, but managing and integrating all these tools can be complex. This suggests a presentation of Smalltalk that permanently displays and synchronizes these tools with example applications.

The system debugger provides a model for this, coordinating an inspector and a class hierarchy browser with the message execution stack. However, particularly for a learner, the debugger can be overwhelming: It is typically encountered in the context of an error; it presents a lot of information about message activity, object states, and implementation; and it is only loosely coordinated with the user's application. This suggests a design that more tightly couples the application with the system tools: incorporating a permanent application view, tailoring the information presented in the tools to be application relevant (e.g., a class hierarchy that displays only classes and methods used by the application), and providing help texts to support integration across tools.

The design of the View Matcher was elaborated by developing specific scenarios of use. The example scenarios in Figure 6 depict learners beginning with high-level goals similar to those of the earlier scenarios, but the user's actions and experiences in pursuing these goals are those afforded by the new system. In "Starting up View Matcher," the coordination among views has been further specified: When the learner opens the blackjack game, the class hierarchy updates to show a list of classes, promoting the inference that these are the objects involved in the application. The start-up scenario also illustrates how the design enhancements combine to support the learner—the availability of a familiar application in combination with the other panes in the tool suggest activities to the learner.[3]

_____

3. Additional scenarios are discussed in Carroll (1990b, pp. 270–272) and in Carroll et al. (1990).

Figure 6. Learning scenarios for the View Matcher. The first three scenarios instantiate the same high-level goals as those presented in Figure 4 but describe pursuit of the goal in the View Matcher environment. The fourth illustrates transfer of learning from the View Matcher to standard Smalltalk.

What Can I Do?

Starting up View Matcher: A new user sees "blackjack" on the View Matcher menu and reasons that this might be a fun and useful way to learn about Smalltalk programs. When the game starts, some classes that seem to be involved in the game (e.g., BlackJack, Card) appear. By selecting these classes, the learner is able to see what appear to be message names and code but is unable to make much progress with it. After playing the game for a while, the user wonders what the empty panes are for, goes back to the menu, and experiments with "Halted Stack Mode." Now when the game is played, it stops in the midst of doing things, and what seems to be a list of messages (the same ones that he or she saw earlier in the browser) is shown. Selecting a message in the list, the learner notes that the other parts of the display are updated, and he or she settles down to try to make sense of all the information.

How Does It Work?

Analyzing card-hand display: The new user wonders how blackjack displays its cards and has noticed that in halt mode the game stops before each display update. The user plays the game to such a point, then looks at the information displayed in the various panes. By selecting the top message on the stack (*showHand*) and reading its associated commentary and code, the learner sees that this message is sent to the card-hand object when it is time to display its cards. The user infers that this message was sent in the course of processing the previous message, just as would be true for a call stack. The remaining messages on the stack seem very confusing, involving a number of non-blackjack classes (e.g., TextPane, Notification Manager), but, using the commentary, the learner is able to see how the responsibility for displaying things is distributed between the window and its components and between the blackjack and its components.

How Do I Do This?

Changing card-hand format: While playing the blackjack game, the learner realizes that the cards are not being displayed effectively; they are listed horizontally, and, for a large hand, this means that not all are visible. A vertical listing would solve this problem. In halt mode, the user discovers that just prior to displaying the player's hand, the game stops, and that the top message on the execution stack is *showHand.* Selecting it causes the browser to locate the associated method code. The user tries making a change to the code, confirms by resuming the game that the format has in fact been changed, and then iterates through this process to get the format to look right.

Using the Full System

Building an address book: After working with blackjack in the View Matcher, the learner decides to try a new application—an address book—using blackjack as a model. Reasoning by analogy, the learner figures that the address list will be like the card hands but that there will be just one of them. The learner goes to the classes used by the blackjack game, borrowing and editing code that seems relevant. When the user first tries to start up the new application, an error message box appears. The user is relieved to see the familiar message execution stack. After looking at the stack, however, the user is distressed to find many messages on the stack that were never encountered in the View Matcher explorations and no commentary to help make sense of them.

The fourth scenario in Figure 6 represents a new kind of concern that inheres in any learning environment: At some point, the learner moves on to work (and continue learning) in the environment. When this happens, the learner may reexperience the first three concerns ("What can I do?", "How does that work?", "How do I do this?"), but now in the context of the target environment and with the View Matcher learning experiences as background knowledge. In this sense, "Using the full system" can be seen as a sort of meta-concern: How do users do the things they were able to do before, and how do they go on from there? An important part of our design hypothesis was that by providing paradigmatic example applications, and by building the coordinated views out of standard system tools, we would ease learners' transition from the View Matcher environment to normal Smalltalk programming.

2.3. Psychological Design Rationale for the View Matcher

The View Matcher system is the end state of the design hypothesis. It was not generated by the design rationale in Figure 5, but then *deductive invention*, more than merely an oxymoron, is probably a technological contradiction (and indeed, here we really do depart from Bacon's optimism about practical science; see Carroll, 1989b). The design of the View Matcher was heuristically guided by the claims analysis of Smalltalk. As the scenarios representing the new design are elaborated and implemented, they then serve as an organizing rubric for analysis of the incipient system (see Figure 7).

The claims embodied in a new artifact can often be seen as evolving from claims analyzed for the precursor artifacts. A case in point is Claim 7a, concerning the role of paradigmatic interactive applications in learning by doing. Figure 8 illustrates how this claim evolved from our analysis of the demos in Smalltalk/V (Claim 5a in Figure 5). Instead of the demos, the View Matcher offers familiar applications that are interactive and that were crafted to exemplify object-oriented decomposition and the separation of application model and user interface. By embedding a paradigmatic example in a tool designed to guide learners' analysis, the View Matcher both exploits the positive aspects of the demos claim and addresses its negative aspects. Of course, it is still possible that our design of the examples was more quirky than we thought or that the View Matcher supports exploration inadequately. However, the design rationale oriented us to these issues, and at the least we made some improvement.

It is typical for design enhancements to introduce new downsides. Thus, we see in Claim 7a that at least one risk of incorporating example applications that are familiar and interactive is that they might be too engaging—learners may get so caught up in their use of the application that their motivation to understand how it works is diminished.

Figure 7. **Psychological claims embodied in the View Matcher for learning. The analysis is organized by the three general learning concerns used in the original analysis of Smalltalk, plus a fourth concern typifying a learning transfer situation.**

What Can I Do?

7a. Exploring paradigmatic interactive applications helps new users learn by doing
     (but users may spend too much time using the application and not enough time learning about it)

7b. Permanent display and coordination of an application's multiple views facilitates learning by doing
     (but the multiple views may provide too complex a representation of the application)

7c. A filtered class hierarchy encourages opportunistic analysis of the objects in an application
     (but users may be frustrated if they want to examine hidden classes)

How Does It Work?

7d. Permanent display and coordination of an application's multiple views support convergent reasoning
     (but does not develop skills for accessing and managing system tools)

7e. The application episodes defined by View Matcher breakpoints are coherent units of analysis
     (but may interrupt normal interaction)
     (but it may be difficult to make sense of an application in the midst of a message send)

7f. The object communication summarized in the message execution stack allows learners to decompose application functionality
     (but learners may be unable to interpret the communication patterns)

7g. The debugger's message execution stack supports learning transfer from procedural programming experience
     (but may encourage learners to rely on a procedural model of computation)

7h. Application-specific commentary for each message in the stack evokes an application-oriented interpretation of object communication
     (but this understanding may not generalize to other applications)

7i. Synchronizing the browser with the method selected in the stack directs attention to the code relevant to the application's current state
     (but it may discourage exploration of methods not on the stack)

7j. Joint updating of the matched views temporally groups information, helping users to parse the display
     (but too much information changing at once might be confusing)

7k. An animated message stack serves as an advance organizer for the halted stack
     (but learners may be frustrated at their inability to explore the animated stack)

How Do I Do This?

7l. Concrete, familiar applications evoke specific subgoals
     (but a particular subgoal may be difficult or impossible to pursue)

7m. Examples that misbehave provide intrinsic motivation for analysis
     (but may encourage too narrow a focus in analyzing the example)

7n. Coordination of an application's multiple views simplifies access to task-relevant information
     (but learners may come to rely on this coordination)

7o. Modifying an application is easier than creating one from scratch
     (but the existing application may not be consistent with the user's current view of object-oriented design)

Using the Full System

7p. Finding and modifying code encourages code reuse as a programming paradigm
     (but it may undermine other reuse strategies like subclassing)

7q. Experience with the system tools in the View Matcher transfers to their use in non-View-Matcher activities
     (but users may be frustrated by the absence of coordinated views)

7r. Paradigmatic examples help users construct principles of good design
     (but learners may overgeneralize the details of particular examples)

7s. Understanding and modifying example applications generalizes well to other programming tasks
     (but may induce too narrow a view of object-oriented programming)

Figure 8. Evolution of demos claim. The figure schematizes an example of our general reasoning heuristic, to capitalize on the constructive aspects of claims while addressing their downsides.

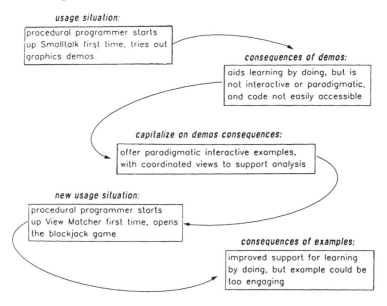

A similar evolution can be seen in the View Matcher's emphasis on the message execution stack. The Smalltalk analyses pointed to the psychological benefits of the stack representation for users pursuing the "How does it work?" concern (for decomposition of application functionality and for learning transfer, see Claims 5e and 5f, respectively, in Figure 5). In the View Matcher, the stack is the controlling view for application analysis: When the application is halted at a breakpoint, the current message execution stack is displayed; selection of a message in this stack causes the class hierarchy browser to display the corresponding method code, the inspector panes to display objects used by that method, and the commentary pane to present an explanation of the method's role in the application transaction being processed (see Figure 3). Increasing the salience of the stack in this way strengthens claims made by Smalltalk (see Claims 7f and 7g in Figure 7). At the same time, problems of stack interpretation are mitigated by providing the explanatory text (see Claim 7h in Figure 7). Of course, encouraging learners to develop an application-specific interpretation of message patterns in the stack increases the risk that the understanding they develop will be too specific to generalize to other situations.

Many of the claims in Figure 7 reflect the consequences of integrating and coordinating multiple views onto an example application. This design feature eases the downsides of many of the original Smalltalk claims—the relative lack of support for exploring the demos, the difficulties in managing the system tools, and the overattention to inheritance relationships (see Claims 5a, 5d, 5e, 5g, and 5h in Figure 5). However, the coordination also introduces new claims with new downsides. Thus, the fact that the tools all provide information about the example application and that they are jointly updated when the user explores the application encourages users to try to make sense of the information in terms of their personal use of the application (see Claim 7b in Figure 7); the downside is that a great deal of information inheres in the multiple views, and this may be overwhelming to a learner.

As another example, consider the use of breakpoints to coordinate the application with the message execution stack: When the user interacts with the application, the application halts at predefined points (user-input events and display updates). This feature addresses the downsides of Claims 5e and 5g in Figure 5; it removes the need for the learner to initiate use of the debugger and inspector information. But it also embodies a new claim, that the application episodes defined by the breakpoints are coherent units of analysis (see Claim 7e in Figure 7). The downside is that natural interaction with the application is interrupted, and this may add to the difficulty of making sense of the application.

In some cases, claims from the original analysis of Smalltalk were dropped from the new analysis. Incorporating a class hierarchy browser as just one of the several view-matched representations reduces the salience of this view and, hence, the emphasis on browsing and on inheritance relationships (see Claims 5c and 5d in Figure 5). Part of coordinating the class hierarchy with the example involves filtering it to display only application-relevant classes and methods. Users can still navigate in the browser themselves, but, because the number of classes is relatively small, the amount of unintentional learning possible is considerably reduced. Indeed, we felt that it was reduced enough to exclude this claim (see Claim 5l in Figure 5) from the View Matcher analysis; our goal was to include only the leading claims of the new artifact. Again, however, the changes to the class hierarchy view introduce a new claim. By limiting the browser to information about application-relevant objects, users may be more likely to engage in opportunistic analysis of these objects (see Claim 7c in Figure 7). The new downside is that they may come upon the names of classes or methods that are not in the filtered view (e.g., generic or inherited functionality) and be frustrated at their inability to explore them.

These examples illustrate how working with a set of claims propagates effects: Claims disappear or change in prominence, new claims appear, and new trade-offs are provoked. As our design work proceeded, claims were continually being recognized, and new trade-offs and design issues caused us to realize new possibilities. For instance, the joint updating of the View Matcher tools temporally groups related events in the interface and helps the learner spatially parse the information display (see Claim 7j in Figure 7). Given the number of simultaneous views in the View Matcher, this is an important psychological consequence, but it was introduced as a side effect of coordinating the multiple views. Another case was the development of an animated view of the message execution stack (see Claim 7k in Figure 7). Our formative evaluations of the View Matcher suggested that some learners interpreted the message stack as a history list of messages sent and not as a list of methods currently suspended at the point of a halt. We introduced an animated view of the execution stack in which the application's message passing is never halted and, therefore, cannot be analyzed in the midst of its execution. Watching the stack grow and shrink in response to user requests prepares learners for their subsequent exploration of the halted stack; the downside is that learners may see something of interest happening in the animated stack and feel frustrated at being unable to stop and analyze it.

The claims listed in Figure 7 under "Using the full system" differ in an interesting way from those we have discussed so far. The relevant scenarios in this case involve the use of the target system, not the View Matcher (see "Building an address book" in Figure 6). Thus, the claims reflect design characteristics of one artifact (the View Matcher) that have consequences for use of another (standard Smalltalk/V). We remarked earlier that such scenarios inhere in any instructional system, but, in fact, any scenario reflecting learning transfer from one artifact to another would have the same character.

The View Matcher provides a good example of Carroll and Kellogg's (1989) point that designed artifacts embody a "nexus of psychological claims" (p. 8), not a list of independent psychological atoms. As in any other view of design, it is not possible to bound the scope of effect of design interventions. The effects of particular design characteristics (e.g., the use of paradigmatic examples and the coordination of the system tools) have wide-reaching psychological consequences. However, the empirical validity of a specific claim can be examined: The permanent display and joint updating of the View Matcher tools either does or does not support the integration of the information displayed in tools (say, relative to the baseline integration supported by the Smalltalk environment); specific error patterns can indicate whether learners become overwhelmed with the amount of information the View Matcher presents or whether experience with the tools as integrated by the View Matcher transfers at all to using the tools as they are presented in the Smalltalk environment.

Because of the interconnectedness of the various claims, such evaluations will always be somewhat inconclusive (although their heuristic value can be leveraged if a diversity of evaluations is jointly interpreted against the rubric of a psychological design rationale). Indeed, this consequence for evaluation clarifies pervasive problems with summative evaluation. Aggregate summative evaluations have served little use in HCI design because they merely order the measured usability of whole artifacts without involving descriptions of the artifacts that could support credit–blame analysis (Carroll, 1989a). The further caveat from design rationale is that this problem may not be mitigated by making analytic summative evaluations (e.g., feature-by-feature assessments of an artifact), because the various features participate in interconnected claims (e.g., evaluating a command name—or even a command language paradigm—may be uninformative unless one also evaluates the interface display metaphor and the semantics of the user's task; Carroll, 1985).

3. A VIEW MATCHER FOR CODE REUSE

When we began to deploy the View Matcher as part of our minimalist curriculum for Smalltalk, several of our users inquired as to whether we could extend the tool for use in routine programming. These users did not make very directive requests; however, they seemed to appreciate something about the View Matcher approach (interactive paradigmatic examples and coordinated views of jointly displayed tools, including interactive execution tracing). We ultimately decided that this was indeed an interesting request to try to respond to, perhaps in part because it required us to discover the details of the request itself.

We decided to focus our redesign effort on facilitating code reuse, because it is so fundamental to Smalltalk programming; Smalltalk offers a rich and extensible hierarchy of reusable classes, and a large portion of any programming project is the discovery and application of existing code. The general reuse task we are addressing is one in which the programmer has a goal in mind and wants to reuse code insofar as this is appropriate. The programmer engaged in this task may have several sorts of concerns (see, e.g., Fischer, 1987; Raj & Levy, 1989). First, he or she may work at identifying a candidate class or classes in the existing hierarchy, pursuing the information retrieval concern of "What can do this?" At some point, a candidate class will have been identified, and at this point the concerns become oriented specifically to that class: The programmer wonders "What can it do?" to see if it does have the desired functionality. If the class does indeed look promising, then the programmer wonders "How is it used?" to reason about how objects of this sort could be used in the project at hand. If problems develop in trying to use the class, the programmer will wonder "How does it work?" so that the problems can be debugged and resolved. Finally, if the programmer determines that the class has some, but not all of the necessary characteristics, the reuse project will encompass

Figure 9. **Reuse scenarios for Smalltalk/V. The scenarios exemplify user concerns that can arise after a programmer has identified a candidate class for reuse.**

What Can It Do?

What Sliders do: A programmer sees Slider in the class hierarchy, and the name sounds like it might be useful to the current project, a color-mixing application. It is a subclass of ControlBox, which seems consistent with this inference. The programmer takes a look at the messages defined for Slider; they sound as if they involve input handling (e.g., *adjustToMouse:*), but it is not clear how they fit together or what the look and feel of the Slider will be. The programmer tries to find an example instance to work with, asking for "instances" of the class, but none exist. Because this is a user interface object, the programmer suspects that it will require considerable set-up to create an example, so he or she finally hunkers down to read through the individual messages.

How Is It Used?

Hooking up a BoardGamePane: A programmer wants to use an instance of BoardGamePane in a chess game being developed and needs to know how to connect it to the current objects. The programmer takes a look at the code for BoardGamePane and sees that it has a *squares* instance variable. He or she asks for "senders" of the *squares:* message, hoping to see situations in which this variable gets set. A long list is returned, but the programmer infers that many of the items in the list are for other implementations of this message. Scrolling through the list, the programmer sees BoardGameWindow. This class sounds promising, so the programmer goes off to look at its code to try to figure out how it uses a BoardGamePane.

How Does It Work?

Analyzing slider scale conversion: The programmer has successfully incorporated three sliders into the color-mixing application, setting them up to have a horizontal layout. As the mouse moves, the slider changes size, but the color output is not right. The programmer puts a halt into the code for *adjustToMouse:*, and when the application halts, he or she inspects the mouse value, which seems correct. Using the debugger stepping functions, the programmer then traces through all subsequent message sends, checking values at each point to see when the error occurs.

---

the "How to extend it?" concern. Our initial design work has focused on the middle three concerns, that is, those involving questions about a target class's functionality, usage protocol, and implementation.

Like the learning concerns, the reuse concerns can be instantiated as specific user scenarios. The scenarios in Figure 9 instantiate these general concerns for the Smalltalk environment and suggest some of the psychological consequences that Smalltalk might have for an experienced programmer interested in reusing a particular class. An analysis of these consequences, combined with our analysis of the previous View Matcher work, formed the starting point for our design of a View Matcher for code reuse.

3.1. Extracting Code Reuse Claims From Smalltalk

The key differences between the learning and the reuse situations are that the goal of the reuse situation is well defined and that the user is assumed to be an experienced programmer fluent at finding and interpreting information in the system. However, analyzing a class in order to reuse it can be seen as a specialized case of learning (the programmer seeks to learn about a particular class or functionality), and the Smalltalk claims we constructed for the reuse situation (see Figure 10) reflect similarities between the two task situations. For example, if an instance of a target class is available, a programmer can send messages to it to explore its functionality, just as a learner might explore a demo application to learn something about how Smalltalk programs work (see Claim 10a in Figure 10). As the first scenario in Figure 9 illustrates, the downside is that the programmer must often create objects from scratch, and the initialization of complex objects (e.g., a user interface object that must be connected to a number of other objects) can be very cumbersome (cf. Gold & Rosson, 1991).

As in the learning scenarios, if programmers can find or create a representative example, Smalltalk encourages them to enlist the system tools in the analysis of the example—in this case, to address concerns of "How is it used?" or "How does it work?" Using the debugger, programmers can trace the message-passing activity of an application that uses a target object to see when and in what context messages are sent to the object (see Claim 10d), as well as to see what messages the target object sends in fulfilling these requests (see Claim 10j); the inspector allows them to examine the target object's changing state over time and to understand how the object manages itself in the process of fulfilling requests (see Claim 10j). But also, as in the learning scenarios, it is up to the programmer to set up an informative situation and to extract and integrate the information relevant to the use or implementation of the target object.

Again, as recorded in our learning analyses, the primacy of the class hierarchy focuses programmers' attention on inheritance relationships. This has consequences for each of the three reuse concerns (see Claims 10c, 10f, and 10k in Figure 10). For programmers wondering whether to use a class, an inheritance context can be quite helpful: If programmers are familiar with the superclass's functionality, they should be able to predict something about the target class's functionality. In contrast, if a programmer wishes to know how an object is used, inheritance relations will help little; what the programmer really needs to know is how other kinds of objects make requests of this object.

The "Hooking up a BoardGamePane" scenario in Figure 9 illustrates a programmer pursuing this question, asking for the "senders" of the squares: message. By looking at the code within which a message is sent, the programmer can determine something about the sending context

Figure 10. **Reuse claims embodied in the design of Smalltalk/V.**

What Can It Do?

10a. Sending messages to an instantiated object supports discovery of its functionality
(but finding or creating a representative instance may be difficult)
(but trying out individual messages may be tedious and distracting to ongoing work)

10b. Class and method names suggest the functionality provided by a class
(but some names are ambiguous or inappropriate)
(but the behavior may be too complex to suggest with a name)

10c. The primacy of the class hierarchy directs attention to inheritance relationships among objects
(but programmers may not be familiar with a class's superclasses)
(but superclass functionality may not predict subclass specialization)

How Is It Used?

10d. Tracing an application's message passing allows programmers to build schemas for use of its components
(but temporal integration of requests to a single object may be difficult)
(but abstraction of communication and control relationships among the objects may be difficult)

10e. Halting an application in the midst of receiving a particular message supports a functional analysis of the message's role
(but it may be difficult to tie this analysis to the visual state of the application)

10f. The primacy of the class hierarchy directs attention to inheritance relationships among objects
(but it may reduce attention to sender–receiver relationships)

10g. Browsing a class's method code helps in understanding how to use the class
(but programmers may be distracted by irrelevant implementation details)

10h. Message-naming conventions allow programmers to predict parts of a class's usage protocol
(but many messages are not covered by convention)

10i. Examining the "senders" of a message supports analysis of the context in which this request is made
(but a sender's listing may include many false alarms)

How Does It Work?

10j. Tracking the messages sent and the state changes produced in the process of evaluating a message allows programmers to analyze message implementation
(but the programmer must initiate, control, and integrate the analysis)

10k. The primacy of the class hierarchy directs attention to inheritance relationships among objects
(but it may reduce attention to other functional relationships, e.g., those provided through instance variables)

(e.g., what kinds of objects send it and when; see Claim 10i in Figure 10). The downside is that, in Smalltalk, a given message name may have multiple implementations (i.e., in different classes), and the implementation invoked in response to a message can only be known at runtime. The senders query returns a list of all code that includes the message string, and this list will often contain many false alarms—code in which this message is sent but not to an instance of the target class.

3.2. Redesigning the View Matcher

Our psychological design rationale for Smalltalk with respect to the code reuse concerns, in concert with that for the View Matcher for learning, focused our consideration of issues for redesign. Again, we worked with these issues by envisioning a design hypothesis. Given our starting goal of applying the View Matcher concept to the code reuse domain, much of our design reasoning was by analogy: What are the analogous representative examples to incorporate into a reuse analysis? What should the multiple views of that analysis convey? How should they be coordinated? To some extent, our ability to reason by analogy in this way indicates that the issues we are concerned with are general—general user propensities and requirements, general limitations of the Smalltalk language/environment, and general opportunities for redesign. But give our a priori commitment to the View Matcher framework, it also probably reflects the bias in our analysis; of course, this is what genre and style in design are all about! (See Newman, 1988.)

The design rationale for learning Smalltalk and for the original View Matcher highlighted the importance of example applications as vehicles for learning and pointed to issues concerning support for analyzing such examples. Similarly, the design rationale for reuse in Smalltalk highlighted the potential of analyzing example applications' usage of a target object, understanding the context of this use, and analyzing the target object's state and activity over episodes of use. By analogy, then, we hypothesized that our new View Matcher should revolve around an example application that makes use of a target object and have consistent views that support an analysis of this use.

Figure 11 illustrates how our design of the reuse examples considered not only the advantages of working with an example instance of the target class (see Claim 10a in Figure 10) but also the claims concerning demos and interactive examples recorded in our prior learning analyses. Users first learning Smalltalk are learning by doing—searching for reasonable goals, pursuing goals opportunistically, trying to make sense of things. Smalltalk's demos encourage this mode of learning, but they are not paradigmatic and do not offer much for learners to do (see Claim 5a in Figure 5); the View Matcher addresses this downside with its provision of paradigmatic interactive applications (see Claim 7a in Figure 7). But the code reuse situation is different. The experienced programmer seeking to

Figure 11. **Evolution of demos and examples in the View Matcher.** The figure schematizes how the usage vignettes in the reuse View Matcher evolved both from the analysis of reuse in Smalltalk and from the earlier analyses of Smalltalk learning.

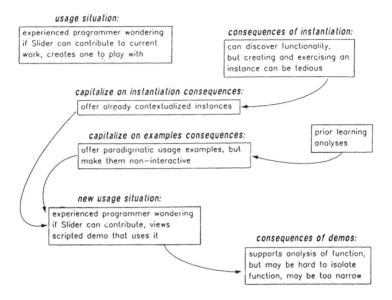

reuse a class is not looking for goals; rather, the programmer is in the midst of a meaningful project and wants a question about reuse answered in a way that distracts as little as possible from the ongoing project. This suggests that the examples for reuse analysis should be scripted (as the Smalltalk demos are)—usage vignettes that exemplify paradigmatic uses of an object but that require no work other than interpretation.

Another key feature of the reuse View Matcher is what we term a *usage view*: The usage vignettes are decomposed into typical usage episodes; these, in turn, are decomposed into messages involving the target class (see Figure 12). This view corresponds to the stack view of the learning View Matcher in that users' interactions with it drive the coordination of the other views. In the View Matcher for learning, the message execution stack supports learners' decomposition of a running application into its basic transactions (via breakpoints delineating interactions between the underlying application and its user interface), helping them to see how the parts cooperate to produce the overall functionality (Claims 7e and 7f in Figure 7). The stack also provides an important path for learning transfer, though perhaps promoting too procedural a view of application activity (see Claim 7g). In the reuse case, we assume experienced programmers as our users, so the transfer consequence becomes less important. Although the debugger does allow programmers to analyze the messages sent to a target object and their subsequent effects (see Claims 10c, 10d, and 10j in Figure 10), it is left to the

Figure 12. **A View Matcher window for the BoardGamePane class.** The usage view is in the upper right; the communication map is in the upper left; the class hierarchy view is in the lower right; and the commentary view is in the lower left. The user has already viewed the Gomoku usage vignette and is exploring the role of the BoardGamePane in the "starting up the Gomoku game" episode. In the communication map, the connection between the window and the subpane has been selected, and it shows the name of the window's instance variable that refers to the instance of BoardGamePane.

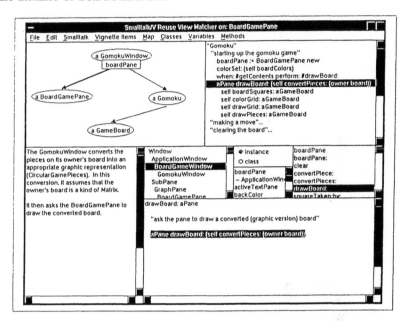

programmer to create, manage, and integrate illuminating tracing episodes. We hypothesized that, by providing an analysis of the vignettes into usage episodes typifying the target object's contribution to the application, the tool would support a decomposition that was both more task relevant and more object-oriented.

As in the case of the View Matcher for learning, many details of our design hypothesis capitalize on the claims of Smalltalk for reuse, while addressing their downsides. Thus, the usage view is an approximation to the message tracing available via the debugger; it enumerates only messages that are sent to the object of interest or that include it as an argument. We added to this message trace a presentation that would make explicit the communication and control relationships among the participating objects (the "communication map" in Figure 12).[4] Finally, we hypothesized

---

4. The antecedents of these representations were developed by Rosson as discussion aids in her object-oriented design workshop at the IBM T. J. Watson Research Center.

that coordinating the inspector with specific requests made to the target object would provide a meaningful structure in which to understand an object's changing state over time.

By including a filtered class hierarchy browser as another coordinated view, we are able to maintain the advantages that the inheritance structure brings to predicting a target class's functionality while addressing the downside that inheritance has little to say about usage. Because the example application exemplifies use of the target object, exploration of it in the browser will necessarily involve assessing true "senders" code.

The design of the reuse View Matcher was elaborated by developing specific scenarios that instantiate the code reuse concerns. As the scenarios in Figure 13 illustrate, a programmer first opening a View Matcher on a target class can view one or more usage vignettes (e.g., a football analyst that uses sliders). Each of these vignettes can then be expanded into its associated usage episodes (e.g., "starting up the game" in the Gomoku vignette, see Figure 12), and episodes can be expanded to display the requests that are either sent to the target object directly (e.g., aPane drawBoard: (self convertPieces: (owner board))) or include the target object as an argument. Finally, each message sent to the target object can be expanded to display messages the object sends to itself in carrying out the request (e.g., self drawGrid: aGameBoard).

The other views of the example application are coordinated with programmers' interactions with the usage view. The communication map provides a graphic representation of the relevant communication context: At the level of a usage vignette, this consists of the relationship of the target object to the controlling objects of the application (e.g., illustrating that boardPane is an instance variable of the BoardGameWindow object, which in turn is an instance variable of the Gomoku object). When a usage episode or a specific message within an episode is selected, the communication map updates as necessary to include other supporting objects. At any time, the user can select one of the objects in the communication map and inspect that object's internal state either before or after the point under analysis in the usage view (e.g., before an episode begins or after it completes, before a particular message is sent or after it is evaluated).

The filtered class hierarchy view is used to show the relevant sender code: When an episode is selected, the browser shows the method code from which this episode was initiated (e.g., the Gomoku newGame method for the "starting up the game" episode); when a particular message in the episode is selected, the browser displays the code within which this message was sent (e.g., drawPieces: is sent within the BoardGameWindow drawBoard: method). The commentary describes the contributions of the target object as a function of the current selection in the usage view either to the application as a whole, more specifically to a usage episode, or, even more specifically, to the evaluation of a particular message.

Figure 13. **User scenarios for the reuse View Matcher. The scenarios instantiate the same high-level goals as those used for analysis of Smalltalk in Figure 9 but describe how the goal might be pursued using the new system.**

What Can It Do?

What Sliders do: A programmer sees Slider in the class hierarchy, and the name sounds like it might be useful to the current project, a color-mixing application. The programmer opens a View Matcher on it and selects the first example, a football player analysis program. A short demo of the football program is shown, and the programmer sees that sliders are being used to manipulate player characteristics that predict several player success measures. The programmer recognizes that this situation is very similar to the needs of the color mixer but goes on to view a second vignette, an economic simulation in which the sliders interact with one another, constraining each other's values.

How Is It Used?

Hooking up a BoardGamePane: A programmer wants to use an instance of Board-GamePane in the chess game being built and needs to know how to connect it to the game objects. The programmer opens a View Matcher on the BoardGamePane class and selects the first example, a Gomoku game. After watching the demo, the programmer examines the object communication map; the map shows that there is a Gomoku object that points to a BoardGameWindow and that the window then points to the BoardGamePane. The programmer wonders where the board comes in and selects the "starting up the game" usage episode; the object communication map updates to show that, in this example, the information about the board comes through the Gomoku and window connections.

How Does It Work?

Analyzing slider scale conversion: A programmer has successfully incorporated three sliders into the color-mixing application, setting them up to have a horizontal layout. As the mouse moves, the slider changes size, but the color output is not right. The programmer opens a View Matcher on Slider, selects the football example, and expands the "increase player speed" episode. The usage analysis updates to show the messages sent to the slider. The programmer expands the *adjustToMouse:* message and sees that it involves an internal request for a scale reading. The programmer examines the code for the scale reading and finds that the conversion algorithm assumes vertical sliders.

3.3. Psychological Design Rationale for the Reuse View Matcher

We have implemented the View Matcher for code reuse (Rosson, Carroll, & Sweeney, 1991a, 1991b). We are continuing to explore the use of psychological design rationale by constructing a claims analysis for the system in order to guide our ongoing prototyping and evaluation work. This interim analysis is presented in Figure 14.

Some of the claims involve familiar features and consequences. For instance, the canned demos are paradigmatic examples of reuse of the

Figure 14. **Psychological claims embodied in the View Matcher for code reuse.**

What Can It Do?

14a. Viewing paradigmatic scripted demos that use an object helps programmers analyze its functionality
(but the concept induced might be too narrow)
(but users may have difficulty isolating the target functionality)

14b. Class and method names suggest the functionality provided by a class
(but some names will be ambiguous or inappropriate)
(but the behavior may be too complex to suggest with a name)

How Is It Used?

14c. Permanent display and coordination of multiple views of an object's usage support convergent reasoning
(but the novelty of some of the views and the amount of information presented may make analysis difficult)

14d. The connections between the target object and other objects in a usage episode provide a template for its incorporation into other designs
(but may convey a fragmented view of the demo application's design)

14e. The requests made of an object during a usage episode provide a template for exercising specific functionality
(but may result in a fragmented view of the usage application's activity)
(but programmers may overgeneralize the details of particular episodes)

14f. The episodes comprising a usage vignette evoke specific subgoals for analyzing how its components are used
(but it may be difficult to connect analysis subgoals to the observed features of the demo)

14g. Directing programmers' attention to the code in which a message is sent facilitates creation of analogous sender code in a new application
(but experienced users may be confused when the browser does not display a message's implementation)

14h. Result-oriented commentary for each message in the usage view promotes a functional analysis of the target object's role
(but may discourage analysis of how a particular request is fulfilled)

14i. Seeing a demo of the vignette provides an advance organizer for the usage view and communication map
(but programmers may forget some features of the demo)

How Does It Work?

14j. Analyzing requests made to self in response to a message supports inferences about message implementation
(but it may discourage a complete decomposition of the request)

14k. Examining the state of relevant objects before and after a message send facilitates analysis of the message's preconditions and consequences
(but some consequences may not be apparent in state changes)

---

target object (see Claim 14a). But they suffer from the downside of any set of examples; they can be incomplete in the functionality or usage protocols that they exemplify. Further, the decision to exemplify target objects from the reuse perspective leads to another downside: Users viewing the vignettes may have difficulty identifying just what functionality is contributed by the target object. Notice, however, that one of the original downsides analyzed for the Smalltalk demos in the learning situation, that scripted demos offer little for the learner to do (see Claim 5a in Figure 5), does not appear in this analysis. As we argued earlier (see Figure 11), experienced programmers seeking to reuse code are not looking for extra things to do, so the scriptedness of the reuse examples is less likely to entrain this negative consequence.

Other claims reflect the consequences of new features—the usage episodes, the communication map, and the sender-oriented class hierarchy view. The usage episodes and the communication map document usage situations that can then serve as templates for reuse in new situations (see Claims 14d and 14e). But these representations are filtered views of the design and message activity of the vignettes, and there may be complex or eccentric situations in which mapping from the template to the new problem is indeterminate. The usage view also suggests a subgoal decomposition of how the target object is used (Claim 14f, which is analogous to Claim 7f concerning the message execution stack). The class hierarchy view supports a copy/edit reuse strategy (Lange & Moher, 1989) by directing attention to the code in which a message is sent (Claim 14g), but it may confuse programmers who expect to see the implementation of the message itself. These risks are salient to us at this stage of design, because we are experimenting with representation and analysis tools that depart significantly from the standard environment.

The evolution of the reuse View Matcher affords a better understanding of application genre. Our analyses capture the similarities between the two View Matcher designs: Both capitalize on the psychological consequences of learning from paradigmatic examples and of reasoning from multiple representations. One can see these similarities as defining a View Matcher genre. Each design also refines the genre as a function of the situation in which the learning occurs: For novices, the examples are interactive and users are encouraged to decompose them into the major transactions between the underlying application and the user interface. For the reuse situation, the examples are noninteractive and users are encouraged to develop object-specific usage analyses.

To this point, psychological design rationale has allowed our View Matcher work to be deliberate and cumulative. We feel that projecting the rationale for our current system as we design additional reuse scenarios, and as we prototype them, provides a powerfully articulate foundation for principled design argumentation. It allows us greater confidence that we are standing on the shoulders of our prior View Matcher design work, and the psychological claims it embodies, and not merely in the vicinity of prior work. But alas, this confidence suggests a final complicating thought: A View Matcher for reuse clearly embodies specific knowledge and strategy claims for programmers who first used the View Matcher for learning. Thus, as we can better articulate and more deliberately emulate and develop the View Matcher approach, we may move on to the psychological consequences of designing species of artifacts.

4. PALEONTOLOGY AS YOU GO

History is an intriguing concept in a design field like HCI. One reason is that the here and now recedes into the distant past with such astonishing speed. One wonders whether people like Dan Bricklin and Bob Frankston, the inventors of VisiCalc, ever expected to be cultural historians, much less dinosaurs. And yet, whatever else they are, they certainly are both of these—and only a decade after they were young turks! (See the interview in Licklider, 1989.) The history of electronic spreadsheet applications for microcomputers dawned just in the fall of 1979, yet it is already rich and valuable, even a bit hoary, to anyone seeking to understand modern technology or to impact successfully the course of future technological evolution. Bricklin and Frankston saw all of that history, but, more than this, they created that history by creating the ancestor of subsequent spreadsheets.

This is another reason why history is so intriguing in HCI: It is deliberately, routinely, and profoundly created by HCI designers in the course of their workaday activities. In the case of the spreadsheet, a few simple but powerful insights into how people work, and how they might work, entrained a substantial revolution in how they do work and, we may be sure, in how they will work in the future.[5] This is weighty stuff: How can we manage our design activities in a satisfying and effective manner when our frames of reference are always radically dynamic and the implications of what we do are oftentimes immediately far-reaching?

Francis Bacon seemingly anticipated that modern science and technology might beget such pressures and responsibilities. Design rationale, taken broadly to incorporate all the approaches described in this volume and others now being pioneered, can be seen as instantiating his "natural history of trades" (Bacon, 1605/1970, p. 10). Because design rationale is

---

5. Dauntingly, we must also consider that, if these developments are allowed to be merely evolutionary, they may also always be optimized too locally (e.g., see Gould, 1989).

developed within the design process itself, it mitigates the paradox of instant history by integrating the analysis and abstraction of a design with the creation and implementation of that same design. Design rationale can help us build a pertinent understanding of the context, the users, the tasks, the technologies, and the situations—as we go. In the 17th century, a lag of 5 or 10 decades in the development of a history of technology turned out to be a decisive limitation; for the spreadsheet, even a single decade would have been prohibitive. Indeed, a reasonable projection is that HCI design history needs to be codified instantaneously in order to be useful.

Our particular approach to design rationale has been to treat it as a vehicle for building contextualized science out of practice. The background for us was the observation that situated artifacts serve as embodied theories in the practice of HCI; the foreground was the question of how to make this observation useful. Design rationale provides a way of getting the implicit theory out of the artifact and its situation of use and into a form that is public and explicit. The representations we have exhibited here are only semiformal, but they are more disciplined and precise than the post hoc case studies we referred to earlier, and they entail a design practice that is more structured and accountable than direct emulation of prior art. This is exactly the kind of consequence we are after: We want to support what is evident, successful, and (therefore perhaps) natural about current design practice in HCI (the transaction between tasks and artifacts in use and the treatment of artifacts as embodied theory), but we want to augment this practice (though not so abruptly or precipitously as to damage or undermine what makes it attractive and efficacious).

Can we do better? We are keenly aware of how difficult it has been to build an applicable and intellectually significant science base for HCI (Carroll & Campbell, 1986). Our strategy now is to let the design material we work with dictate its own analysis to the extent we can. We have adopted a principle of "ontological minimization" (Carroll, 1990a, p. 321): The scientific ontology of a practical domain should add as little as necessary to existent practical ontology. This ecological value system urges us to distinguish between those concepts and techniques that are applicable in understanding and creating tasks and artifacts versus those that are not. On these grounds, again, design rationale appears particularly suitable to us: It impels grounding the scientific interests of HCI research in the practical concerns of HCI design.

We can clearly do better. Psychological design rationale already justifies itself to us in our own design work. Our practice with it has converged rapidly in the past 2 years, and it has also proven to be rich in spawning work on new issues and application domains in our laboratory. For example, we are now applying it to the design of design tools (which has raised

interesting issues about designing scenarios in which scenarios are designed—and evaluating claims and trade-offs at multiple levels of use;
Rosson & Carroll, 1993). We are also exploring how conventional cognitive
descriptions of users (e.g., task-action grammars), which have been of very
limited direct use in HCI design, might find use as tools for developing
psychological design rationale (Payne, 1991).

For us, three sorts of concerns lie in the immediate future: programmatic, methodological, and instrumental. As mentioned earlier, our research program is essentially to exercise and develop our approach through
our own design work and to explore significant theoretical issues within
the context of these design exercises. An important issue in the View
Matcher example is the generalization and abstraction of the design rationale occasioned by our decision to build a View Matcher for code reuse.
This is, of course, a very limited illustration of what we briefly described
as a psychology of tasks (see Figure 2). Our work showed how general
concepts such as paradigmatic examples, view coordination, and filtering
could be abstracted and generalized from one design project to another
(see also Singley & Carroll, 1995 [chapter 8 in this book]).

More significantly, it showed that detailed psychological consequences
could be generalized and adduced within a subsequent design argument:
The usage vignettes in the View Matcher for reuse are a specific development
of the example applications in the View Matcher for learning; both exemplify
typical functionality, but the learning examples are interactive in order in
engage the user where the reuse examples are scripted in order to disrupt
minimally the user's work. Similarly, the usage episodes in the View Matcher
for reuse were a development from the application transactions defined by
breakpoints in the View Matcher for learning; both provide decompositions
of the example, but the former delineates the typical roles (patterns of
messages and behavior) of a particular object, whereas the latter delineates
a series of message sends involving a variety of objects but over a more limited
range of function and course of time. This suggests how prior HCI design
work can provide a basis for subsequent design, but it also raises many
questions. For example, how will the exercise go when there are a thousand
design rationales in our inventory instead of one or two? Confronted with
this, designers may feel that it is better just not to know!

At the same time, we are poignantly aware of how small a piece of the
design rationale problem we are developing. Our analysis concentrates,
roughly speaking, on the cognitive psychology constituted in HCI artifacts.
Clearly, however, other domains of analytic discourse should be brought
to bear: By purporting to have value, social utility, and internal structure,
designed artifacts embody claims pertaining to economics, sociology, software engineering, electrical engineering, and so forth. Indeed, Harrison,
Roast, and Wright (1989) specifically suggested that our notion of psycho-

logical design rationale be complemented by claims pertaining to software engineering and systems design issues in a more comprehensive design rationale. But even granting limitation to the domain of cognitive psychology, one could ask whether the vocabulary of that domain is uniquely appropriate or appropriate at all: That is, will designers ever bother to penetrate this vocabulary? It is an open question whether the distinctions and generalizations supported by cognitive psychology compensate for the potential insularity of using its jargon.

Methodologically, we feel that we have demonstrated the feasibility of psychological design rationale: We can do what we wanted to do, and it seems to add value. This is compelling to us, in part, because our work was directed at a realistic HCI design problem and, in fact, has produced one system already in use outside our laboratory. Of course, this is only the first and the smallest step, although it justifies the effort of bothering to worry further. We need to explore the role that psychological design rationale might play in developing analytic summative evaluation in HCI design (evaluation directed at collections of interconnected psychological claims, not at whole artifacts or isolated features of artifacts). We need to investigate the reliability and transferability of our method to other designers and analysts as we continue to investigate and improve its efficacy (we have had some encouraging preliminary experiences with this; see Bellamy & Carroll, 1990). To some extent, we are addressing this by publishing worked examples, that is, concrete models that can be critiqued and adapted. All our work, however, pertains to exploratory systems design (small groups implementing powerful tools on powerful platforms) in contrast to product design (large groups working within existing and diverse constraints; see Grudin, 1995 [chapter 16 in this book]; Sharrock & Anderson, 1995 [chapter 15 in this book]). We have begun to expand our designer–user set by providing courses and conference tutorials (e.g., Carroll & Rosson, 1992b; see also Carey, McKerlie, & Wilson, 1995 [chapter 13 in this book]). We have a long way to go.

Finally, and directly related to the issue of managing a science base of design rationales, is the instrumental concern—tools. The tabular summaries we used in this article may be the stuff of research discourse and may even be useful to researcher-designers like us, but we have no illusion that these presentations could be generally suitable. One approach we imagine is designing a hypermedia browser for exploring claims, artifacts, and tasks, both as targets and as mutual contexts, presented as text, spoken narration and testimonials, graphics, video episodes, and so forth. We are only at the very beginning of this effort (cf. Conklin & Begeman, 1988; Fischer, Lemke, McCall, & Morch, 1991 [chapter 9 in this book]; Gruber & Russell's, 1995 [chapter 11 in this book], for discussion of more "generative" tool strategies).

Francis Bacon's "natural history of trades" is an intriguing idea. In this electronic welter of HCI, this idea is alive in an exciting and multifaceted *Zeitgeist* directed at realizing the deliberated evolution of useful and usable tools and environments. We believe that this work will transform practice in HCI design by rendering the evolution of new tasks and artifacts more deliberative, even science based. Beyond this, we believe it may transform our understanding of the nature and role of science in practice, perhaps providing a paradigm for a psychology of design.

NOTES

Background. This chapter first appeared in the *Human–Computer Interaction* Special Issue on Design Rationale in 1991.

Acknowledgments. This work was made possible by the rich context provided by our colleagues at the IBM Thomas J. Watson Research Center during the period 1988–1991. In particular, we thank Rachel Bellamy for collaboration since 1988 on psychological design rationale of Smalltalk, Rachel Bellamy and Janice Singer for collaboration in 1989 on the initial claims analysis of the View Matcher for learning, Eric Gold for discussion of reifying the usage context of Smalltalk instances, and Christine Sweeney for collaboration in the development of the reuse View Matcher system. Our general framework and its application in this chapter owe much to Rachel Bellamy, Robert Campbell, Wendy Kellogg, Steve Payne, and Kevin Singley. We also got important suggestions from John Bowers, Jeff Conklin, Tomasz Ksiezyk, Clayton Lewis, Tom Moran, and Randy Smith.

Authors' Present Addresses. John M. Carroll and Mary Beth Rosson, Computer Science Department, 562 McBryde Hall, Virginia Tech, Blacksburg, VA 24061. Email: carroll@cs.vt.edu, rosson@cs.vt.edu

REFERENCES

Bacon, F. (1970). *The two books of the proficience and advancement of learning* (Book 2). New York: Da Capo. (Original work published 1605)

Basalla, G. (1988). *The evolution of technology.* New York: Cambridge University Press.

Bellamy, R. K. E., & Carroll, J. M. (1990). Redesign by design. In D. Diaper, D. Gilmore, G. Cockton, & B. Shackel (Eds.), *Human–computer interaction: Interact '90* (pp. 199–205). New York: North-Holland.

Bellamy, R. K. E., & Carroll, J. M. (1992). Structuring the programmer's task. *International Journal of Man–Machine Studies, 37,* 503–527.

Buckingham Shum, S. (1995). Analyzing the usability of a design rational notation. In T. P. Moran & J. M. Carroll (Eds.), *Design rationale: Concepts, techniques, and use.* Hillsdale, NJ: Lawrence Erlbaum Associates. [Chapter 6 in this book.]

Carey, T., McKerlie, D., & Wilson, J. (1995). HCI design rationale as a learning resource. In T. P. Moran & J. M. Carroll (Eds.), *Design rationale: Concepts, techniques, and use.* Hillsdale, NJ: Lawrence Erlbaum Associates. [Chapter 13 in this book.]

Carroll, J. M. (1985). *What's in a name? An essay in the psychology of reference.* New York: Freeman.

Carroll, J. M. (1989a). Evaluation, description and invention: Paradigms for human–computer interaction. In M. C. Yovits (Ed.), *Advances in computers* (Vol. 29, pp. 47–77). Orlando, FL: Academic Press.

Carroll, J. M. (1989b). Feeding the interface eaters. In A. G. Sutcliffe & L. A. Macaulary (Eds.), *People and computers V* (pp. 35–48). London: Cambridge University Press.

Carroll, J. M. (1990a). Infinite detail and emulation in an ontologically minimized HCI. *Proceedings of the CHI '90 Conference on Human Factors in Computing Systems,* 321–327. New York: ACM.

Carroll, J. M. (1990b). *The Nurnberg Funnel: Designing minimalist instruction for practical computer skill.* Cambridge, MA: MIT Press.

Carroll, J. M., & Bellamy, R. K. E. (1989, April). *Smalltalk as a theory of programming.* Paper presented at the Third Annual Conference on Empirical Studies of Programming, Austin, TX.

Carroll, J. M., & Campbell, R. L. (1986). Softening up hard science: Reply to Newell and Card. *Human–Computer Interaction, 2,* 227–249.

Carroll, J. M., & Campbell, R. L. (1989). Artifacts as psychological theories: The case of human–computer interaction. *Behaviour and Information Technology, 8,* 247–256.

Carroll, J. M., & Kellogg, W. A. (1989). Artifact as theory-nexus: Hermeneutics meets theory-based design. *Proceedings of the CHI '89 Conference on Human Factors in Computing Systems,* 7–14. New York: ACM.

Carroll, J. M., Kellogg, W. A., & Rosson, M. B. (1991). The task-artifact cycle. In J. M. Carroll (Ed.), *Designing interaction: Psychology at the human–computer interface* (pp. 74–102). New York: Cambridge University Press.

Carroll, J. M., & Rosson, M. B. (1990). Human–computer interaction scenarios as a design representation. *Proceedings of the 23rd Annual Hawaii International Conference on Systems Sciences,* 555–561. Los Alamitos, CA: IEEE Computer Society Press.

Carroll, J. M., & Rosson, M. B. (1992a). Design by question: Developing user questions into scenario representations for design. In T. W. Lauer, E. Peacock, & A. C. Graesser (Eds.), *Questions and information systems* (pp. 85–99). Hillsdale, NJ: Lawrence Erlbaum Associates.

Carroll, J. M., & Rosson, M. B. (1992b). Getting around the task-artifact cycle: How to make claims and design by scenario. *ACM Transactions on Information Systems, 10,* 181–212.

Carroll, J. M., Singer, J. A., Bellamy, R. K. E., & Alpert, S. R. (1990). A View Matcher for learning Smalltalk. *Proceedings of the CHI '90 Conference on Human Factors in Computing Systems,* 431–437. New York: ACM.

Carroll, J. M., Thomas, J. D., & Malhotra, A. (1979). A clinical-experimental analysis of design problem solving. *Design Studies, 1,* 84–92.

Conklin, E. J., & Begeman, J. L. (1988). gIBIS: A hypertext tool for exploratory policy discussion. *ACM Transactions on Office Information Systems, 6,* 303–331.

Conklin, E. J., & Burgess-Yakemovic, KC. (1991). A process-oriented approach to design rationale. *Human–Computer Interaction, 6,* 357–391. Also in T. P. Moran

& J. M. Carroll (Eds.), *Design rationale: Concepts, techniques, and use*. Hillsdale, NJ: Lawrence Erlbaum Associates, 1996. [Chapter 14 in this book.]

Ferguson, E. S. (1977). The mind's eye: Nonverbal thought in technology. *Science, 197*, 827–836.

Feyerabend, P. (1988). *Against method* (rev. ed.). New York: Verso.

Fischer, G. (1987). Cognitive view of reuse and redesign. *IEEE Software, 4*, 60–72.

Fischer, G., Lemke, A. C., McCall, R., & Morch, A. I. (1991). Making argumentation serve design. *Human–Computer Interaction, 6*, 393–419. Also in T. P. Moran & J. M. Carroll (Eds.), *Design rationale: Concepts, techniques, and use*. Hillsdale, NJ: Lawrence Erlbaum Associates, 1996. [Chapter 9 in this book.]

Gibson, J. J. (1979). *The ecological approach to visual perception*. Boston: Houghton Mifflin.

Gold, E., & Rosson, M. B. (1991). Portia: An instance-centered environment for Smalltalk. *Proceedings of OOPSLA '91*, 62–74. New York: ACM.

Gould, S. J. (1989). *Wonderful life: The Burgess shale and the nature of history*. New York: Norton.

Gruber, T. R., & Russell, D. M. (1995). Generative design rationale: Beyond the record and replay paradigm. In T. P. Moran & J. M. Carroll (Eds.), *Design rationale: Concepts, techniques, and use*. Hillsdale, NJ: Lawrence Erlbaum Associates. [Chapter 11 in this book.]

Grudin, J. (1995). Evaluating opportunities for design capture. In T. P. Moran & J. M. Carroll (Eds.), *Design rationale: Concepts, techniques, and use*. Hillsdale, NJ: Lawrence Erlbaum Associates. [Chapter 16 in this book.]

Guindon, R. (1990). Designing the design process: Exploiting opportunistic thoughts. *Human–Computer Interaction, 5*, 305–344.

Harrison, M. D., Roast, C. R., & Wright, P. C. (1989). Complementary methods for the iterative design of interactive systems. In G. Salvendy & M. J. Smith (Eds.), *Designing and using human–computer interfaces and knowledge-based systems* (pp. 651–658). Amsterdam: Elsevier.

Kuhn, T. S. (1970). *The structure of scientific revolutions* (2nd ed.). Chicago: University of Chicago Press.

Lange, B. M., & Moher, T. G. (1989). Some strategies of reuse in an object-oriented programming environment. *Proceedings of the CHI '89 Conference on Human Factors in Computing Systems*, 69–74. New York: ACM.

Lewis, C., Rieman, J., & Bell, B. (1991). Problem-centered design for expressiveness and facility in a graphical programming system. *Human–Computer Interaction, 6*, 319–355. Also in T. P. Moran & J. M. Carroll (Eds.), *Design rationale: Concepts, techniques, and use*. Hillsdale, NJ: Lawrence Erlbaum Associates, 1996. [Chapter 5 in this book.]

Licklider, T. R. (1989, December). Ten years of rows and columns. *Byte*, pp. 324–331. [Includes interview of Dan Bricklin and Bob Frankston by Janet Barron on pp. 326–328]

Mack, R. L., & Nielsen, J. (1987). *Software integration in the professional work environment: Observations on requirements, usage, and interface issues* (IBM Research Rep. No. RC 12677). Yorktown Heights, NY: IBM T. J. Watson Research Center.

MacLean, A., Young, R. M., Bellotti, V. M. E., & Moran, T. P. (1991). Questions, options and criteria: Elements of design space analysis. *Human–Computer Interaction, 6*, 201–250. Also in T. P. Moran & J. M. Carroll (Eds.), *Design rationale:*

Concepts, techniques, and use. Hillsdale, NJ: Lawrence Erlbaum Associates, 1996. [Chapter 3 in this book.]

Newman, W. M. (1988). The representation of user interface style. In D. M. Jones & R. Winder (Eds.), *People and computers IV* (pp. 123–144). Cambridge, England: Cambridge University Press.

Payne, S. J. (1991). Interface problems and interface resources. In J. M. Carroll (Ed.), *Designing interaction: Psychology at the human–computer interface* (pp. 128–153). New York: Cambridge University Press.

Polanyi, M. (1958). *Personal knowledge: Towards a post-critical philosophy.* Chicago: University of Chicago Press.

Raj, R. K., & Levy, H. M. (1989). A composition model for software reuse. *ECOOP '89: Proceedings of the Third European Conference on Object-Oriented Programming,* 3–24. London: Cambridge University Press.

Roberts, T. L., & Moran, T. P. (1983). The evaluation of text editors: Methodology and empirical results. *Communications of the ACM, 26,* 265–283.

Rosson, M. B., & Alpert, S. R. (1990). Cognitive consequences of object-oriented design. *Human–Computer Interaction, 5,* 345–379.

Rosson, M. B., & Carroll, J. M. (1993). Extending the task-artifact framework: Scenario-based design of Smalltalk applications. In H. R. Hartson & D. Hix (Eds.), *Advances in human–computer interaction* (Vol. 4, pp. 31–57). Norwood, NJ: Ablex.

Rosson, M. B., Carroll, J. M., & Bellamy, R. K. E. (1990). Smalltalk scaffolding: A case study in minimalist instruction. *Proceedings of the CHI '90 Conference on Human Factors in Computing Systems,* 423–429. New York: ACM.

Rosson, M. B., Carroll, J. M., & Sweeney, C. (1991a). Demonstrating a View Matcher for reusing Smalltalk classes. *Proceedings of the CHI '91 Conference on Human Factors in Computing Systems,* 431–432. New York: ACM.

Rosson, M. B., Carroll, J. M., & Sweeney, C. (1991b). A View Matcher for reusing Smalltalk classes. *Proceedings of the CHI '91 Conference on Human Factors in Computing Systems,* 277–284. New York: ACM.

Schön, D. A. (1982). *The reflective practitioner: How professionals think in action.* New York: Basic Books.

Sharrock, W., & Anderson, R. (1995). Organizational innovation and the articulation of the design space. In T. P. Moran & J. M. Carroll (Eds.), *Design rationale: Concepts, techniques, and use.* Hillsdale, NJ: Lawrence Erlbaum Associates. [Chapter 15 in this book.]

Simon, H. (1981). *The sciences of the artificial* (2nd ed.). Cambridge, MA: MIT Press.

Singley, M. K., & Carroll, J. M. (1995). Synthesis by analysis: Five modes of reasoning that guide design. In T. P. Moran & J. M. Carroll (Eds.), *Design rationale: Concepts, techniques, and use.* Hillsdale, NJ: Lawrence Erlbaum Associates. [Chapter 8 in this book.]

Smith, D. C., Irby, C., Kimball, R., Verplank, W., & Harslem, E. (1982, April). Designing the Star user interface. *Byte,* pp. 242–282.

Suchman, L. A. (1987). *Plans and situated actions: The problem of human–machine communication.* New York: Cambridge University Press.

Wexelblat, A. (1987). *Report on scenario technology* (IMCC Tech. Rep. No. STP-139-87). Austin, TX: Microelectronics and Computer Technology Corporation.

5

Problem-Centered Design for Expressiveness and Facility in a Graphical Programming System

Clayton Lewis
John Rieman
Brigham Bell
University of Colorado

ABSTRACT

This chapter presents a case study in the use of problems in design. Problems—concrete examples of user goals whose accomplishment a system is intended to support—were used to describe the intended function of a graphical programming system and to manage the growth of the space of design alternatives for the system. Problems were also used to evaluate alternative designs: They served as bench marks for comparing both the solutions offered by differing designs and the work required of users to reach these solutions. The problem-centered design process includes a representation of design rationale in which the strengths and weaknesses of design alternatives in dealing with specific problems, rather than abstract connections among design issues, are central.

Clayton Lewis is Professor of Computer Science and a Member of the Institute of Cognitive Science at the University of Colorado; his interests include methods in user interface design and especially the design of programming systems. **John Rieman** is a computer scientist and cognitive scientist with interests in user interface design and cognitive architecture; he is currently a postdoctoral fellow at the Medical Research Council Applied Psychology Unit in Cambridge, England. **Brigham Bell** is a computer scientist with interests in visual programming and intelligent tutoring systems; he is a Member of the Technical Staff of the Advanced Computing Technologies group at US West Advanced Technologies in Boulder, Colorado.

CONTENTS

1. INTRODUCTION

This chapter presents a design case study that illustrates a new approach to a design question common in human–computer interaction: How can the design rationale for large numbers of complex, interacting design alternatives be managed? Because the case study examines the design of a programming system, the chapter also describes an approach to dealing with usability considerations in programming systems, rather than in end-user systems with closed functionality.

Both of these issues are approached by placing *problems* at the center of the design process. Here we mean problem not in the sense of something wrong, like a health problem, but in the sense of a statement of a user goal, like a problem in algebra or physics. Design alternatives are linked to problems that demonstrate concretely how well or how poorly an alternative supports some relevant user goal. Design rationale is expressed in these linkages. The expression may be explicit, in cases where a designer composes a statement of the abstract issues involved in applying an alternative to some problem, or it may be implicit, when the description of an alternative and an associated problem is left as a concrete example of an advantage or disadvantage of the alternative.

Problems are also central in addressing usability within the design process. We identify two aspects of the usability of a programming system, which we term *expressiveness* and *facility*, and we use problems to evaluate both aspects. A programming system shows adequate expressiveness on a problem if there exists a program of reasonable size within the scope of the system that solves the problem. A programming system shows adequate facility on a problem if a user can easily write a program that solves the problem, without the need for extensive problem solving or rarely available background knowledge.

We develop these points and their background in the balance of this introduction, beginning with an overview of the major phases of the case study and the main ideas illustrated in them. The introduction also relates the approaches we developed in the case study to earlier work. The second section of the article describes the case study in detail. The third and final section presents the lessons we draw from the case study.

1.1. Overview of the Case Study

The problem-centered design process emerged during our efforts to design a programming environment for animated simulations, the project described in detail in Section 2. Here we preview the case study, painting the major phases of the design process with a broad brush and highlighting the roles played by problems in these phases.

Representing Design Goals as Target Problems. The proposed programming environment, called ChemTrains, had the general goal of allowing nonprogrammers to create animated graphical simulations. From our own experience and through discussions with potential users, we identified a small number of simulation problems that seemed to typify what a user would want to program with our system. We called these *target problems.* For example, one target problem was to create a simulation in which the user could manipulate the gas control on a graphic of a Bunsen burner, causing the flame of the burner to change size and the material in a beaker over the burner to change from solid to liquid to gas.

Evolving the Space of Design Alternatives. We had a general idea of the computational model ChemTrains should present, based partly on experiences with other simulation environments and partly on intuition.

But as we began to define the details of the system, the target problems became the primary focus of the design effort. The need to address aspects of the target problems drove our search of the space of possible designs, and design ideas that applied to none of the targets were dropped from consideration.

In the course of our working discussions, the original target problems were joined by problems of a different kind. The target problems were all complete, in that they described something a user might want to simulate using ChemTrains. In discussing approaches to these problems, however, there often emerged particular aspects of one of these simulations that raised issues of special significance. For example, as part of a simulation of document routing in an office, there arises the particular problem of modeling distribution lists. These interesting pieces of the original target problems often cropped up repeatedly in our discussions and, like the target problems, came to be known by individual names.

Besides these problem fragments, another kind of small problem joined the original target problems in our discussions. When a new design alternative was proposed, we needed to compare it with others. Sometimes the difference between two alternatives was captured by framing a new problem on which the alternatives clearly diverged. These problems, like the fragments of the target problems, were not complete simulation problems but, rather, picked out just some key part of a complete problem. They too cropped up repeatedly and acquired names as we discussed alternatives. We call all of these small problems, whether they arose as fragments of target problems or were created in order to compare design alternatives, *microproblems*.

Capturing Design Rationale. It was necessary for us to keep track of the rationale for our various design decisions, partly for reasons extrinsic to the design project and partly as a way of managing design ideas that would drop out of consideration at one point and be revived at another point. We needed to know if an idea being brought back for discussion had had something wrong with it that had led to its being dropped or whether it had just been displaced by some other seemingly attractive alternative. It became apparent that the problems we had discussed, both the target problems and the smaller problems we derived from them, formed the basis of our design rationale. Rather than remembering abstract statements of what was wrong with a design idea, we found ourselves remembering what problem the idea had failed to solve. We further found that we often could not formulate satisfactory abstract statements of these difficulties even if we tried.

These observations led us to formulate a problem-centered representation of design rationale. Design alternatives are linked to problems that they solve poorly or well, rather than to abstract statements of their weaknesses or strengths.

Collapsing the Design Space. Even though the target problems acted to constrain the design space, the number of design alternatives brought into consideration became quite large, and of course the number of possible combinations of design alternatives became astronomical. We called a halt to the exploration of design alternatives and described three complete alternative designs for ChemTrains. The three designs reflected different overall approaches to the design problem.

Evaluating Competing Designs Using the Target Problems. We analyzed each of these designs, again using our target problems as a focus. We first considered expressiveness: Could each design support a solution to every problem, and were the solutions compact and relatively simple? We then evaluated facility: What problem-solving efforts and special knowledge, beyond the basic syntax of the language, would a user require to bridge the gap between the problem statements and the solutions?

Checking Generality Using Reserved Problems. This analysis of the three candidate designs identified a clear winner, but we needed to compare the designs on a broader basis than just the target problems used in the design process. After all, one or all of the designs might have been narrowly tuned to work well only on these problems. To guard against this possibility, we had held in reserve a further set of target problems that we did not consider during the design phase. We compared the designs on these *reserved problems*, using the same methods of evaluation.

User Testing. In the final phase of the case study, we implemented a prototype of the most promising ChemTrains design and asked test users to solve one of the target problems with the prototype. The results provide a partial check on the evaluation methods we used in the design process.

1.2. Related Design Approaches

Our use of problems is an outgrowth of earlier work suggesting the value of detailed analysis of examples of use in user interface design (Carroll & Rosson, 1990; Gould, 1988; see also Young, Barnard, Simon, & Whittington, 1989). Because of the complexity of the sequences of events that constitute successful use of a user interface and because of the cost of using live user testing to explore the advantages and disadvantages of differing designs, these authors advocated the preparation of detailed scenarios that included each step a user would take in using a design. By preparing a scenario, a designer could verify that a given task could actually be accomplished with a design, and by examining the scenario he or she could identify parts of the interaction that were complicated, imposed high memory demands, or were awkward in other ways. The cognitive walk-through procedure (Lewis, Polson, Wharton, & Rieman, 1990) is an example of the use of scenarios to examine the mental processes required in performing a task with a given design.

We use the term *problem* rather than *scenario* in describing our approach to bring out an important distinction. A scenario, as the term is used in earlier work, usually includes not only a description of a user's goal but also a description of the steps the user would execute to accomplish the goal using a particular design. In our adaptation of these ideas, we use the term problem to refer to a task description that is independent of any system that might be used to solve it. Our approach focuses on problems, not scenarios, because a problem can be used to compare and contrast different design alternatives, a crucial task within the design process, whereas a scenario provides evidence only about the design alternatives used to generate it.

Our approach is also closely related to the use of tasks as a basis for comparing systems, as in Card, Moran, and Newell (1983), and to a long tradition of task analysis in human factors. We use the term problem rather than task because the microproblems that play a crucial part in the evo-lution of the design space are too fragmentary, and often too closely tied to particular design alternatives, to be comfortably called tasks. Neverthe-less, the logic of our method is essentially the same as that in Card, Moran, and Newell: Systems can be compared by examining what users must do to accomplish representative tasks.

1.3. Designing for Usability in a Programming System

As Newell and Card (1985) pointed out, the field of programming language design has been untouched by any explicit consideration of usability. Although the work of Brooks (1977), Soloway and colleagues (Soloway & Ehrlich, 1984; Spohrer, Soloway, & Pope, 1985), Green and colleagues (Green, 1986; Green, Bellamy, & Parker, 1987; Sime, Greene, & Guest, 1977), and others has produced useful insights into mental processes in programming and how language features help or hinder them, language designers seem to be unaware of this work. Language designs are presented and evaluated mainly on the basis of the characteristics of programs rather than the characteristics of the mental processes needed to produce programs. In an excellent text, MacLennan (1987) provided a clear summary of guiding principles of language design; it contains no reference to mental processes.

To bring users' mental processes into our design considerations, we adopted two criteria for evaluating designs—expressiveness and facility—influenced by Mackinlay's (1986) analysis of expressiveness and effectiveness in graphical representations. A language scores high on expressiveness for a given problem if a program that solves that problem is short and simple. A language scores high on facility for a problem if the process of creating a program that solves that problem is simple. We have deferred attention to another important evaluation criterion, comprehensibility.

How We Assess Facility. For assessing facility, we adapted the cognitive walkthrough method (Lewis et al., 1990; Polson, Lewis, Rieman, & Wharton, in press). In our approach, one outlines the mental steps required to develop a program from a problem statement and looks for steps that are unlikely to be made successfully, given the user's knowledge. In making this assessment, the designer can choose any sequence of mental steps he or she chooses. If all the steps are straightforward, this sequence is probable and the language gets high facility marks. If any steps do not seem likely to be made, however, the rating suffers.

In applying this idea, it immediately became clear that no programming language is likely to show high facility without assuming some background knowledge on the part of users. This can be illustrated by Soloway's work on learning Pascal (Spohrer et al., 1985). Knowledge of the statements of Pascal is not adequate to write programs. Rather, one must know about *plans* that show how to combine statements into groupings that perform a function that can be linked to the problem. For example, the accumulator variable plan shows how to use iteration to calculate the sum of a collection of values. The plans can be seen as bridges that connect what the user wants to do (calculate the average of some rainfall measurements) to the statements of Pascal (assignment, iteration), often via knowledge of some mathematical abstractions (sum). A Pascal programmer who knows about sums, iteration, and assignments, but not about accumulator variables, has a long, frustrating problem-solving episode ahead.

Doctrine and Facility. We call the knowledge about a language that must be available to use it effectively *doctrine.* Our use of this term is borrowed from military science, in which, for example, *tank doctrine* would refer to a body of ideas about how to use tanks effectively in a battle. We include in doctrine only knowledge specific to a language or its use, not commonly available general knowledge.

Evaluating a language for facility really means evaluating the language and its doctrine. A language can show poor facility in two ways. First, it can require a lot of doctrine, because its features are hard to relate to the kinds of problems it is supposed to solve. Second, on a particular problem, even with its doctrine, it may be impossible to develop a solution without extensive problem solving. Accordingly, our evaluation of competing designs included the development of appropriate doctrine followed by an analysis of how easily the particular problems could be solved given each design and its accompanying doctrine.

2. THE CASE STUDY

In this section, we present a more detailed discussion of the ChemTrains design effort, the work that brought our ideas of problem-centered design into focus. We describe the role of problems in all phases of the design process.

The presentation here is based on copies of design descriptions generated during the project, notes that we took during the design process, and written analyses carried out in the latter phases. During preparation of this chapter, we used this material to construct a graphical representation of the overall space of design alternatives we developed and the problems we considered; we draw on this retrospective reconstruction in summarizing the evolution of the design space.

2.1. The ChemTrains System and Its Goals

The ChemTrains system is a tool intended to permit users without programming background to construct animated graphical models of systems for which they have a qualitative behavior model, such as a Turing machine or document flow in an organization. The interactions among the graphical objects are specified by means of production rules, and their movements are defined by predrawn paths. Changing the initial state of a simulation can cause different production rules to fire; different objects to be created, deleted, or modified; and different paths to be selected and followed.

Four designers (the three authors of this study and Victor Schoenberg) were involved in the design project at different times. The initial design sketch, prepared by two of the designers, provided an outline still recognizable despite the intervening design work described later in this section. ChemTrains models were to show objects moving among places on the screen along visible or invisible paths that were specified graphically. Objects participated in reactions, like chemical reactions, which could modify them or delete them. Like chemical reactions, ChemTrains reactions were thought of as occurring only when the objects involved appeared in the same place. The name ChemTrains was suggested by the role of these reactions and the role of paths, thought of as railroad tracks. Figure 1 shows a simple simulation in an early version of the system.

Figure 1. Solution to the office problem in an early ChemTrains. The simulation window shows four places: three offices and a copy room. Places are connected by paths, at the ends of which are triangular filters. The people in the offices, the memo "To Dept," and the copier are all objects. When the simulation is run, the filters pull objects matching their names onto the paths, causing the "To Dept" memo to take the path to the copy room. In the copy room, the rule shown in the reaction window will fire, changing the single memo into three memos, addressed to the office holders. The filters attached to the copy room will direct those memos to the appropriate offices, where additional reaction rules will simulate how each office occupant disposes of the memo.

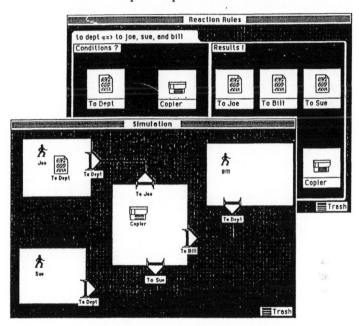

2.2. The Source and Function of Target Problems

The early ChemTrains design was developed using two problems as illustrations of the intended functionality. The office problem, the source of the simulation in Figure 1, showed how documents might flow in an organization, including copying and the use of distribution lists. The Bunsen burner problem was a trivial qualitative physics problem showing how varying the control of a gas burner could cause the burner's flame to vary in size, which in turn would cause the phase of water in a beaker to be ice, water, or steam.

The office and the Bunsen burner problems were the first examples of what we came to call target problems. These are complete descriptions of things someone might want to do with the system, things which the system is intended to support. Initially, these problems arose as tools for deliberation and discussion of the proposed design. Asked what the ChemTrains project was all about, we could describe the class of user who might have one of these problems, then explain how ChemTrains would support a solution. The description gave a solid foundation from which the discussion could have been extended to more general or abstract terms. However, perhaps because of their concreteness, problems typically remained the focus of early discussions. A colleague might ask, "What if the user wanted to simulate a more intelligent distribution of memos, like my secretary does?" Or, "What if I want to show the gas flow to the Bunsen burner?" Such questions gave us the opportunity to explore more of the design's strengths and weaknesses, in terms of the original problems.

Similarly, staying within the spirit of problems, we sometimes considered how ChemTrains would handle a problem from a colleague's domain. "Can ChemTrains simulate the running score of a pair of squash players?" Or, "Can ChemTrains simulate a Petri net?" Some of these problems seemed to fall within our general goals for ChemTrains, goals that were becoming better defined by, and in terms of, the problems. Simulations of a Petri net and a Turing machine were added to our set of target problems because they seemed to be appropriate applications for Chem-Trains and because they posed interesting difficulties for the design. In other cases, problems seemed to fall outside of our immediate goals for ChemTrains. Our abstract design rationale was often insufficient for us to explain why this was the case; we could only point to the problems that ChemTrains should be able to handle.

The role of problems in defining the functionality of ChemTrains became especially clear when the fourth designer joined the team. It was the new designer's contention that the initial design was too limited, but his abstract arguments did not persuade the other designers. Because he could not show the value of his proposed features on the trivial office and Bunsen burner problems, he posed the more difficult problems of simulating a winning tic-tac-toe player and simulating a mouse exploring a maze. The other designers were persuaded that these challenges were within ChemTrain's scope, so they were added as additional target problems, forcing us to give serious consideration to the new designer's proposed features.

Taken together, then, the target problems came to represent the functional objectives of the ChemTrains system. They served to document the system's goals and to document implicitly the rationale for the features included in the system. As we describe in the next subsection, they also served to drive our investigation into the design space. Brief descriptions of the six target problems are shown in Figure 2. (The complete problem statements, together with complete documentation of other aspects of the case study, are in Rieman, Bell, & Lewis, 1990.)

Figure 2. **Brief description of the six target problems. (The descriptions used in the walkthroughs were more detailed.)**

Bunsen burner: Show how the flame of a Bunsen burner responds when the user moves a control knob to the off, low, or high position. Show water in a beaker above the flame changing from ice to water to steam as the flame changes.

Office: Show three offices and a copy room. A memo addressed to any one of the office holders travels to that person's office and is destroyed. A memo addressed "to office" travels to a copy room, where copies are made that travel to each of the office holders.

Tic-tac-toe: The simulation should play a nonlosing game of tic-tac-toe against the user.

Maze: Simulate an intelligent mouse searching a maze for cheese. A technique is described by which a real mouse can keep track of its progress through the maze, using crumbs and bits of string. The maze is specified, but the simulation should work if the maze is changed.

Petri net: Show the operation of Petri net transitions, which fire and generate a new token on their output place when tokens are found on both of their input places. A specific network of transitions is specified, but a general solution is required.

Turing machine: Show the operation of a Turing machine, which includes a moving read/write head, a tape, an internal state variable, and a set of rules describing the head's actions. A specific task, changing a string of bits on the tape to even parity, is specified, but a general solution is required.

2.3. How the Design Space Evolved

The initial design ideas for places, objects, paths, and reactions obviously left many points to be investigated, including how movement along paths was to be controlled, how reactions would be specified, the nature of objects, and more. Among ourselves, we attempted to describe in detail how a ChemTrains implementation would allow a user to solve the target problems, and the design space became both larger and more detailed. Obviously, the problems caused us to propose design features, but proposed features also had an effect on problems. The interaction was roughly as follows.

Target Problems Dissolve Into Microproblems. In attacking a given problem with some version of a design, some aspects of the problem were easily dealt with whereas others caused trouble. We developed microproblems by pulling out the key parts of target problems that illustrated trouble in the design. For example, matching a memo "to office" with a distribution list called "office" is a microproblem derived from the office problem.

Microproblems Expand the Design Space. Consideration of microproblems led to the proposal of design alternatives. Where one design had trouble, another design could be suggested that could deal with the trouble.

Design Alternatives Engender New Microproblems. Comparison of design alternatives led to formulation of new microproblems. Design Alternative B might outdo Design Alternative A on Microproblem 1, but what about this new situation, Microproblem 2? Microproblem 2 shows what is wrong with B and good about A.

Microproblems Merge Into Existing or New Target Problems. Usually these new, emergent microproblems could be embedded in an existing target problem, either by focusing attention on some hitherto neglected aspect of one of them or by extending the scope of one. Sometimes, however, the new microproblems seemed to bring out something that the system should be able to do but that no existing target problem required. This could (as in the case of the fourth designer's proposed features) lead to the addition of new target problems.

Each of the processes described is related to problems. We did, on occasion, voice concerns over abstract design issues, and the features or solutions proposed by individual designers sometimes reflected preferences for abstract goals in both a general sense (e.g., simplicity) or a more specific one (e.g., locality of reference). But abstract concerns in themselves never provided sufficient rationale for adding a feature. It was problems that drove the search further into the design space.

2.4. Capturing Design Rationale

The chapters in this book develop the arguments for giving design rationale an important place in the management of design. In the ChemTrains project, two reasons predominated. First, the ChemTrains project was part of a larger project seeking to strengthen the role of cognitive theory in system design (Doane & Lemke, 1990), and as part of this we needed to keep track of the extent to which cognitive considerations did or did not play a role in the design. Second, and of more general relevance, we needed to manage a large, shifting set of design alternatives. We needed to recall why, for example, Idea A had been chosen over Idea B at some stage in design, because Idea B had resurfaced and looked attractive in connection with some new issue.

Our efforts to keep track of our design decisions and the reasons for them led us to give problems a key role. We found that problems nearly always provided the basis for rejecting design alternatives. We seldom abandoned an idea on abstract groups; rather, we composed a concrete problem that showed why the idea would not work. Recording the reasons for such decisions was easy if we just described the problems but was very hard if we instead tried to present abstract characterizations of the design issues. In addition, we found problems easy to think about, remember, and discuss. Design debates were punctuated with references to problems rather than abstract design issues.

2.5. Examples of the Relationship Between Problems and Design Alternatives

To give a flavor of the interaction among target problems, micro-problems, and the design space, this subsection includes examples of the design space evolution described in the previous subsections. Two points are worth noting at the outset. First, some of the problems and design alternatives require considerable explanation, even in the concrete terminology of problems; to reformulate them in abstract terms would be difficult. Second, we were able to describe these and other examples, some of them months after the relevant discussions had taken place, with little difficulty. Because the problems were the rationale for the design decisions, a few brief notes provided a powerful memory cue into a tightly interconnected space of decisions and deliberations. Our ability to recall this material may be related to its concreteness: Concrete materials are easier to learn in verbal learning studies (see Crowder, 1976, for discussion).

Example 1: Target Problem Yields a Microproblem, Which Spawns a Design Alternative. The original ChemTrains design prescribed that reaction rules could only operate when all participating objects were in the same place. This limitation was motivated by one designer's intuition that it would make rules simpler and easier to understand than if objects situated anywhere could freely interact.

This restriction on interaction caused immediate difficulty in handling the Petri net raw problem. Figure 3 shows part of that problem. The circles in the diagram represent input places (not to be confused with ChemTrains places!), and the horizontal lines represent transitions. The rules of operation for a Petri net specify that, when one or more tokens are in each input place for some transition, one token is absorbed from each input place and a single token is passed out of the transition (along the paths leading to the bottom of the diagram). Notice that a token in the middle input place may trigger the firing of either transition, depending on where other tokens are.

In considering what ChemTrains places should be set up to model the Petri net, there is obvious trouble. One would like to have a place enclosing both input places for each transition, because tokens in the input places

Figure 3. **Petri net problem.**

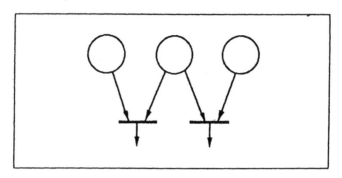

need to interact to trigger the transition. But because the middle input place is shared by two transitions, these places would have to overlap, something not contemplated in the original ChemTrains design. This shared input place microproblem, part of the Petri net target problem, directly spawned a new design alternative: permit ChemTrains places to overlap.

Example 2: Design Alternatives Spawn a Microproblem. Alternative ways of specifying rules, termed the *before–after* and the *operations* alternatives, arose from consideration of how rules might most naturally be written, an issue that cut across the target problems. In the before–after approach, a rule would be specified by describing the relevant state of the model before the rule applied and then by showing the state that would be produced by the rule. In this approach, one would not say explicitly that an object was to be deleted, for example, but rather would simply include it in the "before" description and omit it from the "after" description. In the operations approach, the "before" conditions would be described as for before–after, but the action of the rule would be described as a collection of operations on objects, like "delete memo." This approach would permit the user to state explicitly what is supposed to happen.

Although considerations of these contrasting approaches was sparked by concern for natural expression rather than by a specific problem, the comparison spawned the formulation of a new microproblem, the *modification* problem, which was framed as a part of the office target problem. In the modification problem, a memo is to have its destination changed for rerouting but carry the same content. In the operations approach to rules, this is easy to arrange. By specifying a modification operation to be applied to the memo, one can easily indicate that only the destination is to be changed, whereas other aspects, including the content, should be left alone. But in the before–after approach, there is trouble. The crux of the problem is the need to indicate that the memo appearing in the "after" description is really the same one that appears in the "before" condition, that is, to indicate that the original memo is being changed rather than that the original memo is to be deleted and a new, unrelated memo is to be created. The simplest before–after approach cannot make this distinction; some machinery must be added, such as explicit links between objects in "before" and "after" or the inclusion of variables in the "before" descriptions whose bindings can be used in the "after" description.

2.6. Overall Character of the Development of the Design Space

The previous examples illustrate the typical patterns of growth of the design space, but they do not capture all of the influences at work. To develop a more complete picture, we constructed retrospectively a list of the design alternatives we considered and how they entered the design discussion. Of 49 design alternatives considered, 16 arose directly from microproblems. Two more arose directly from target problems, with no intervening microproblem being identified. A further 15 alternatives arose as refinements of alternatives spawned from problems. The remaining 16 alternatives arose not from target or microproblems but from knowledge of other systems, such as Hypercard, or from abstract considerations the designers brought to bear on the ChemTrains design, such as the comprehensibility of representation with local, as opposed to global, definitions of terms. Thus, two thirds of the design alternatives arose from consideration of problems.

The same retrospective representation of the design process allows us to see where microproblems came from. Of a total of 22 microproblems, 13 arose from target problems or as parts of other microproblems, and the remaining 9 were spawned to help evaluate design alternatives.

A design alternative can be linked to a problem in one or two ways: It can arise as a solution to a problem, or, as with the 9 microproblems just noted, a problem can be created as a way of indicating a strength or weakness of the alternative. Of the 49 design alternatives, 43 were linked to problems in one of these ways, and 8 were connected only to abstract considerations.

As might be expected, some of these design alternatives that were not linked to problems were ones not considered very fully in the design discussions. Four of the eight alternatives not linked to problems were considered extensively, and all of these represented general design approaches, such as the use of graphics to represent rules. Thus, all design alternatives not linked to problems were either ones not examined closely or ones capturing considerations that cut across all of the problems.

2.7. Collapsing the Design Space

It became apparent that development of new microproblems and design alternatives could go on indefinitely. It was also clear that, although we could produce designs that were expressive enough to solve the target problems using many combinations of the features we had proposed, we could not assess the facility that different combinations of features would afford without examining the combinations in detail. Individual design features in themselves did not promise facility or the lack thereof; only by examining the process required to solve a problem with a complete design could we determine this. For a combination of reasons, including the need to report on our work and the feeling that further investigation of individual features was an effort with diminishing returns, we decided the time had come to close down exploration of the design space and to evaluate one or more complete ChemTrains designs.

We elected to define three complete designs (ZeroTrains, ShowTrains, and OpsTrains), which represented distinct ways of choosing alternatives within the space. In ZeroTrains, the simpler alternative was chosen whenever possible. Thus, for example, ZeroTrains retained the restriction that rules could apply only to objects in a single place. OpsTrains chose alternatives for power, influenced strongly by experience with the OPS family of rule-based systems (see Forgy, 1984). Thus, OpsTrains views places as a kind of object, allows objects to be nested, and permits variables in rules. ShowTrains, taking a middle path between simplicity and power, differs most noticeably from the other two designs along a different dimension—concreteness. In ShowTrains, rules are not expressed in any notation but rather are specified by demonstration, a technique suggested by Maulsby's work on specifying procedures by example (Maulsby & Witten, 1989). The ShowTrains user always works within the context of the model, pointing out the objects that figure in the condition of a rule and then manually making the changes that should occur if the rule applies.

The three designs we chose come nowhere near exhausting the possible combinations of design alternatives in the design space, but the designs did permit us to compare the consequences of designing by three distinct intuitions that we found increasingly in conflict in design discussions. We could not settle by discussion whether the power of OpsTrain would make it easier to apply than ZeroTrains or whether the concreteness of ShowTrains would make it easier to understand and use than the other two designs.

2.8. Evaluating Expressiveness

The first step in evaluating the three designs was to attempt to find solutions to all six target problems using each design. We were confident from discussion of the various design features that this would be possible, but some last-minute adjustment to the designs was permitted in preparing the solutions. As expected, solutions were found for all the problems for each design. Although we had no specific criteria for expressiveness, all solutions fell within our general expectations as to the number of programming elements (objects, places, paths, and rules) required.

2.9. Developing the Doctrine

The solutions to the target problems, along with the descriptions of the three designs, were the target material we required to develop doctrine for each design, that is, the ChemTrains-specific knowledge that a user would need in order to produce a solution to a given problem using a given design. Optimally, the user solving a problem should perceive a series of unambiguous matches between elements of the problem, on the one hand, and items of doctrine, on the other, with each item of doctrine prescribing the next action the user should take. We developed the doctrine iteratively, using the following procedure:

- Prepare an initial list of items of doctrine, representing all the special knowledge that seems to be needed to solve the target problems.
- Step through the user actions that lead to a solution to each target problem. For each action, consider the problem statement, the current state of the program, all the doctrine, and the user's current beliefs and expectations. If the user has insufficient reason to take the required action, add or modify an item of doctrine that will cause the actions to be taken in this situation. If doctrine seems to suggest more than one action, make the doctrine more specific to prevent confusion.
- Step through the solution again, until a complete and unambiguous set of doctrine is ensured.

In keeping with our goals that ChemTrains be easy to learn and generally applicable, we strove to keep our doctrine short, and we avoided doctrine that was overly specific to a target problem.

This process of tracing out a path from a problem statement to a solution, using the doctrine at each step to determine what to do next, is similar in spirit to the cognitive walkthrough used for user interface evaluations by Lewis et al. (1990). In the cognitive walkthrough, the focus is on the adequacy of cues provided by the user interface to guide choices the user must make. In our analysis, the focus is on the adequacy of the doctrine. We call our variant a *programming walkthrough*. A programming walkthrough evaluates the facility of the underlying design indirectly: A design has high facility if there is a body of doctrine for it that guides each step in the solution process and is not large or complex.

Evolution of Solutions as Doctrine Was Developed. As the doctrine developed, it sometimes happened that a new solution to a problem was adopted. Some of the original solutions had reflected insight into a design that could not be conveyed readily in doctrine or had resulted from a good deal of trial-and-error problem solving. Some of these could be replaced by solutions that could be derived more readily from a reasonable body of doctrine.

The Bunsen burner solution in ZeroTrains illustrates this joint evolution of doctrine and solutions. In the original solution, the place containing the flame was expanded to include both the beaker and the various places for the control. This is an economical solution to the problem of allowing the parts of the burner setup to interact under the ZeroTrains restriction that interacting objects must be in the same place. But it proved hard to formulate doctrine that could reliably indicate which of two or more places to expand in situations like this. After examining the requirements of this and other problems, the following item of doctrine was developed:

> D17. The objects in a trigger condition for an event must all be in one place. If they are not, create a new place that encloses the placed where the objects are found, and put a catalyst object C in the new place but outside all the original places. Include C in the trigger for the event.

This doctrine leads to a less economical solution, because a catalyst object is needed, but the doctrine seems to provide direct guidance in a wide range of cases.

Revision of Problem Statements as Doctrine Was Developed. We also found the need to refine the statements of the target problems during this phase of the process. The original problem statements often permitted different approaches to the problems, which in turn led to different solutions. For example, the maze problem as originally stated could be solved by following walls or by marking routes that have been explored. Had these differences in approach to the problems reflected differences in what was natural in the designs or had they been driven by doctrine, they would have been informative. But in some cases it seemed that the differences simply reflected arbitrary choices, and they complicated comparison of the designs. In these cases, we selected one approach for each problem and specified it in the problem statement.

The doctrine for the three designs is excerpted in Figures 4, 5, and 6. Because we had no experience in describing doctrine, the designers used different presentations, as can be seen. Although differences in form and content make precise comparison impossible, it appeared that the three designs required about the same amount of doctrine. Taking an item of doctrine to be a statement of a choice to make or action to take, together

Figure 4. **Excerpts from OpsTrains doctrine.**

RO1:	IF	starting,
	THEN	draw a picture of envisioned interface as it would initially appear to the user.
RO2:	IF	an object can move or can be placed in a particular area of the interface that is not drawn,
	THEN	draw an object that is big enough to contain the objects and specify that the object is to be "hidden in simulation."
RO5:	IF	a place object is to be used to hold different types of objects but never more than one at a time and the place object is empty,
	THEN	create an object to denote that the object is empty, place it in the empty place object, and specify that the object is to be "hidden in simulation."
RR1:	IF	an object should be moved or deleted or a new object should be created based on specific conditions that may exist in the picture,
	THEN	enter the replacement rule editor, copy all of the objects and paths relating to the conditions and all the objects and paths to be modified from the main picture to the pattern picture of the rule, copy these objects and paths onto the result picture, modify the objects in the result picture appropriately, and give the rule a name that is appropriate for the task it does.
RR2:	IF	an object or path is to be deleted when a rule is executed,
	THEN	remove that object or path from the result picture.
RR3:	IF	an object or path is to be added when a rule is executed,
	THEN	create a new object or path, or retrieve an existing one and add it to the result picture.
RR5:	IF	an object in the pattern of a rule may match any object regardless of its display,
	THEN	specify that this object is a variable. (a big V will be placed over the variable object in the pattern)

with some statement of the conditions in which it is relevant, OpsTrains has 21 items, ShowTrains has 24, and ZeroTrains has 26.

2.10. Evaluating Facility and Expressiveness: The Walkthroughs of Record

We concluded this phase of the design effort by producing "walk-throughs of record" for each of the six target problems in all three designs. In these walkthroughs, the doctrine and design features were fixed across all problems. Each walkthrough took as its starting point a written description of a target problem. The walkthrough then described the sequence of problem-solving activities and system interactions that the programmer would need to follow to solve the problem. At every point, the programmer's critical decisions were justified by citing specific items of doctrine and features of the problem. Because the designs and doctrine had been

Figure 5. Excerpts from ShowTrains doctrine. An introduction to the doctrine describes the simulation involving, a cat, bird, and worm, which is referred to in some of the points of the doctrine.

D1. *Sketch a snapshot:* On a blank piece of paper, sketch a "snapshot" of the running simulation—something like Figure 1. Do not spend a lot of time on this, but produce a rough graphic that shows how you expect the screen to look. Do not worry about marking things as objects, places, or paths, but spend a few moments thinking about how you want to interact with the simulation as it is running. For example, will you want to have a button to click that creates more worms for the bird?

D2. *Identify objects:* Decide what is going to move or be created, deleted, or modified. These will be objects. Create the objects that should exist when the simulation begins to run. Give them meaningful names and graphics.

D6. *Create object-changing rules:* For each place, decide what conditions will cause objects to be created, modified, or deleted in that place. Set up exactly those conditions, highlight the relevant items, and demonstrate to ShowTrains what should occur. Be sure to give each rule a meaningful name—this will make it easier to revise the simulation.

Figure 6. Excerpts from ZeroTrains doctrine.

D1. First draw a sketch of roughly what you want to see on the screen for the model.

D2. Things that can move around in the picture or that can be added to or removed from the picture or that can change as the model runs will represented by objects. Begin a list of objects for the problem, including different versions of objects that change. Things that can be moved or changed by the user are also represented by objects.

D3. Draw outlines on your sketch to show places where objects can appear, disappear, or change or to which objects can move.

D4. Draw paths to connect places between which objects can move. Put names on the ends of the paths in such a way that the paths leaving any place all have different names. If a path will be used only one way, you only need to give a name to the end objects will enter.

D7. Things that can happen in ZeroTrains are creations of new objects, deletion or modification of existing objects, or objects moving along a path. One or more of these things that should happen together is called an event. Make a list of events for your model. For each event, say when it should occur, that is, what situation in the model triggers it.

D17. The objects in a trigger condition for an event must all be in one place. If they are not, create a new place that encloses the places where the objects are found and put a catalyst object C in the new place but outside all the original places. Include C in the trigger for the event.

Figure 7. Statement of Bunsen burner problem used in walkthroughs.

Purpose: Show how changing the position of the control on a Bunsen burner affects the state of the water in a beaker.

Task description: Show a Bunsen burner with a beaker of water on top of it. Also on screen, separated from the burner, show a control with positions high, low, and off. Allow the user to manipulate the control. When the control is in the high position, a large flame should be visible between the burner and the beaker and the water should be shown as a cloud of steam (the cloud should stay in the beaker). When the control is in the low position, a small flame should be visible and the water should look like plain water. When the control is in the off position, no flame should be visible and the water should be shown as a cube of ice.

---

developed with these six problems as targets, the walkthroughs offered a chance to exhibit each design in the best possible light.

The solutions described by the walkthrough of record allowed us finally to nail down our claims as to the expressiveness and facility of each design, at least insofar as the target problems were concerned. Facility was a function of the length of the doctrine and the amount of problem solving required to produce the solution. Expressiveness was judged by the number of ChemTrains places, objects, paths, and rules that the solution required.

In the following paragraphs, we describe a walkthrough of record for the OpsTrain design, using a single target problem—Bunsen burner. We then briefly discuss how the Bunsen burner walkthroughs for the other designs differed from OpsTrains.

The Bunsen Burner Problem. The target problem asks the programmer to simulate the operation of a Bunsen burner, showing how the flame changes as a user drags a control knob into one of three positions: off, low, and high. The simulation must also show how water in a beaker above the flame changes phase—from ice to water to steam—as the flame changes. Figure 7 shows the full problem statement. All items of OpsTrains doctrine used during the walkthrough were given in Figure 4.

Summary of the OpsTrain Walkthrough. In OpsTrains, objects and places are not different elements: Any object may contain other objects. An object is identified by its shape and by the objects it contains. The doctrine for OpsTrains tells the programmer what to do as a first step, but no further order of action is specified. Instead, doctrinal rules suggest appropriate actions to take in various situations. For the Bunsen burner simulation, Doctrine RO1 (see Figure 4) tells the programmer to start by drawing the visible objects of the simulation. The programmer draws the Bunsen

Figure 8. OpsTrains solution to Bunsen burner problem.

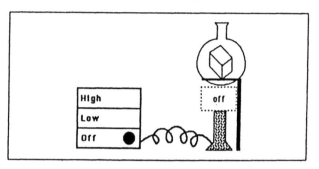

burner, the beaker, and the control panel, which is made up of three separate rectangles in which the control may be located, marked by the text strings "Off," "Low," and "High" (see Figure 8). Drawing an object in Ops-Trains causes the object to be created. Additional items of doctrine induce the programmer to draw a place for the flame to appear (Doctrine RO2), enclosing an object whose graphic is the text string "off" (Doctrine RO5). In compliance with the doctrine, both the "off" object and the flame-place object are defined to be invisible when the simulation runs.

Doctrine RR1 then suggests that the programmer specify the OpsTrain reaction rules that describe the activity in the simulation. The same item of doctrine describes the general procedure for specifying a reaction rule: Set up the simulation window the way it will appear when the reaction is to occur, then copy and paste the relevant patterns of objects and paths into both the "pattern" and "result" sides of a new reaction rule. Finally, modify the result side of the reaction rule to show appropriate changes, which may involve deleting objects, creating new objects, or sending objects along paths. Two additional items of doctrine, RR2 and RR3, map the general idea of reactions into the specific needs of the problem for deletion or creation.

Guided by this doctrine, the programmer begins to specify the reaction rule that causes the flame to change from off to a low-flame appearance when the control is moved into the low position. As the programmer specifies the reaction rule, another item of doctrine becomes applicable. The appearance of the flame when the control is moved into the low position is irrelevant; whether it is a high flame or the invisible "off" object, it should be transformed into a low-flame graphic. In this situation, Doctrine RR5 advises the programmer to make the flame's appearance a variable in the condition of the rule. The programmer follows this advice, marking the flame with a graphic *V* that indicates its variability.

For the Bunsen burner simulation, a total of six reaction rules are required. Three reaction rules change any flame object to a low flame,

Figure 9. Two of the six reaction rules in the OpsTrains solution for the Bunsen burner problem.

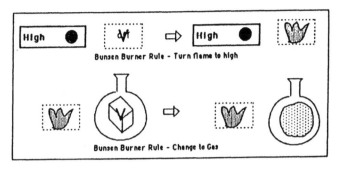

high flame, or the invisible "off" marker when the control is in the low, high, or off position, respectively. Three more reaction rules are needed to transform any phase of the water to ice, liquid, or steam, depending on whether the object in the flame-place is "off," low flame, or high flame. The walk-through shows that the programmer enters each of the reaction rules as a direct result of the application of doctrine to the problem statement. The graphic versions of two of the reaction rules are shown in Figure 9.

Summary of the ShowTrains Walkthrough. The final ShowTrains solution is similar to OpsTrains. However, the walkthrough shows a solution path that requires many more decisions on the part of the programmer. In ShowTrains, elements of the simulation may be objects, places, or paths. Whereas the OpsTrains programmer merely had to decide that some real-world thing would be an object, the ShowTrains programmer has to decide whether it will be a place or an object. ShowTrains is further complicated by the fact that objects, places, and paths have both a name and a graphic and that either or both may be significant in the condition part of a reaction.

ShowTrains claim to ease of use is concreteness. What the OpsTrains programmer must do in a separate rule window, the ShowTrains programmer can do on the main screen. However, it takes only a few items of OpsTrains doctrine to describe the extra steps, and the actual work saved in ShowTrains is far overshadowed by ShowTrains complexity of decision making.

As with OpsTrains, a total of six reaction rules are needed: three to set the graphic of the flame, depending on the position of the control, and three to set the graphic of the water, based on the graphic of the flame. In the second three rules, it is the graphic of the flame that is significant, whereas in the first three it is the location of the control knob, a distinction

Figure 10. **ZeroTrains solution to the Bunsen burner problem.**

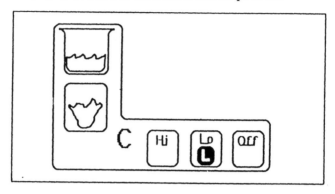

that the programmer must indicate by appropriate highlighting during rule creation.

Summary of the ZeroTrains Walkthrough. In the ZeroTrains design, the elements from which a simulation must be built are places, objects, and paths. These elements are simple graphics, with no names. Places may be nested or overlapped, but objects cannot contain other objects. Reactions have fewer options in ZeroTrains than in OpsTrains or ShowTrains, and the ZeroTrains final solution is significantly different from the solutions in the other designs.

The walkthrough is complicated by ZeroTrains limitations, which must be overcome by tricks that are guided by doctrine. For example, rules can only be specified in terms of objects, not places, so extra objects have to be created to mark relevant places. ZeroTrains cannot test for an object's absence, so a no-flame object has to be in place when the flame is off.

ZeroTrains is also limited by the principle that things can interact only when they are together in the same place. When places overlap, the meaning of "together in the same place" depends on an information-hiding convention that we referred to as the *membrane rule,* and making the Bunsen burner simulation work involves using a catalyst object, an action directed by doctrine (see D17 in Figure 6). The catalyst is the large *C* in Figure 10.

The final section of the walkthrough finds the programmer translating the events list into reaction rules in the ZeroTrains graphic reaction-editor window. Because nothing can be variabilized in a ZeroTrains reaction, events specifying "a or b" in their condition must be entered as two separate reaction rules. Twenty-four reaction rules are required for the final solution.

2.11. Comparing the Programming Walkthroughs

Comparison of the walkthroughs for the three designs revealed a consistent pattern of differences, clearly attributable to particular design decisions. The comparison does much to resolve the conflicts of intuition that bedeviled our efforts to evaluate these design decisions earlier in the development of the design.

Overall, the power-oriented OpsTrains design showed the greatest facility. The central ideas of the problem statement could be translated quite directly into OpsTrains rules guided by doctrine of about the same size and complexity as for the other designs. ShowTrains placed second, with many solutions reachable about as directly as OpsTrains and others requiring considerable problem solving even with the doctrine supplied. ZeroTrains placed last, with each solution requiring work directly attributable to getting around ZeroTrains "simplifying" assumptions. Overall, the comparison provides no support for the intuition that the simplicity of the ZeroTrains design would pay back in simple, clear doctrine what it lost in power. The concreteness emphasized in the ShowTrains design also had limited positive effect, and the design's middle course between simplicity and power seemed to combine the worst rather than the best traits of each.

It is not surprising that variables permitted more economical solutions in OpsTrains. It is perhaps more surprising that the use of variables did not require substantial doctrine. The OpsTrains doctrine guides the user in an approach similar to that used in Query by Example (Zloof, 1975), in which the user first writes a rule with no variables that handles a specific case. The user then replaces some constants with variables to produce a rule that handles other cases. As in Query by Example, this approach appears to allow the OpsTrains user to create rules that use variables in a fairly complex way, for example, to require that designated parts of two different objects must be the same, without really understanding much about how the variables work. As we describe later, this prediction from the walkthrough analysis was not borne out in tests with users.

It is important to note the difference between the expressiveness evaluation of the designs, based just on the characteristics of the solutions supported by them, and this facility analysis, which examines the process of arriving at solutions. Consider as an example the ZeroTrains requirement that rules involve only one place. It might appear that simply examining the solutions available with and without this restriction would suffice to establish its cost. However, the restriction does not make solutions much bigger. Only in the programming walkthroughs does it become obvious that meeting the ZeroTrains restriction requires a great deal of extra work. This can be seen clearly in the Bunsen burner example.

The success of OpsTrains challenges the widely held intuition (see, e.g., MacLennan, 1987) that languages should contain a minimum number of concepts. Until the programming walkthroughs were complete, it seemed very likely to us that the power of OpsTrains would exact a cost and that understanding how to select among and to apply its larger number of capabilities would require either more doctrine or more problem solving. This did not happen, because the features of OpsTrains matched the naturally arising statements of problems more directly than did the limited features of, in particular, ZeroTrains. If added features in a language make it easier to relate the language to real problems, they confer a significant advantage.

2.12. Evaluating the Designs for Generality Using Reserved Problems

The whole design process up to this point was strongly shaped by the set of target problems. The growth of the design space and the evaluation of alternative complete designs for expressiveness and for facility relied on these problems. The economy and coherence gained by concentrating on a few problems are accompanied by a risk: What if the resulting designs were narrowly tuned to look good on these problems but would fail on other problems that were within the intended scope of ChemTrains but were not included in the target problem set?

To address this question, we held a number of target problems in reserve, not using them in design discussions or in the evaluation process. The reserved problems were the explicit stand-in for the wide class of problems for which we hoped ChemTrains would be useful. Knowing that they were on the agenda, we were encouraged to define features in general terms, even though the features were initially proposed to solve specific problems.

Source of the Reserved Problems. These reserved problems included one of the original motivating problems for ChemTrains, modeling sunspot development, which was discussed briefly at the very start of the design process and then withdrawn from consideration. To this, we added other problems that we encountered in the literature or in other work that seemed within the intended scope of ChemTrains. Brief descriptions of the reserved problems are shown in Figure 11.

After completing the expressiveness and facility evaluation of the three designs on the original target problems, we did walkthroughs for each design on each reserved problem. Neither the designs nor the doctrine used in the earlier evaluation was allowed to change, in keeping with the goal of detecting narrow focus in design or doctrine resulting from our concentration on the original target problems earlier in the design process.

Figure 11. Brief descriptions of the four reserved problems. The descriptions used in the walkthroughs were more detailed.

Doorbell: Simulate an electromechnical doorbell. At the push of a button, current generates a magnetic field, pulling on an arm that rings a bell and opens the circuit. A spring pulls the arm back, and the cycle repeats.

Grasshoppers: Grasshoppers hatch, consume resources, change to adults, lay eggs, and die. Show how availability and importance of resources affect the population of adult grasshoppers over many generations (after Varley, Gradwell, & Hassell, 1973).

Sunspots: Show how a tube of magnetic field lines bends and rises above the surface of the sun, causing sunspots to appear. Then show how the tube rises even higher, the field lines separate and reconnect, and a solar flare is produced (after Patrick McIntosh, personal communication, June 19, 1990, and Tandberg-Hanssen & Emslie, 1988).

Copier: Show how an office photocopier operates. Actions to simulate include transfer of charge, transfer of toner particles conditional on charge, and coordinated motion of the paper and an internal belt (after Schrager, Jordan, Moran, Kiczales, & Russell, 1987).

Results of the Walkthroughs With Reserved Problems. The analyses using the reserved problems produced results generally consistent with those from the original target problems. The advantage of OpsTrains over the other designs in both expressiveness and facility was even greater than in the original target problems because the reserved problems tended to be bigger. The lack of variables in ZeroTrains and the limitations on variabilization in ShowTrains meant that many more rules were required under these designs than in OpsTrains, a fact already apparent in the earlier analyses but of more consequence for the bigger reserved problems.

Although the analyses of the reserved problems did not markedly change our assessment of the three designs, they did reveal that some potentially important design issues were not adequately raised by the original target problems. Three of the reserved problems—doorbell, sunspots, and grasshopper—involved time-dependent processes, which had not figured in any of the original target problems. None of the three designs had adequately direct ways of representing such processes.

The reserved problems also exercised the ability to represent structured objects more fully than the original problems. In the grasshopper problem, for example, it was natural to think of the state of a grasshopper as combining several components reflecting food and water needs, stage of maturation, and the like. This was difficult to represent in ShowTrains and ZeroTrains.

The ability in OpsTrains to move paths around was useful in the doorbell problem to open and close a circuit. This feature was not used in any of the original target problems.

None of the designs provided rotation of the graphics for objects. This is needed to portray the motion of segments in a moving belt in the copier problem.

The doctrine for the three designs stood up quite well to the reserved problems. Even more than with the designs themselves, we had been concerned that our doctrine might have been tailored to the requirements of the original target problems, but there were few indications that different doctrine would have worked better for the reserved problems. Both Show-Trains and ZeroTrains had in fact included some "advanced" doctrine that was not needed in the original target problems, but this was not needed for the reserved problems either.

All in all, the reserved problems confirmed the conclusions drawn from the original target problems, but they also revealed that the original problems did not cover all aspects of the intended design space. This difficulty could be met in part by including more complex target problems in the starting set for design development.

2.13. User Testing of a ChemTrains Prototype

Our programming walkthroughs convinced us that OpsTrains was the design of choice. We considered incorporating additional features into the design and performing further walkthroughs but decided first to test the walkthrough results empirically, using the OpsTrains design on which the walkthroughs had been performed. We used a minimal OpsTrains prototype, including an interactive graphic editor, a pattern matcher and rule executor, and sufficient features to allow the Bunsen burner problem to be programmed.

Method and Subjects. Our claim from the programming walkthroughs was that a nonprogrammer could use this version of ChemTrains to solve the problems we had considered, relying only on the knowledge described by the doctrine. To test this claim, we asked subjects to program one of our original six problems, a Bunsen burner simulation, using the prototype. The results of that empirical work are described in detail in Bell, Rieman, and Lewis (1991) and are summarized here.

Six subjects performed the task, working individually. Four had some traditional programming experience, one was an experienced programmer, and one had no programming experience. The instructional material given to the subjects was essentially the same doctrine used in the programming walkthroughs. We rewrote the doctrine in complete sentences and made some word substitutions to avoid jargon, but we did not try to find a really good way to present the doctrine. Half the subjects read the doctrine and began the task; the other half had the doctrine read and explained to them before beginning the task. When subjects reached a point in the task where they seemed unable to proceed, we intervened and gave sufficient help to allow them to continue. Because we were interested in the adequacy of the information contained in the doctrine, we called subjects' attention to pertinent sections of the doctrine, if there were any, before providing any new information.

Results. All six subjects were able to apply the more straightforward items of doctrine and make initial progress toward a solution. The experienced programmer completed the entire task with only minor problems. There were several areas, however, in which subjects had problems that we had not anticipated in the walkthrough analysis.

All subjects had some problem understanding the general way in which ChemTrains operated. All subjects failed to appreciate the need to draw objects on screen to define the location where certain reactions, such as the change in the shape of the flame, occurred. This was a problem even though it was explained in the doctrine.

Three Ways the Walkthroughs Fell Short. We place subjects' problems into three categories, each of which has implications for the programming walkthrough methodology and for ChemTrains. First, some of the problems were caused by "holes" in the doctrine. Subjects' inability to comprehend how a user would interact with the system and how the firing of a rule would affect the main screen fall into this category. In writing the doctrine, we were simply too close to the problem to appreciate the need to state all the basic facts.

A second category of problems occurred when subjects' real-world knowledge interfered with doctrinal prescriptions. This explained the failure to apply the doctrine that described the need for a place in which the flame would be located. Every subject initially assumed that ChemTrains could distinguish a pattern by evaluating how close two objects were to each other, just as there is a real-world distinction between a flame directly above a burner and a flame floating in space.

Predicting interference from real-world knowledge during the walkthrough is difficult, and adjusting the doctrine to trap all possible misconceptions would be impossible: The doctrine simply cannot be extended to describe everything that the language cannot do. This may be an area in which empirical testing is required to identify particularly attractive false paths that the doctrine should block.

The third category of problems involves situations for which the doctrine prescribed programming procedures but did not convince the subject that the recommended procedures would work. The notable example here is variables. Because the doctrine stated exactly where and how a variable should be used, the walkthrough predicted that a programmer familiar with the doctrine should be able to write programs using variables.

Indeed, some of the subjects recognized where variables should be used and even described how to use them. Then, having determined how to apply the doctrine, they balked at taking the appropriate action because they could not understand how it would affect the program. Thus, doctrine that is informationally adequate may fail because users do not feel that they understand it.

This analysis suggests that doctrine should describe the results of each action it prescribes so the programmer can confidently select and take the action when applicable. Describing the effect of variables raises the issue of variables' implications for program comprehensibility. As noted earlier, we have not addressed this important issue, although others have (see Gilmore & Green, 1984; Green, Sime, & Fitter, 1981).

3. METHODOLOGICAL LESSONS FROM THE CASE STUDY

There are obvious limitations in what can be concluded from any single case study, especially when (as here) the study is done in an academic setting, free of the pressures of schedule, budget, and users. ChemTrains is a small project, in scope and staffing, although not smaller than many "real" projects. Any one study involves just one group of designers and one design problem, both of which may differ importantly from what will be encountered elsewhere. Nevertheless, we venture some observations that may or may not be found to apply in other situations.

3.1. The Role of Problems in Developing Design Alternatives

Problems played three key roles in the growth of the design space for ChemTrains. First, target problems defined the functional scope of the intended design. Second, microproblems served to promote the formation of new design alternatives to deal with the difficulties they represented. Third, microproblems were used to evaluate or compare design alternatives by capturing difficulties associated with the use of some design alternative.

In these roles, the concreteness of problems, both target problems and microproblems, was crucial. We do not think we could have prepared an abstract specification of the functional requirements of ChemTrains. Or rather, we could have, but only by beginning with a list of target problems and trying to describe abstractly the characteristics apparently required in these problems. We simply lacked abstract concepts and categories adequate to describe what ChemTrains should do; Buckingham Shum (1995 [chapter 6 in this book]) discusses this problem. Collecting target problems was easy and natural.

The concreteness of microproblems was equally important. Occasionally, an issue arose in abstract form (e.g., conflict resolution) because it was familiar from previous work, but usually an issue arose as a difficulty whose nature needed to be clarified before we could describe it. A concrete microproblem was the natural way to clarify a difficulty and to communicate it.

In reporting the dominance of concrete problems in our design deliberations, we are not arguing that abstractions are in some way bad. To the contrary, we recognize abstractions as valuable, when they are available. But our experience, in this poorly understood domain of design, was that few useful abstractions were available to us at the start, and it was very difficult to develop them as we worked.

Another key point about the problems we used, as noted by Ray McCall (personal communication, May 7, 1990), is that they all represent pieces of the intended scope of the ChemTrains design. This may seem trivial, but we found it to be significant in keeping the design process focused. In a discussion of abstract issues, it is easy to stray from matters that are important to matters that are interesting but not really relevant. Our problems, however, were things we really needed to work on.

This is obvious for the target problems: We could not accept failure on any of the target problems without changing the scope of the design. For microproblems, the relationship is conditional but still strong. As noted earlier, some microproblems were spawned by consideration of a current design alternative and embodied some difficulty for that alternative. Because microproblems were nearly always parts of target problems, they could not be neglected without neglecting a target problem or abandoning a design alternative.

3.2. The Role of Problems in Design Rationale

As noted before, the overall goals of the ChemTrains project required us to capture design rationale. We initially attempted to record the issues under debate and the relationships between them in a manner similar to issue-based information systems (IBIS; see Fischer, Lemke, McCall, & Morch, 1991 [chapter 9 in this book], and Conklin & Burgess-Yakemovic, 1991 [chapter 14 in this book]). We tried to show issues deriving from issues, as in that framework, but we got bogged down. One difficulty was that it was not easy, as already noted, for us to give abstract statements of the issues with which we dealt. Indeed, even preparing the tabulation of the design space that we used in Section 2.6 required us to develop abstract descriptions of design alternatives that we did not use in the original design discussions and, hence, did not have readily at hand.

A second, related difficulty was that the problems kept getting in the way. We found that we thought about design alternatives and their strengths and weaknesses in terms of the microproblems they were connected to. The description of the evolution of the design space in Section 2.6 makes clear that the consideration of almost all of the design alternatives included one or more problems and that most of the development of the design space can be described in terms of the relationships between problems and alternatives. But there was no natural way to incorporate this organizing material, the problems, into the issue structure. The same difficulty would arise in the questions, options, and criteria framework described by MacLean, Young, Bellotti, and Moran (1991, [chapter 3 in this book]). As soon as we recognized the organizing role of problems, it became very easy to map out our design discussions; it was not always as easy, however, to think of terms that would be meaningful outside the context of the specific problems with which we dealt.

In hindsight, problems could have been incorporated into other representations of design rationale, including the IBIS family. Issues would be of two kinds: "What design alternative should be chosen for such and such a design choice?" and "How should such and such a problem be solved?" Problems would either be embedded in issues of the second kind or would appear as arguments for or against design alternatives that appeared as possible answers to issues of the first kind.

This embedding of the problem-design alternative structure in an IBIS-like framework does not avoid the need to develop abstract statements of design issues, which we found difficult, as noted earlier. Further, some problems appear in the design rationale both in the role of arguments for or against alternatives and as issues to be resolved, and it is not clear how to frame the statement of a problem so that it can play these roles simultaneously. As an issue, a problem would be framed as, "How can such and such a problem be solved?" As an argument, it would appear as, "Such and such an alternative cannot deal with such and such a problem." Flexible interpretation of the issue structure might accommodate these role clashes, but at the least we can observe that the issue-centered view did not suggest to us a natural way of capturing our design deliberations, even if in hindsight we could have found a way to work within that view.

The structure of problems and design alternatives we created contains the design rationale for ChemTrains in a way that is both implicit and concrete. We wrote no statements explaining that we chose alternative A for reason R. However, by examining the microproblems associated with a design alternative, one can reconstruct these reasons readily, in the form of concrete, specific examples of situations in which a specific alternative deals with difficulties encountered by other alternatives.

Despite the naturalness of capturing our design rationale in this form, we did not keep any up-to-date record of it. Besides the obvious overhead associated with any recording process, we see two reasons for this. First, we found we could keep the rationale in our heads during design discussions. Even though these discussions took place over a long period of time and with many interruptions, we did not feel that we were losing our grasp of the issues or were revisiting already plowed ground. The concreteness of the problems probably was important in allowing us to remember them and the alternatives tied to them. We do not know whether our ability to reconstruct rationale in the relatively small ChemTrains project would extend to larger projects.

A second reason we did not keep our rationale up to date was that we became confident of our ability to reconstruct the problem structure if needed. This is closely related to the point just made, but it meant that, even though we intended to produce a record of the design rationale, we felt little pressure to produce it as we worked.

The fact that, except for the very incomplete presentation in this chapter and the more extensive presentation in Rieman et al. (1990), we represented design rationale only for our own use during the design process means that we have not addressed some additional serious issues. How can problems be described fully enough so that someone not involved in the original design discussions can reconstruct them adequately? How could we ourselves reconstruct the problems at a much later time? These are serious matters for any representation of rationale, but we do not feel they would be more difficult to address for problem-centered rationale than for another approach.

3.3. The Role of Problems in Evaluating Expressiveness and Facility in Programming Languages

The use of concrete problems in the ChemTrains design meant that we could not only compare the solutions afforded by different designs but also the processes involved in thinking of the solutions. We could address the facility as well as the expressiveness of the designs.

Programming Walkthroughs as a Preview of Prototype Usability. Like other walkthrough methods (Lewis et al., 1990), the programming walkthrough can be seen as a partial substitute for, or preparation for, prototype evaluation. It can give an indication of probable weaknesses in a design without the expense and effort of building a prototype and testing it.

As the results of our empirical testing make clear, evaluating facility by programming walkthrough is fallible. We nevertheless feel that the walkthrough approach, even if not followed by empirical testing, moves design debates about programming language design a step away from intuition and taste, where they have been. A designer can try to demonstrate that his or her language supports a natural approach to a task by anatomizing the specific sequence of decisions needed to accomplish the task. A critic can point to a particular decision as unmotivated or unguided by knowledge plausibility possessed by users.

Our work also emphasizes the importance of Soloway's concept of plan (Spohrer et al., 1985), not only to the analysis of the use of existing languages but also to the design of new languages. A language design should be viewed as including not just the features of the language but also the doctrine needed to put it to use. It appears to be practical to evaluate both of these elements of a design together and, indeed, is of limited value to evaluate the features without the doctrine.

Limitations of Programming Walkthroughs. A limitation of our evaluation is that it examines only the process of writing programs, without attention to understanding them, testing them, or modifying them, each of which has an important cognitive component. Certainly language design influences these other tasks as well. Further, in considering programming, we have dealt only with what might be called *shallow* programming, where sufficient knowledge is available (encoded as doctrine, in our case) to guide the writing of programs with very little search. Surely much programming is *deep* programming, where appropriate choices of representation and procedure are not obvious up front, and search and backtracking are required. Perhaps more limited designs would be easier to reason about and would show an advantage in deep programming that does not appear in shallow programming.

3.4. On Beyond ChemTrains

In subsequent applications of the method, with the benefit of the experience described in the case study, we are changing two aspects of the method:

1. *Use bigger target problems:* The target problems we chose to use in the design discussions were small enough so that we could discuss possible solutions quickly. They allowed us to uncover important design alternatives with little effort. However, the larger target problems held in reserve pointed out serious shortcomings in all three designs. The larger target problems not only showed that previously acknowledged weaknesses were more serious than we anticipated but also pointed out some unseen weaknesses. See the discussion of scenario generation in Singley and Carroll (1995 [chapter 8 in this book]).

2. *Start programming walkthrough analysis earlier:* We did a lot of design before considering facility explicitly and before doing any programming walkthroughs. As a result, new design alternatives, motivated by problems of facility, arose late in the process.

Although the ChemTrains design was done as part of a research activity, rather than as a "real" development project, we are finding that the problem-centered approach is useful in more realistic settings. Indeed, the use of problems has further advantages not shown in the ChemTrains study, when designers must communicate with sponsors or potential users. The same attributes that made problems easy for us designers to discuss make them a good basis for discussion between designers and users who may find it difficult to interpret abstract statements of requirements or abstract arguments for or against design alternatives.

3.5. Conclusions

As we grappled with the complexities of designing a usable graphical programming system, the ChemTrains design process became a study in the use of concrete problems, as opposed to abstract principles and analysis, in design. Concrete problems entered the process as a way to define objectives and focus our efforts. Once on the scene, they proliferated and penetrated all aspects of design and evaluation. They helped to organize a complex design space. They also made it possible for us to evaluate not only the solutions afforded by the design but also the mental processes required to realize solutions.

NOTES

Background. This chapter first appeared in the *Human–Computer Interaction* Special Issue on Design Rationale in 1991.

Acknowledgments. Victor Schoenberg helped get the ChemTrains design started in early discussions and prototyping. Robert Weaver helped develop the programming walkthrough approach and clarified the role of doctrine in language evaluation. Ray McCall provided many suggestions, welcome encouragement, and keen insight in our efforts to understand and capture our design process. We thank Tom Moran, Phil Barnard, and two anonymous reviewers for helpful suggestions on an earlier version.

Support. This work was supported by Grant IRI-8944178 from the National Science Foundation. Related research was sponsored by US West Advanced Technologies. Clayton Lewis was a sabbatical visitor at the Center for Advanced Decision Support in Water and Environmental Systems during final preparation of this chapter.

Authors' Present Addresses. Clayton Lewis, John Rieman, and Brigham Bell, Department of Computer Science, Campus Box 430, University of Colorado, Boulder, CO 80309. Email: clayton@cs.colorado.edu, rieman@cs.colorado.edu, and bell@cs.colorado.edu

REFERENCES

Bell, B., Rieman, J., & Lewis, C. (1991). Usability testing of a graphical programming system: Things we missed in a programming walkthrough. *Proceedings of the CHI '91 Conference on Human Factors in Computing Systems*, 7–12. New York: ACM.

Brooks, R. (1977). Towards a theory of cognitive processes in computer programming. *International Journal of Man–Machine Studies, 9*, 737–751.

Buckingham Shum, S. (1995). Analyzing the usability of a design rationale notation. In T. P. Moran & J. M. Carroll (Eds.), *Design rationale: Concepts, techniques, and use*. Hillsdale, NJ: Lawrence Erlbaum Associates. [Chapter 6 in this book.]

Card, S. K., Moran, T. P., & Newell, A. (1983). *The psychology of human–computer interaction*. Hillsdale, NJ: Lawrence Erlbaum Associates.

Carroll, J. M., & Rosson, M. B. (1990). Human–computer interaction scenarios as a design representation. *Proceedings of the 23rd Annual Hawaii International Conference on System Sciences*, 555–561. Los Alamitos, CA: IEEE Computer Society Press.

Conklin, E. J., & Burgess-Yakemovic, KC. (1991). A process-oriented approach to design rationale. *Human–Computer Interaction, 6*, 357–391. Also in T. P. Moran & J. M. Carroll (Eds.), *Design rationale: Concepts, techniques, and use*. Hillsdale, NJ: Lawrence Erlbaum Associates, 1995. [Chapter 14 in this book.]

Crowder, R. G. (1976). *Principles of learning and memory*. Hillsdale, NJ: Lawrence Erlbaum Associates.

Doane, S., & Lemke, A. (1990). Using cognitive simulation to develop user interface design principles. *Proceedings of the 23rd Annual Hawaii International Conference on Systems Sciences, Software Track*, 547–554. Los Angeles: IEEE Computer Society Press.

Fischer, G., Lemke, A. C., McCall, R., & Morch, A. I. (1991). Making argumentation serve design. *Human–Computer Interaction, 6*, 393–419. Also in T. P. Moran & J. M. Carroll (Eds.), *Design rationale: Concepts, techniques, and use*. Hillsdale, NJ: Lawrence Erlbaum Associates, 1995. [Chapter 9 in this book.]

Forgy, C. L. (1984). *OPS5 user's manual* (Tech. Rep. No. CMU-CS-81-135). Pittsburgh, PA: Carnegie Mellon University, Computer Science Department.

Gilmore, D. J., & Green, T. R. G. (1984). Comprehensive and recall of miniature programs. *International Journal of Man–Machine Studies, 21*, 31–48.

Gould, J. (1988). How to design usable systems. In M. Helander (Ed.), *Handbook of human–computer interaction* (pp. 757–789). New York: North-Holland.

Green, T. R. G. (1986). Computer languages: Everything you always wanted to know, but no-one can tell you. In F. Klix & H. Wandtke (Eds.), *Man–computer interaction research: MACINTER-I* (pp. 249–259). Amsterdam: North-Holland.

Green, T. R. G., Bellamy, R. K. E., & Parker, J. M. (1987). Parsing and gnisrap: A model of device use. In G. Olson, S. Sheppard, & E. Soloway (Eds.), *Empirical studies of programmers: Second workshop* (pp. 132–146). Norwood, NJ: Ablex.

Green, T. R. G., Sime, M. E., & Fitter, M. J. (1981). The art of notation. In M. J. Coombs & J. L. Alty (Eds.), *Computing skills and the user interface* (pp. 221–251). London: Academic Press.

Lewis, C., Polson, P., Wharton, C., & Rieman, J. (1990). Testing a walkthrough methodology for theory-based design of walk-up-and-use interfaces. *Proceedings of the CHI '90 Conference on Human Factors in Computing Systems*, 235–242. New York: ACM.

Mackinlay, J. (1986). Automating the design of graphical presentations of relational information. *ACM Transactions on Graphics, 5*, 110–141.

MacLean, A., Young, R. M., Bellotti, V. M. E., & Moran, T. P. (1991). Questions, options, and criteria: Elements of design space analysis. *Human–Computer Interaction, 6*, 201–250. Also in T. P. Moran & J. M. Carroll (Eds.), *Design rationale: Concepts, techniques, and use*. Hillsdale, NJ: Lawrence Erlbaum Associates, 1996. [Chapter 3 in this book.]

MacLennan, B. (1987). *Principles of programming languages*. New York: Holt, Rinehart & Winston.

Maulsby, D., & Witten, I. (1989). Inducing programs in a direct-manipulation environment. *Proceedings of the CHI '89 Conference on Human Factors in Computing Systems*, 57–62. New York: ACM.

Newell, A., & Card, S. (1985). The prospect for psychological science in human–computer interaction. *Human–Computer Interaction, 1*, 209–242.

Polson, P. G., Lewis, C. H., Rieman, J., & Wharton, C. (in press). Cognitive walkthroughs: A method for theory-based evaluation of user interfaces. *International Journal of Man–Machine Studies*.

Rieman, J., Bell, B., & Lewis, C. (1990). *ChemTrains design study supplement* (Tech. Rep. No. CUCS 480-90). Boulder: University of Colorado, Department of Computer Science.

Shrager, J., Jordan, D. S., Moran, T. P., Kiczales, G., & Russell, D. M. (1987). Issues in the pragmatics of qualitative modeling: Lessons learned from a xerographics project. *Communications of the Association for Computing Machinery, 30*, 1036–1047.

Sime, M. E., Green, T. R. G., & Guest, D. J. (1977). Scope marking in computer conditionals—A psychological evaluation. *International Journal of Man–Machine Studies, 9*, 107–118.

Singley, M. K., & Carroll, J. M. (1995). Synthesis by analysis: Five modes of reasoning that guide design. In T. P. Moran & J. M. Carroll (Eds.), *Design rationale: Concepts, techniques, and use*. Hillsdale, NJ: Lawrence Erlbaum Associates. [Chapter 8 in this book.]

Soloway, E., & Ehrlich, K. (1984). Empirical studies of programming knowledge. *IEEE Transactions on Software Engineering, SWE-10*, 595–609.

Spohrer, J., Soloway, E., & Pope, E. (1985). A goal/plan analysis of buggy Pascal programs. *Human–Computer Interaction, 1*, 163–207.

Tandberg-Hanssen, E., & Emslie, A. (1988). *The physics of solar flares*. New York: Cambridge University Press.

Varley, C. G., Gradwell, G., & Hassell, M. (1973). *Insect population ecology: An analytic approach*. Berkeley: University of California Press.

Young, R. M., Barnard, P., Simon, T., & Whittington, J. (1989). How would your favorite user model cope with these scenarios? *ACM SIGCHI Bulletin, 20*, 51–55.

Zloof, M. (1975). Query by example. *AFIPS Conference Proceedings, 44*, 431–432.

EMPIRICAL STUDIES
OF DESIGN RATIONALE

6

Analyzing the Usability of a Design Rationale Notation

Simon Buckingham Shum
University of York

ABSTRACT

Semiformal, argumentation-based notations are one of the main classes of formalism currently being used to represent design rationale (DR). However, using them makes particular demands on the designer, which cannot be ignored by proponents of such approaches. One way to tackle the challenge posed by DR's representational overheads is to understand the relationship between designing, and the idea-structuring tasks introduced by the DR notation. This can then inform the design of the notation in order to support the process of externalizing ideas within the notational vocabulary, and turn the structuring effort to the designers' advantage. Whereas our understanding of the demands on designers of using such representations has to date been drawn largely from informal and anecdotal evidence, in recent years, studies have demonstrated that the close study and characterization of design activity "in the natural" is a powerful way to define requirements for subsequent support technology. It is within this paradigm that the current work has been conducted.

Two studies of DR in use are reported, in which designers used the QOC (Questions, Options, and Criteria) notation (MacLean, Young, Bellotti, & Moran, 1991 [chapter 3 in this book]) to express rationale for their designs. In the first study, a substantial and consistent body of evidence was gathered, describing the demands of the core representational tasks in using QOC, and the variety of strategies that designers adopt in externalizing ideas. The second study suggests that an argumen-

Simon Buckingham Shum is a Research Fellow at the Knowledge Media Institute, The Open University, UK, studying implications and applications of interactive media and the internet for learning and design.

CONTENTS

tation-based design model based around laying out discrete, competing Options is inappropriate during a depth-first, "evolutionary" mode of working, centered around developing a single, complex Option. In addition, the data provide motivation for several extensions to the basic QOC notation. The chapter concludes by comparing the understanding of QOC that emerges from these studies, with reports of other argumentation-based approaches in use.

1. THE NEED TO STUDY DESIGN RATIONALE IN USE

Semiformal, argumentation-based notations are one of the main classes of formalism currently being used to represent design rationale (DR). Although, from a notational perspective, graphical formalisms are well suited for recording design arguments as they arise, doing so also introduces representational overheads for the designer. This chapter is concerned with understanding the nature of the extra cognitive work introduced by argumentation-based DR. This is obviously important in the context of a fast-flowing, time-pressured activity such as software design, in which "documentation" is already a bad word.

One way to tackle the challenge posed by DR's representational overheads is to understand the relationship between designing, and the idea-structuring tasks introduced by the DR notation. This can then inform the design of the notation in order to support the process of externalizing ideas within the notational vocabulary, and turn the structuring effort to the designers' advantage. Whereas our understanding of the demands on designers of using such representations has to date been drawn largely from informal and anecdotal evidence, in recent years, studies have demonstrated that the close study and characterization of design activity "in the natural" is a powerful way to define requirements for subsequent support technology (e.g., Minneman, 1991; Olson & Olson, 1991; Tang, 1991). It is within this paradigm that the current work has been conducted. However, "usability" as addressed by analyses of this sort is of course but a partial contributor to the general acceptability of a design representation or method. Accounts of the organizational context in which DR must survive make sobering reading (Grudin, 1995 [chapter 16 in this book]; Sharrock & Anderson, 1995 [chapter 15 in this book]), and should ensure that claims for usability at the individual and small-group design levels, crucial though this is, are set in context.

This chapter reports the first step of an investigation into usability at the cognitive representational level, namely, characterizing the interaction between designing and using DR for expressing and rationalizing ideas. Attention focuses on the cognitive properties of a particular argumentation-based DR notation, *QOC* (*Questions, Options,* and *Criteria*), used within the *design space analysis* perspective (MacLean, Young, Bellotti, & Moran, 1991 [chapter 3 in this book]). In this earlier work, MacLean et al. focused on the "rationale" underlying the design of QOC as a representation for DR. Subsequently, some preliminary observations about DR usability were made within an analytic framework (Shum, 1991b). This chapter develops the story, focusing on empirically based analyses of the QOC authoring process, and its relationship to different modes of software design activity. The analyses of the data gathered in these studies address issues relating to QOC's cognitive compatibility and utility for software design, with broader implications for other DR approaches and their associated representational schemes.

2. THE STUDIES: DESIGNERS, TRAINING, AND TASKS

Analyses of two empirical studies are presented in this chapter. All of the data reported are drawn from video-based observational analyses of design problem solving. In Study A, 12 pairs of software designers (16 professionals/8 students) spent an hour using QOC to redesign and rationalize the user interface to a bank's automated teller machine (ATM). The task was based on that employed by MacLean et al. (1991 [chapter 3 in this book, Figure 4]).[1] In Study B, in which different modes of design are considered, an electronics doctoral research student used QOC while designing Smalltalk-80 data structures over three 1.25-hour sessions.

In Studies A and B, the designers were taken through a tutorial that introduced DR as a general concept, and QOC specifically. In accord with the more retrospective design space analysis approach, emphasis was placed on the importance of developing coherent rationales that would communicate clearly to an outsider the key issues and reasoning behind the designs. The QOC tutorial tasks were intended to give the designers practice in structuring natural discourse semiformally. Designers were required to translate into QOC the key aspects of several fictional design discussions, such that a third party could understand what had been discussed. By varying the length of these design discussions, and the medium in which they were presented (as transcripts plus sketches, or as a video recording of a discussion) the representational task became steadily more demanding. Details can be found in Shum (1991a).

In all of the studies, designers used pens and large sheets of paper as opposed to a software tool. Under these conditions, the authoring process could be studied with minimal interference from extraneous factors, while preserving or even enhancing properties of the online medium such as display space, resolution, and ease of local editing. Although computational tools can alleviate some of the mechanical overheads of the task, the core tasks of deciding how to express reasoning as structured argumentation remain essentially unchanged.

Throughout this chapter, extracts from the design transcripts are used to illustrate points. Most of the examples in this section are from Study A's ATM design problem, which centered around reducing customer queues without sacrificing the number of services offered. Other extracts are from Study A tutorial exercises: One task was to design the remote control for a video recorder intended for the elderly, and the other was to design an airport public information symbol to indicate a "1-hour left-luggage office."

1. Two tasks were in fact used, in a between-subjects experimental design. The first described user steps for a Standard-ATM (SATM) interface, and requested a new design and DR. The second additionally described the Fast-ATM interface, and several SATM usability problems. Half of the designers were required to evaluate the FATM, and if necessary, devise and rationalize changes. The data from both tasks are combined for the analysis presented here.

3. CORE REPRESENTATIONAL TASKS IN USING QOC

In order to translate ideas into QOC, the designer is faced with three basic cognitive tasks: deciding what *kind* of an idea one has (classification), how to *label* it meaningfully (naming), and how it *relates* to other ideas (structuring). Before these tasks are illustrated, however, it is necessary to emphasize their nonlinear relationship, that is, the exploratory, opportunistic nature of the process.

3.1. QOC Construction as an Opportunistic Activity

When studying designers using QOC, it soon becomes clear that externalizing ideas as structured argumentation is not a smooth, top-down process. Continual revision and switching from one task to another characterize QOC authoring as an opportunistic mode of working (Guindon, 1990). The QOC evolves through multiple, sometimes embedded, represent-and-evaluate cycles, switching between different parts of the structure.

Various approaches to representing QOC were adopted, demonstrating that the process of developing QOC analyses is quite different from the orderly structure of the final product. For instance, in the following extract in which the designers discuss how the ATM should dispense different kinds of output, it was most natural to generate Options, then Criteria, and the Question last of all[2]:

[Study A: Pair 2]
P: Halifax machines drop everything into a little drawer . . . the Question here is . . . well the ideas *[i.e., Options]* are as is, and everything from one place. The Criteria are . . .
D: What are you going to call the Question though?
P: Hmm, I get caught on the Questions. . . . the Criteria are natural feel to it—getting it from different holes doesn't feel natural
D: Actually, it's more like a teller, more human
P: What? If you get it all from the same hole?
D: The same kind of thing like when you go the counter, and the guy gives you it through the little slot
P: *[writes]* security in mind (everything from a draw—feels secure). *[Linking Options to Criteria]*—as is—it doesn't have a natural feel to it, everything from different slots.
    What's the Question here? *[frustrated tone]*.
D: erm . . . I suppose physical layout of . . .

---

2. In longer transcript extracts, the key points are shown in **bold**, and ideas recorded as QOC like this. Dialogue that has been cut from extracts is indicated by (. . .). Additional explanatory notes are placed in square brackets, e.g., *[points to sketch]*.

Figure 1. [Study A tutorial exercise: Pair 2] Moving opportunistically to a new Question and Options as they arise, and then back to complete the original (numbers show sequence of ideas). In QOC, boxed Options indicate a decision, or at least a working commitment.

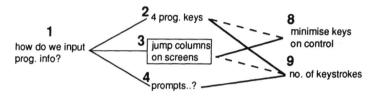

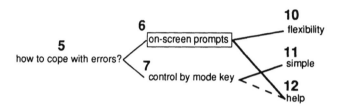

P: Layout of holes *[starts to write]*
D: Physical layout of input/output stuff *[P. writes* `layout of I/O for`
     `cash/card/receipt`*]*

Figure 1 shows the order in which QOC was constructed in another situation, illustrating switching between Questions to capture new ideas as they suggest themselves.[3]

In some cases, subjects explicitly adopted "strategies" to represent the QOC, as ways of imposing some structure on their task. For example, in order to control the tendency to pursue new ideas as they arose opportunistically, one pair declared:

[Pair 12: Study A]
M: Let's try to write down some of the Questions of concern here, and then some Criteria, and then I think some of the Options will come from that—possibly . . . 'cos Questions and Criteria are related a lot aren't they
N: So if we have one heading Questions and another here for Criteria, which might not be related
M: Right—I'm going to concentrate in terms of Questions, and then the Criteria might come from the Questions. `length of queue is a Question` . . .

About a minute later, they started to discuss the details of a possible Option but stopped themselves intentionally, aware that at this early stage it might be a tangent:

---

3. Apart from "cleaned-up" graphic appearance, all QOC examples used in this chapter are copied directly from designers' QOC drawings, unless otherwise stated.

N: So could you have a machine that behaves like a regular ATM or a fast ATM depending on where you put the card in?

N: You could . . . but would that be . . . ? *[stops herself from exploring this idea]*—right, so

M: Yeah—so *[writes]* `variety of machines` . . . *[no further discussion on that Option until later]*

It is important to note, however, that strategies of this sort seemed to be short-term, flexible modes of working; that is, they did not govern the structure of the whole session, and within them there was flexibility to attend to different parts of the QOC and to switch strategies, such as: (a) listing Questions in advance, before elaborating them; (b) generating Options and Criteria, and then the Question; and (c) generating Questions and Options first, and then evaluating with Criteria.

QOC authoring is not only opportunistic when used during conceptual problem solving of the sort required by the ATM task. Even when the decisions have been made and all the arguments are known (i.e., *retrospective DR*), working out how best to represent them is a separate task. Having recognized that the authoring process is far from a tidily sequenced activity, let us now turn to its constituent tasks.

3.2. Learning to Classify Ideas

This section focuses on the normal process of classifying ideas. When QOC is being used, the vocabulary and orientation of discussion inevitably changes, with regular references to the new constructs, for example, *that's an Option; could that be a Question?; this Criterion keeps coming up,* and so forth. As with any language, fluency increases with use, such that arguments are more smoothly and accurately translated over time. In Study A, it was found that normally subjects classified ideas without spending much time discussing what type they should be. However, as the examples of restructuring QOC show (see Section 3.4), the first translation that springs to mind is not necessarily the optimal representation. Even relative "experts" with QOC (e.g., QOC's developers) still engage in restructuring, reclassification, and renaming, revision activities that are dealt with later. The following examples illustrate classification difficulties characteristic of early use of QOC:

1. Figure 2 shows a typical error in classifying an idea (`natural order` is initially recorded as an Option, but then corrected to be a Criterion).

2. In the following extract, an Option is first represented as a Question:

[Study A: Pair 2]

D: What does this do then?—the Fast ATM do—if you press the cash amount and not *[put in]* the card?

Figure 2. [Study A: Pair 2] Correction of a typical error in classifying an idea (a Criterion as an Option).

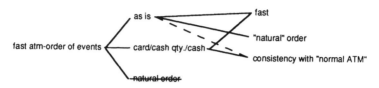

P: I guess it just goes "Ha Ha," and clears itself. It'll have a t i m e o u t, or c l e a r k e y.

D: Yeah, but that's an Option though—a Clear key.

P: On where?

D: Well, on the normal ATM. You could change the order of the events, but you have a clear key and a timeout

P: f a s t A T M o r d e r o f e v e n t s *[writes this as a new Question]* **. . . that's an Option?**

D: Yeah that's an Option

P: Oh, it's an Option that addresses this security issue isn't it *[deletes Question:* c l e a r k e y *and* t i m e o u t *on* A T M, *and writes* c l e a r k e y + t i m e o u t *as an Option].* **What's the Question? It's sort of a vandalism issue . . .**

3. This extract shows the pair generating ideas in discussion, and then striving to represent those ideas as QOC:

[Study A: Pair 5]

R: That's a design decision *[pointing to the keypad sketch].* Deciding that Cancel is going to be the emergency get-out, and that we should stick with that as its mode—the mode of that key is "get out of this, now."

J: Well, it certainly is . . . it goes back to this question of b u t t o n s—is that what the Cancel . . . I mean we've added the Cancel key, and consider that to be a design conclusion—I like that because it's intuitive. I suppose that could be a Criterion—in fact it's a very important user-Criterion. *[sigh]* How do we record that?

(. . .)

J: We've not got much time left. We could do with identifying this Cancel key more specifically. But I don't know what the Question is.

R: In fact the Question is "How does the card get returned?"

J: You can put it under the question of b u t t o n s . . . . . . I think the Enter key and Cancel key are part of the question of b u t t o n s. That's the Question *[indicates* Q. *of* b u t t o n s *]*—now there are probably other Questions that feed into this, and I don't know that you can actually . . . Hah! *[laughs]* When is something a Question, and when is it a Criterion?

In the aforementioned extract, the decisions and arguments are clear in J.'s mind, but neither he nor his partner are sufficiently fluent with QOC to translate them. J. notes in particular the informality inherent in QOC (e.g., how to classify ideas).

3.3. Naming and Renaming

Naming entities is often a process of renaming. The renaming of nodes was a prevalent activity for every pair of designers in Study A. Renaming reflects the problem-solving process of developing ideas; if a QOC is constructed as the problem is explored, it is inevitable that node-names that do not reflect current understanding of the problem must be updated.

Naming in QOC takes up a significant amount of time for several reasons. First, a node's name must be succinct, and convey the idea it represents. Second, to aid interpretation, a further constraint on Criteria is that they be expressed positively, for example, easy to learn, low error rate, low cost, high speed.[4] Third, a particularly important characteristic of names is *focus*. Focus refers to the level of generality at which the idea is expressed: A Question may address several issues; an Option may embody several key features that differentiate it from others, but not along the dimension that is addressed by the Question; a Criterion might be expressed so generally (e.g., intuitive; simple) that it is hard to see how it relates to an Option. A fourth property of a name is its relationship to others of its type: It should be *distinctive*. An Option may really be an example of another, or two Criteria might really be reexpressions of each other, for instance, each side of a trade-off, which depending on the context, might be useful or redundant. Both distinctiveness and focus in naming are characteristics of "well-formed" QOC, insofar as they force designers to be precise about the nature of an idea.[5]

Although these requirements were not made explicit to designers in Study A, they still appreciated the importance of finding good names for ideas. The next two extracts illustrate the cooperative process of refining names:

[Study A tutorial exercise: Pair 5]
R: So how are we going to . . .
T: "Keys to what kind of functions . . ."
R: That's not a very good way of putting it . . .
T: It's like the "classes of functions . . ."
R: Classes! That's the way to put it.
T: What classes of function keys?

R: . . . You see teletext is the only thing you read—you don't read other things—you don't read the picture.
T: What we have are two negative reasons

---

4. With this constraint, supports Assessments (solid links) to Options can always be interpreted as "pros," and (dashed) objects-to links as "cons"; because Criteria have different weights (Section 5.3), decisions clearly cannot be made on the basis of how many supports links Options have, but they provide an initial visual indication.

5. Principles for well-formed structures were collated as a "QOC styleguide."

R: We have to make them positive though . . . ok, so, "easy to read?"
T: Well, that wasn't the point was it? um . . . it was like that they couldn't
    actually . . .
R: They can't see it, so they don't need it.
T: *[laughs]* Yeah—it's like the Criterion is that you're providing a function
    that they can actually make use of, and they can't make use of the
    teletext because it's too small to read.
R: OK—useful function?
T: Yeah, ok.

The following comment summarizes the experience of many subjects
in learning to express Criteria positively:

[Study A: Pair 5]
R: . . . I mean, I really struggled on that first exercise, and found that very
    awkward and very difficult. In fact the thing I found most difficult was
    negating everything, so that the attribute was a positive attribute
J: Yes
R: I just couldn't get my brain to pick out the right word to describe that
    attribute.

3.4. Structuring and Restructuring

The dominant organization that presents itself to someone browsing a
QOC diagram is the Question structure, and it is under Questions that
design ideas must be eventually placed. For this reason, the problems that
designers choose to address through the Questions are important: Questions reflect the way in which a problem is currently understood, and guide
the direction of future deliberation. Many examples of Question structuring and restructuring were collated in this and other studies, three of
which are reproduced next. Note that renaming of Questions is covered
here (rather than in the previous section) because of their importance in
shaping the macrostructure of the QOC:

1. Working out the Question together:

[Study A: Pair 12]
G: *[new sheet]* Our first Question—"Do you want different kinds of ATMs
    for the different—you know, a fast ATM and a fully functional one—or
    do you want to do it all at once?"
J: OK, so what's the Question—'cos those are the Options aren't they?
G: Do you want . . um . . well the Question is . . . em . . .
J: Single machine
G: Yeah, kinds of . . . do you want to have just one machine or do you
    want to have . . .
J: Well, those are Options
G: Yeah, well I know those are Options *[laughs]*, but the Question can kind
    of beg the question

Figure 3. [Study A: Pair 2] Refocusing a general Question to capture the issue subsequently addressed by the Options.

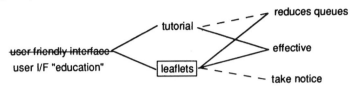

J: Well, it could be a Question—"do you want a variety of machines?" Yes or No

G: Well, *[Question]* number of ATM designs: one and more than one—typically two: *[Options]* fast and fully functional

2. The subjects return to their first general Question, and realize that they have now more precisely identified the real problem:

[Study A: Pair 10]

T: Oh no. *[pause—returns to Question]* Except that this isn't really how to develop user interface—it's [really to do with] the first screen isn't it?

A: What do we show initially?

T: Yeah

A: *[changes first Question]* what do we display on 1st screen?

3. Similar to the last example, Figure 3 shows how a Question is refocused to express the problem which the generated Options now seem to be addressing (user interface education).

As Bellotti, MacLean, and Moran (1991) emphasized, asking the right Questions is critical to developing a useful design space representation, and avoiding particular mental sets or design fixations (Jansson & Smith, 1991). The data collected in these studies in fact demonstrate that Question revision is the natural process that designers follow when using QOC, even though the Study A designers (from whom all of the previous examples are taken) were not explicitly told to work on refining Questions. Design space analysis, with its particular emphasis on asking good Questions, encourages and builds on this activity.

The aforementioned examples showed that the reformulation of Questions often takes place in response to the Options generated. The relationship is reciprocal, however, because an insight into the nature of a Question can lead to restructuring of those Options, by moving them to new or existing Questions; another example would be making an implicit criterion embedded in a Question explicit as a Criterion. This two-way

Figure 4. [Study A tutorial exercise: Pair 8] **Generating a new Question from an
existing one, when it is realized that the last Option plays a different role from
the others.**

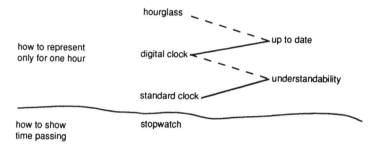

Figure 5. [Study A: Pair 8] **Making a requirement explicit (as a Criterion) that was
formerly implicit (embedded in the Question).**

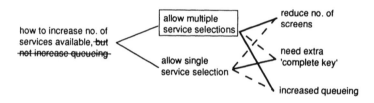

process is often tightly interwoven. Several examples of restructuring are
presented next:

1. In Figure 4, Options to a Question are moved when it is realized
that the Question breaks down into two separate Questions (the Options
are separated to indicate that stopwatch and subsequent Options now
respond to a second Question).

2. Questions sometimes embody an important assumption (an implicit
requirement) that the Options subsequently generated do not satisfy equally.
In Figure 5, the designers make the Criterion increased queueing
explicit, rather than leaving it embedded in the Question. It can then be
used to highlight a difference in quality between the two Options.

3. In another incident (Study A: Pair 10), two designers initially recorded
several ideas as Options (reduce response time, minimize key
depressions, minimize no. screens) in response to a high-level
Question, How to reduce queues? They then realized that they really
wanted to choose all of them, which was a clue that they could serve as
Criteria. The Question was restructured accordingly, and the Criteria reused
in subsequent Questions. This pattern has been observed on other occasions,

and reflects the process of defining goals (or requirements) as the first step to formulating and evaluating solutions. Recognizing and making regularities such as these explicit is a valuable way in which to share expertise among QOC users. A regular pattern such as this also serves as a form of *cognitive task scenario*, which a future QOC tool should support, given that this is a common way in which designers reason with the notation.

4. In Study B, a design session involved the gradual refinement of Options. The designer redrew his QOC structure in order to make this explicit and went on to develop the hierarchy further, as shown schematically in Figure 6.

3.5. Summary: The "Nuts and Bolts" of Using QOC

Open-ended, ill-structured, "wicked" problems (Rittel & Weber, 1973) are rendered manageable only through the exploratory process of framing and reframing views in order to better understand constraints on the solution space. Being able to succinctly "name" the key issues in such problems is much of the work in solving it. Designing the ATM user interface engendered such a mode of working, which led to extensive revision of QOC names and structure as ideas developed.

Although the amount of effort devoted to classifying, naming, and structuring was not documented quantitatively, the resources each takes depends on the users, task, and familiarity of domain. Ongoing experience in using QOC does, however, suggest that these tasks persist as features of "expert" QOC use, although experts are able to draw on strategies for advancing the QOC in situations that might "stall" a less experienced user (cf. Section 3.4, Example 3, and the heuristics proposed by MacLean et al., 1991 [chapter 3 in this book], see Appendix). One would not in fact expect such features of the task to disappear because a claim made explicitly by proponents of semiformal notations is that the discipline of expressing ideas within a constrained vocabulary encourages a dialogue with the representation, which can "talk back" to the designer and expose weaknesses in thinking.

The previous analysis of authoring behavior was based on data from a task intentionally selected to allow QOC to be studied—the problem domain was novel, with many issues left open, and the alternative ATM designs (presented to half the designers) focused attention on trade-offs between Options. In contrast, the evidence described next, from Study B, points to a possible boundary to QOC's scope of application. Specifically, the study suggests that QOC's focus on arguing about design spaces of competing Options is poorly suited to work dominated by the evolution of a single Option.

Figure 6. [Study B] Overview of QOC analysis to show how it was restructured in order to refine the Options (Option numbers added to show the transition). Note that the process of making the Option hierarchy explicit prompted the designer to change Option 1.3 to Option 1.2.3, and to decompose Option 1.2.2 one level further (as the designer was using pen and paper and renamed and restructured extensively, the original QOC was littered with changes, which have been omitted for clarity). Reproduced from Buckingham Shum and Hammond (1994) with permission of Academic Press, Ltd.

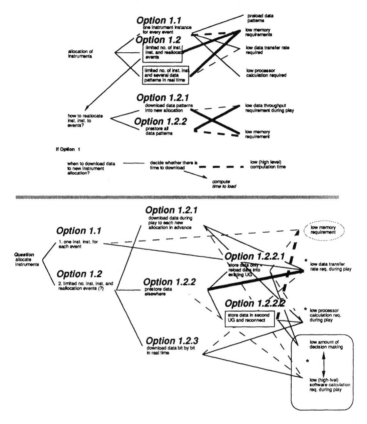

4. PROBLEMS USING QOC IN EVOLUTIONARY DESIGN

Three sessions were spent with a designer who was working on developing a music composition system in the Smalltalk environment. In Session 1, it became clear that many of his ideas were already quite well developed, as a lot of thinking had been invested in the problem beforehand; the main task to which QOC was put was therefore rationalization and decision making. The designer was very positive about QOC's role in this context, and it was clear from the data (Shum, 1991a, Case Study 1) that QOC had assisted in drawing out existing but vague ideas, and clarified relationships between Options and Criteria that would have otherwise remained unarticulated (see Figure 6 for an extract from Session 1). In Sessions 2 and 3, however, serious difficulties were encountered in using QOC, and no explicit DR was constructed. It is on these sessions that attention now focuses.

Let us begin by characterizing what is termed the *evolutionary* mode of working, that is, the iterative development of what the designer conceptualized as one, complex design Option. The designer spent Sessions 2 and 3 developing representations of two Smalltalk data structures, respectively, a hierarchy of data types, and a table of data types such that each column progressively refined the previous one.

The designer described the method of developing the hierarchy in Session 2 as follows:

[Study B]
What I'm doing is a sort of consistency check—thinking through the implications of what I'm doing—this draft suggestion here. And I'll incrementally alter things *[i.e., the data structure]*—I mean I've already done that many times to get to this stage . . .
[points out that he's refining an earlier sketch from his notes] . . . "Gradual refinement" is the phrase. I don't know the Options until I test the previous Option.

In Session 3, the content of the problem was different but the mode of working very similar:

I'm postulating a structure, going round and round testing it, and drawing a few example diagrams of applications of that structure to a real situation— getting it into some concrete familiar objects . . . checking that the abstract structure fits that, and changing it if it doesn't.

To summarize, the design problem solving in Sessions 2 and 3 was characterized by the following: (a) opportunistically driven generate–evaluate cycles to refine the form of the message-passing hierarchy; (b) use of complex, concrete examples as test cases for the abstract structure; (c) management of numerous constraints within the message-passing hierarchy; and (d) application of much implicit Smalltalk programming knowledge.

Given this mode of working, attention now focuses on its apparent incompatibility with the tasks demanded by QOC's explicit, argumentative mode of design. The designer's mode of working should also become clearer through the additional extracts presented.

4.1. Difficulties Encountered With QOC Constructs

Problems using each of the three main constructs (Questions, Options, and Criteria) are illustrated by extracts from Sessions 2 and 3, and then discussed.

Questions.

[Study B]
What are the decisions that I've made? *[tries to formulate Question]* "Do you have . . ." it's difficult to put it in terms of that kind of Question . . . "Is a device configuration a separate class or just one type of category?" Now I don't know how I answer that—it just fits in that it's . . . there's no real doubt about that, it just fits in consistently.
E: So you can make up a Question if you have to?
Yes, in some cases. Let's see if I can make up any more then . . . this idea of having a category of submodules which it's allowed to have—that just arose as a solution to a problem—and what was the problem? The problem was that you have a structure where each object can't just have any old object that's available as its child, only selected ones. And there may be many children . . . *[describes relationships between types . . .]*

Generally, there are two ways in which Questions can be used: either by posing an extremely general Question such as what is the best data structure for a primitive event? or through a long series of Questions each of which addresses the problem of the current iteration. Whereas, on the one hand, a very general Question offers no analytical power to the designer and no insight to someone else trying to understand the design, on the other, the implicit nature of the designer's expertise would make explicit recording of numerous iterations as context-specific Questions unrealistic—the consequences would be enormous DRs, with a corresponding increase in authoring overheads. In sum, because Questions in design space analysis are meant to pick out generic or important dimensions of the design, it was difficult to apply them usefully in this context.

Options.

[Study B]
. . . but as for articulating possibilities—they only arise consecutively. I couldn't have initially said, "We've got two ways of doing it: like I've done it for *those* messages, or *that [i.e., a hypothetical alternative]*—now let's think which is the better one." The only way you arrive at the second one is by having the first one there and thinking, "Now, still what's wrong with that?" You go more in a linear way than a bifurcated way being implied by your DR scheme.
. . . As it's a linear, iterative, refining kind of design, the Options are less useful.
. . . Options are hard to parcel up—often very similar but for one detail.

> It's almost a problem of notation: how to record each stage, as it's an iterative process.
> . . . There's no set of Options—it's just one big mass you have to sort out into categories.

In sum, for the designer, discrete Options were impossible to identify because the design of the final structure was treated as the evolution of one Option over time; the difference between each version was only one, or a few fine details; there were effectively tens of versions marking the path of the design. Thus, even had he wished to, there was no obvious way to express Option evolution.

Criteria. Criteria were the only elements of QOC that could be easily made explicit during Sessions 2 and 3. In the extracts shown next, the designer tries to identify Criteria that he has been using up to that point, and is able to identify the main tradeoff between simplicity and non-repetition of data. This incident may illustrate the benefits of explicitly considering Criteria: Having enlarged on each of the Criteria, the designer concludes that in fact nonrepetition could be easily satisfied within Smalltalk's environment (through "pointer references"). Although there is little doubt that he knew of this facility, and that nonrepetition is a good principle in object-oriented programming, it is not clear if up to that point the connection had been made in his mind, such that he no longer worried about nonrepetition of data as a problem (i.e., as a relevant Criterion):

> [Study B]
> I want independent instrument and event. The event's got a signal list, and I don't want them to be tied together. I suppose I'm trying to have the simplest data structure possible—simplicity. I don't want repetition—nonrepetition of data—I'm trying to express all these positively *[i.e., as positive Criteria]*. On the other hand there is a slight conflict between nonrepetition and simplicity of data: To have nonrepetition you have to have lots of references to things, which can make it a lot less simple.
> *E: Do you have a general policy on that for the whole design, or does it depend on each situation?*
> In a way I'm still learning about it. As Smalltalk has pointer references anyway, it's actually quite efficient. I can put a new object in, and unless I do a deep copy of the object, it will just be a pointer anyway, so it won't be that . . . so I think really simplicity I'm coming down to is the key element, and I don't care about data apparently being repeated, because it won't actually place much overhead on the system. . . . So simplicity's more important than nonrepetition.

However, despite being able to articulate the main Criteria, the designer made the following comments about making them explicit:

I don't know *how* I'm making half of these decisions. I think it's a whole block of expert knowledge—well experience—that I've built up of object-oriented programming; having seen examples, it's very difficult to articulate every reason for everything.
[turns to Criteria noted during the session] Here we've come up with some Criteria—again useful to have those down, but a lot more difficult to go back over this and explain to people why I've done it this way. All I can say is "well, this one works at this stage." Difficult to go back to another branch and say, well, this didn't work because . . . everything *[i.e., earlier versions of the design]* would work, but this simplicity idea is a difficult one to then give alternatives to—it's a subtle one.

Similar comments were made in Session 3. In sum, although it was useful to have Criteria recorded, the difficulty in introducing any additional reasoning structure (Questions and Options) meant that Criteria could only be referred to in general terms, applying to the whole structure that constituted the "Option." Later, the designer commented that it was impossible to explain why a structure was "good" at the level of local decisions. In the final analysis, the only Criterion was "did it work?":

[Study B]
. . . Now, there's a very complex relationship as to why that's better—I sort of just hit on it 'cos it seemed to fit in—it just sort of happened. I mean I know how I got it—by working round the problem, drawing examples, and you just get a feel . . . you abstract from the concrete examples into the structure that will . . . *[It's]* sort of inductive, whereas I think DR is deductive.
. . . the only Criterion is, "Does it enable, or not?" There's not a set of things it could fulfill—it either does or it doesn't. If it doesn't enable you to create a detailed concrete structure, then it's no good—that's the only Criterion!

To conclude, clearly, for any subproblem in design the ultimate Criterion is does it work?, but for the purposes of QOC, this needs to be reexpressed as more focused Criteria that bridge from that goal to specific design features. The designer did make Criteria such as consistency, flexibility, nonrepetition of data, and reduce real-time calculation explicit during Sessions 2 and 3, but the difficulty in breaking the problem down into discrete subproblems, and the tight interdependencies, meant that Criteria like these remained useful only at a global level of application, rather than for assessing alternatives to subproblems within the design space.

4.2. Characterizing the Relationship Between QOC and the Two Modes of Designing

The picture emerging from this study describes the interaction between the way of working that QOC engenders, and two different modes of design. In order to clarify this picture, let us start by considering an extract from the designer's own characterization of QOC's relation to different modes of designing:

[Study B: Session 1]
I think almost there's two different . . . the activity I was doing here was different from what I was doing yesterday, the actual low-level structures *[i.e., in work prior to Session 1]*. This *[Session 1]* is more about big decisions—policies—whereas this *[refers to own notes]* is about implementation. So this *[QOC]* is certainly very useful for strategies.
. . . it was quite tricky to think in that way *[with QOC]* then *[for Session 1]* but it was useful . . . but I think I'd find it almost impossible to wrench myself into thinking in that way *[for Sessions 2 and 3]*; it would be unnatural almost, 'cos it's not that kind of path that you take when you're doing this sort of thing.

On two occasions he drew analogies to the way of working in the latter sessions:

Have you ever watched someone design a circuit board?
It's more . . . like painting a picture of something, and you ask why did you put the trees there and not there? There are so many Criteria and they all interrelate.

These images characterize a mode of designing for which the mode of working engendered by QOC is ill-matched, given the present evidence. The primary difference is that argumentation-based DR is based on a model for representing competing alternatives to substantive problems, rather than the evolution of a single alternative over time.

The designer's characterization of the sessions as "strategic" versus "low-level" design captures to some extent the contrast between laying out and deliberating over alternatives at key decision points, and pursuing one Option in detail to better understand its properties. However, this should not be interpreted as "conceptual versus implementational" design; the distinction may be better characterized as breadth-first versus depth-first design. This then allows for either mode of working (laying out of multiple Options, or exploration of single Options) at any point between upstream, highlevel conceptual design and downstream, low-level coding.[6] In reality, of course, there will often be an interplay between these modes of working, which suggests that designers intending to use QOC need to recognize when the formalism is likely to be of most use.

A point also worth considering is that most decisions made during evolutionary modes of working (whether high- or low-level) may in fact not be of interest to other domain experts, because they are in some sense

---

6. I am grateful to Tom Moran and an anonymous reviewer for drawing out this distinction more clearly.

Figure 7. The relationship between a QOC Design Space Analysis, the analytic "breadth-first" mode of designing, and the evolutionary "depth-first" mode, as contrasted in Studies A and B.

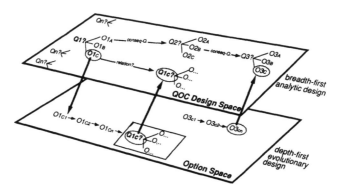

"routine"—the decisions are not contentious given a certain level of expertise. If this is indeed the case, then the work in evolving *Option X* has value only to the extent that it clarifies for the designer what is meant by "*X*" in relation to *Y* or *Z* as Options to a superordinate ("strategic") Question. It is in working through and laying out the relative merits of each for which structured argumentation may be best suited and most useful, both for the designer and others subsequently.

In the light of this analysis, the relationship between these two modes of working can be thought of as streams of activity occupying different "spaces" in the DR domain (represented as parallel layers in Figure 7, after Lee & Lai, 1991 [chapter 2 in this book]). These spaces are associated with breadth-first and depth-first modes of designing to different degrees.

The *design space* is what QOC represents explicitly; its development is associated most strongly with an analytic frame of mind in which issues are delineated, and Options are laid out for consideration in a breadth-first fashion. Study A and Study B Session 1 demonstrated that design space analyses emerge most naturally when the problem demands this mode of working (although, as reported, coherent rationale is not "waiting to be externalized" from designers' minds or "captured" from ongoing discussion, but is *constructed* in dialogue with colleagues, the problem, and the QOC structure). In terms of Lee and Lai's (1991 [chapter 2 in this book]) analysis, QOC design spaces primarily address the *issue, alternative,* and *criterion spaces* (see chapter 2, Figure 2); these have been merged within the design space layer of Figure 7.

The alternative space, by Lee and Lai's (1991 [chapter 2 in this book]) definition, allows relations between alternatives to be made explicit. However, in QOC's vocabulary, Option transformation is not explicitly supported, indicated by the separate *Option Space* in Figure 7. The Option Space illustrates how an evolutionary mode of working—focusing on developing individual Options through iterative redesign and testing—relates to the Design Space with which QOC is primarily concerned. The arrows linking the two spaces reflect the fact that Options may evolve both before they are represented within the Design Space (Option O3c), and after (O1c).

From the point of view of representing a concise, intelligible design space analysis, it is often unnecessary, tedious, or impossible to make successive versions explicit—indeed, a feature of a good design space analysis is that it clarifies the *boundaries* of a given possibility space, rather than detailing the minor differences between features of a local space. However, although minor transformations to Options will remain to a large extent undocumented, on occasion it may be useful to record (in the Design Space) a significant issue in the design of a particular Option (e.g., Question Q1c).

A representational issue that then arises is how such rationale is integrated with the main QOC analysis, given that it derives from a different space to other Questions, and may not make sense if simply added (as shown by the queried relation between O1c and Q1c). At present a general *consequent-Question* relationship is used in QOC, which could be refined if necessary to indicate *what kind* of consequent Question it is. However, the vocabulary for a lightweight DR scheme should be extended with care, because declaring a new relationship has implications on two fronts. First, users must ensure that all *future* instances of that relationship use the new link type (introducing more choices at the point of creation), and second, in order to maintain the integrity of the existing QOC, all existing instances of that relation (currently implicit), need to be converted to take advantage of the new more precise link type. Whether or not this is judged to be worthwhile will depend on factors such as the current complexity of the notation, and the anticipated future benefits that may be accrued.

The analysis summarized in Figure 7 informs, and is informed by other reports of DR in use. However, before relating this to other work, the next section considers further the issue of when and how to extend QOC.

5. QOC'S EXPRESSIVENESS

The *expressiveness* of a representation describes its coverage of the target domain to be represented: Is the vocabulary sufficient to represent all of the important concepts, relationships, and scenarios of use? Moreover, even if certain classes of information are logically representable, the extent to which the representation *eases* access (visually, or computationally) to important information, and hides irrelevant detail, should be taken into account. Highlighted next are several points of notational weakness that

Figure 8. [Study A tutorial exercise: Pair 8] QOC can be misused: a poor way to capture the evolution of Options, as devised by two designers. Each Option was rejected in favor of a better version, starting from the top. Each Criterion serves only to object to the "current" Option in order to show the next improvement. For instance, the third Option, stopwatch with 1 at top was devised because stopwatch failed to show time passing; however, because it was judged to be misleading, stopwatch with solid line pointing to 1 was devised, and so forth. Expressively, this is a weak representation: (a) The absence of Assessment links to other Options implies that the Criterion is irrelevant, which is clearly not the case (e.g., showing time passing applies to all Options); (b) it is not shown that subsequent Options in fact satisfy all Criteria more fully than their predecessors.

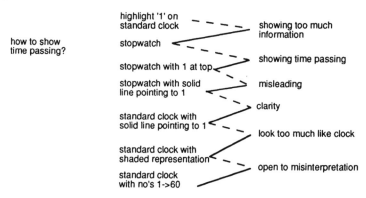

became evident in the QOC studies, and illustrations of the ways in which designers overcame these. The minimalist conditions in which designers were studied (using a very concise notation, and with only pen and paper) helped to make requirements for more support all the more obvious.

5.1. Representing Evolution Within QOC Structures

Much has been said about Option evolution in previous sections. QOC's vocabulary at present cannot easily express how Options are transformed, preferring instead to express the *results* of those transformations within the design space—a product as opposed to process orientation. One response to data such as presented in Study B is therefore to leave such deliberation unsupported, the position being that design space analysis is not intended to support such design tasks, and the evidence indicates that it is a rapid, intuitive process that should not have to bear the weight of explicit representation. This approach may indeed prove to be the most pragmatic.

The alternative position is to provide constructs such as the special-izes, generalizes, and simplifies relations in gIBIS. Through selective use, significant developments in Options could be made explicit,

Figure 9. In the absence of QOC conventions for expressing one Question's contingency on an earlier decision, two designers express the relationship informally through annotation under the first Question, and context-dependent phrasing of the second Question.

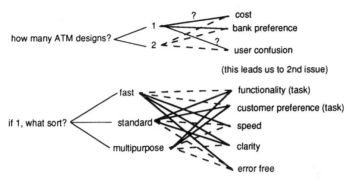

and later retrieved, perhaps with changing design requirements or constraints. Isolated incidents in Study A suggest that in the absence of notational conventions, "pseudorepresentations" may be invented instead (e.g., Figure 8). As is always the case, the cost of tool support, and any cognitive benefits that the notation provides, is the added overhead of explicit structuring.

An alternative to notational support for version control is support from the environment in which the DR tool runs. A tentative conclusion is that because QOC was developed with simplicity and minimal overheads as a goal, the most promising way to introduce QOC evolution may be through environmental support from generalized hypertext version control mechanisms, rather than through extensions to the notation itself, which inevitably introduce user overhead.

5.2. Expressing Constraints and Dependencies

In software design on any significant scale, keeping track of dependencies becomes a major task. In one sense, design is all about discovering and subsequently controlling important dependencies. In terms of QOC, dependencies manifest as constraints on the selection of Options in different parts of the space about a design. Not surprisingly, in the course of the QOC studies, there were a number of incidents in which dependencies between Options were encountered. With only the core QOC notation and pen and paper, there was little support for representing dependencies and constraints elegantly. As the designers were not taught any notational conventions, links across the QOC were verbally noted, or ad hoc notational devices invented (e.g., Figure 9).

5.3. The Subtleties of Expressing Options, Criteria, and Assessments

If QOC is to be used during design deliberation, it needs to be sensitive to the exploratory nature of the process. Commitment is often delayed as alternative routes are partially developed to assess their potential as solutions. Consequently, it was found that binary choices in the status of entities frustrated designers in a number of respects. For instance, there was not enough expressive power in the *selected/rejected* distinction for Options, leading to the invention of visual devices to reflect indecision (e.g., a dashed box drawn around two Options to indicate tentative selection).

Many comments were also made about the lack of a more sensitive scheme for *evaluating* Options. First, designers pointed to the need to prioritize Criteria, given that simply counting supports links was manifestly an inadequate way to make decisions. Given the evidence that this facility is needed, the issue of representational overheads is perhaps not at stake here: Designers will make use of different weightings as the need arises.

Although Criterion weighting was important, the majority of incidents and comments on QOC's representation of the evaluation space related to the need to show relative strengths of *Assessment* links. Within QOC, Assessments are relative to each other for a given Question, and no semantics can be inferred between Assessments made for different Questions. However as the following examples demonstrate, the formalism runs into problems even in expressing within-Question Assessments.

Designers often described Assessments as being "more positive" than others, and as with Option selection, invented their own graphical conventions for encoding link strengths such as double-links or thicker links to represent gradations of the supports and objects-to relations. One pair also made extensive use of what they called "neutral" Assessments, signified by *omitting* to link an Option to a Criterion. Neutral links were used 11 times in three design exercises, but under several different circumstances. Generally, they served as Assessments for Options that were judged to fall "in between" other Options that had positive and negative Assessments, that is, Assessments not judged to be sufficiently good or bad to merit a +/– link (Figure 10). Elsewhere, neutral Assessments were used to mean still other things (*not yet decided* and *not yet discussed*).

Another pair of designers placed "?"s over links to represent undecided status as they worked. However, this occurred under two circumstances, again leading to ambiguity for an outsider: (a) when an Assessment could not be made until further design details had been finalized, and (b) when they were not confident that an Assessment was correct (e.g., the designers did not feel qualified to make evaluative decisions about psychological Criteria such as display clarity or attentional requirements).

Figure 10. [Study A: Pair 6] An example of the ambiguous use of "blank links" to express "neutral Assessments." The Criterion speed of other functions "neutrally" assesses the first two Options, but for different reasons: current atm is neutrally assessed as it was judged to be better than proposed atm, but not to the extent that it merited a *supports* link. However, the FATM (FastATM) Assessment against that Criterion was also left blank, but in this case because the Criterion was simply irrelevant (the FATM only offers a cash service, so has no "other functions").

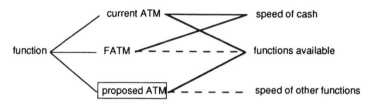

As indicated, a variety of visual codes can be used to convey relative Assessment strengths. In whatever way it is implemented, the results reported here are evidence that a more sensitive Assessment scheme would be a welcome notational extension, and would not add unnecessary overheads.

6. CONNECTIONS WITH OTHER RESEARCH

This final section considers reports of other approaches to DR in use, focusing specifically on the interplay between design and representing DR, as expressed in Figure 7. The purpose is not only to draw attention to relevant aspects of other work, but to assess the sufficiency of the present QOC analysis, and argumentative DR more generally. More comprehensive reviews are presented elsewhere (Buckingham Shum, in press; Buckingham Shum & Hammond, 1994).

Lewis, Rieman, and Bell (1991 [chapter 5 in this book]) have argued that design is inherently problem centered, and moreover that it is too distracting to abstract from concrete to more general characterizations of those problems as required by most DR notations. In a design project of their own, they expressed DR informally as a series of problems and alternatives. Several patterns were observed in their own deliberations as to how problems, subproblems, and alternatives interacted to move the design forward (e.g., *"micro problem derived from raw problem, design alternatives spawned by micro problem"*).

The observation that design is often problem centered provides insight into the ways in which Questions, Options, and Criteria can productively interact to move a design forward. Depending on the nature of the problems, the patterns reported by Lewis et al. (1991 [chapter 5 in this book]) can be easily reexpressed as dynamics operating in either the Design or Option Spaces (e.g., *focused Question derived from general Question, and Options generated for this consequent Question*). However, the key question is clearly whether problems can be easily reexpressed at the level of QOC. The evidence of the QOC studies indicates that they can, but that reformulation is often necessary as the ill-structured design space is explored. Ultimately, it is the importance that the designers place on having a clear, reusable DR resource that will dictate the extent to which this effort is judged worthwhile.

Fischer, Lemke, McCall, and Morch (1991 [chapter 9 in this book]) report several studies of design students trying to use PHI (Procedural Hierarchy of Issues—McCall, 1986, 1991). It was found that PHI was extremely difficult to use during the actual "construction" phases of design (i.e., development of the solution, as opposed to preparatory design discussion), although the problem was being decomposed hierarchically into subissues by the designers, which in principle mapped well to the PHI method. Fischer et al. interpret these results in terms of Schön's (1983) conception of the expert's design process. Schön asserted that design comprises several mutually exclusive modes of activity, termed *knowing-in-action*, *reflection-on-action*, and *reflection-in-action*, as described next.

Parallels between Figure 7, McCall's (1986, 1991) data, and Schön's (1983) concepts can be identified as follows. Identification and comparison of Questions and Options corresponds to reflection-on-action—a "stepping back" in order to consider the nature and scope of the local design space. PHI, like QOC, was well suited to such deliberation. In terms of Schön's framework, instructions to use PHI as the dominant design model prolonged rationalistic reflection-on-action, sidetracking the students into "talking round" the actual design (i.e., perpetuating *discussion about* the design space).

Study B clearly showed the difficulty of representing useful DR while engaging in "artifact construction" (the data structures). This rapid testing and changing of the artifact, coupled with a reluctance or even inability to interrupt and articulate one's process is aptly characterized by the concept of knowing-in-action: ". . . the knowing is *in* the action. We reveal it with our spontaneous, skillful execution of the performance; and we are characteristically unable to make it verbally explicit" (Schön, 1983, p. 25).

DR's interest in the concept of reflection-in-action has already been noted, and this continues to hold in the current analysis: "In an *action present*—a period of time, variable with the context, during which we can still make a difference to the situation at hand—our thinking serves to reshape what we are doing while we are doing it. I shall say, in cases like this, that we reflect-*in*-action" (Schön, 1983, p. 26).

In Study B, use of knowledge in this way was intrinsically embedded in the construction/evolution process in the Option Space. Many of the problems encountered were not worth recording—indeed, could not be succinctly expressed (cf. Lewis et al.'s, 1991 [chapter 5 in this book], perspective). However, reflection-in-action may on occasion encounter issues worth noting as rationale (e.g., several potentially workable alternatives, or a decision made under extenuating circumstances); reflection-in-action might then become reflection-on-action, moving the attentional frame to the broader design space (represented by Q1c in Figure 7). Two questions that are as yet unresolved are first, whether designers are sufficiently aware of what makes useful DR to recognize such situations, and second, whether breaking out of construction to reflect via QOC would be unacceptably disruptive. If both of these are the case, one solution might be to briefly record a word or two to capture the problem (i.e., during the "action-present"), to be returned to later for more deliberate reflection.

Turning to another key issue, Conklin and Burgess-Yakemovic (1991 [chapter 14 in this book]) focus on the contrast between "structure-oriented," retrospective DR, which emphasizes the capture and communication of the logical content and structure of design reasoning, and "process-oriented" DR, which preserves the form and order of ideas in the deliberation process, analogous to a "narrative" with less rationalization.[7] Conklin and Yakemovic explicitly state that reusability is a secondary concern for narrative DR, because process-orientation biases toward minimizing the overheads by minimizing the amount of DR rationalization.

The evidence provided by the QOC studies clarifies certain implications of this approach. If gIBIS Issues are simply recorded one after another as they arise, with little retrospective analysis, there may be representational problems to overcome in understanding how argumentation of different sorts holds together; that is, the content of the different spaces in Figure 7 will be merged. It is almost inevitable that additional structuring will be necessary to manage the growing DR. A DR environment could therefore support users by differentiating DR from different spaces.

Work related to gIBIS has addressed the development of a real-time collaborative version, called rIBIS. Rein and Ellis (1991) briefly reported on users' experiences with the tool over 16 meetings. All but one of these meetings was described as "mostly unsatisfying and frustrating" by their participants, with significant difficulties encountered in using the IBIS method to structure discussions. It was concluded that the main causes of the problems were participants' inexperience with IBIS notation, and the complexity of the rIBIS user interface. There was no analysis of the nature of the difficulties with IBIS, so these findings cannot be related to the QOC studies in detail; the comments on gIBIS also apply to rIBIS DRs if

---

7. The expressions "capturing" versus "constructing" DR describe points on a continuum of how rationalized the DR is. In principle of course, all DR is constructed to some extent (J. C. Tang, personal communication, June 26, 1992).

they are constructed as narrative argumentation. However, the rIBIS results further demonstrate that without sufficient training even the simpler DR notations like gIBIS and QOC can be unnatural to use.

Last is the claims-based approach to DR, which has been directly concerned with the use of DR in design evolution. Carroll and Rosson (1991 [chapter 4 in this book]) describe the redesign of a View Matcher support tool, from its original use for novices learning Smalltalk, to a View Matcher to aid more expert users in *code reuse*. They demonstrate that, in combination with their scenario-based design methodology, the structuring of claims by a high-level task analysis produced a representation for DR that helped them to build in a principled manner on the claims analysis of the original design, leading to, "greater confidence that we are standing on the shoulders of our prior View Matcher design work . . . and not merely in the vicinity of prior work."

In terms of Figure 7, the evolution of the View Matcher corresponds to the exploration of and commitment to portions of a new design space with a particular relationship to the original. Understanding the differences and commonalities between the two spaces is what Carroll and Rosson (1991 [chapter 4 in this book]) refer to as understanding the "species" of an artifact. In order to transform one design space into another, new constraints (derived from the tasks to be supported) are taken into account, leading to the development of new portions of the space. The use of claims-based DR to support redesign, as described by Carroll and Rosson, can therefore be seen as operating at the level of the design space (evolution of whole Option sets), rather than the transformation of individual Options, as reported in Study B.

As a particular approach, psychological DR through claims extraction provides a more systematic way in which to identify Criteria relevant to user's tasks; it helps to answer the question often asked of QOC, "Where do the Criteria come from?" This question is also addressed by Singley and Carroll (1995 [chapter 8 in this book]), who distinguish several modes of reasoning that in their experience served as sources of psychological design constraints.

7. CONCLUSION

Is representing design rationale as structured argumentation too much work? This chapter demonstrates that the answer is not a simple yes or no; QOC played both facilitatory and obstructive roles, depending on the mapping between the design process and the QOC representational process. However, more studies are needed to add weight to the case that expressing design rationale as semiformal argumentation can assist the reasoning process, improve decisions, and is pragmatic in the context of design practice.

This chapter, and other analyses of design activity (e.g., Olson et al., 1995 [chapter 7 in this book]), contribute toward establishing empirical foundations on which to develop not only appropriate support technology, but also training and methodological guidelines (e.g., MacLean, Bellotti, & Shum, 1993). In Carroll and Rosson's terms (1991 [chapter 4 in this book]), "artifacts" such as tools, training courses, and methodologies make assumptions or "claims" about cognitive and group design processes. This chapter clarifies the usability claims that QOC makes as a notation, and thus serves as a point of departure in the design of tools and training intended to facilitate its use.

NOTES

Background. Please note that my name has changed from Shum to Buckingham Shum (and thus this chapter should be referenced under "B"). However, my previous papers are still under the name of Shum.

Acknowledgments. I am grateful to Jack Carroll, Nick Hammond, Allan MacLean, Andrew Monk, Tom Moran, and an anonymous reviewer for critical comment and discussion that helped to shape this chapter. My thanks also to the Impact Project at Nestlé Rowntree, Logica Cambridge Ltd., the University of York Music Technology Research Group, and Warehouse for providing designers.

Support. This research was funded by Rank Xerox EuroPARC, Esprit Basic Research Action 3066 (The AMODEUS Project), and SERC CASE Award 88504176.

Author's Present Address. Simon Buckingham Shum, Knowledge Media Institute, The Open University, Milton Keynes, MK7 6AA, U.K. Tel. +44 1908-653800, Fax: +44 1908-653169. E-mail: S.Buckingham.Shum@open.ac.uk, WWW:http: //kmi.open.ac.uk/

REFERENCES

Bellotti, V. M. E., MacLean, A., & Moran, T. (1991). Structuring the design space by formulating appropriate design rationale questions. *SIGCHI Bulletin, 23*(4), 80–81 (Extended version available as: Working Paper RP6/WP6, AMODEUS Project). Cambridge, UK: Rank Xerox EuroPARC.

Buckingham Shum, S. (in press). Design argumentation as design rationale. To appear in: A. Kent & J. G. Williams (Eds.), *The Encyclopedia of Computer Science and Technology.* New York: Marcel Dekker, Inc.

Buckingham Shum, S., & Hammond, N. (1994). Argumentation-based design rationale: What use at what cost? *International Journal of Human-Computer Studies, 40*(4), 603–652.

Carroll, J. M. & Rosson, M. B. (1991). Deliberated evolution: Stalking the View Matcher in design space. *Human–Computer Interaction, 6,* 281–318. Also in T. P. Moran & J. M. Carroll (Eds.), *Design rationale: Concepts, techniques, and use.* Hillsdale, NJ: Lawrence Erlbaum Associates, 1996. [Chapter 4 in this book.]

Conklin, E. J., & Burgess-Yakemovic, KC. (1991). A process-oriented approach to design rationale. *Human–Computer Interaction, 6,* 357–391. Also in T. P. Moran

& J. M. Carroll (Eds.), *Design rationale: Concepts, techniques, and use*. Hillsdale, NJ: Lawrence Erlbaum Associates, 1996. [Chapter 14 in this book.]

Fischer, G., Lemke, A. C., McCall, R., & Morch, A. I. (1991). Making argumentation serve design. *Human–Computer Interaction, 6*, 393–419. Also in T. P. Moran & J. M. Carroll (Eds.), *Design rationale: Concepts, techniques, and use*. Hillsdale, NJ: Lawrence Erlbaum Associates, 1996. [Chapter 9 in this book.]

Grudin, J. (1995). Evaluating opportunities for design capture. In T. P. Moran & J. M. Carroll (Eds.), *Design rationale: Concepts, techniques, and use*. Hillsdale, NJ: Lawrence Erlbaum Associates. [Chapter 16 in this book.]

Guindon, R. (1990). Designing the design process: Exploiting opportunistic thoughts. *Human–Computer Interaction, 5*, 305–344.

Jansson, D. G., & Smith, S. M. (1991). Design fixation. *Design Studies, 12*(1), 3–11.

Lee, J., & Lai, K-Y. (1991). What's in design rationale? *Human–Computer Interaction, 6*, 251–280. Also in T. P. Moran & J. M. Carroll (Eds.), *Design rationale: Concepts, techniques, and use*. Hillsdale, NJ: Lawrence Erlbaum Associates, 1996. [Chapter 2 in this book.]

Lewis, C., Rieman, J., & Bell, B. (1991). Problem-centered design for expressiveness and facility in a graphical programming system. *Human–Computer Interaction, 6*, 319–355. Also in T. P. Moran & J. M. Carroll (Eds.), *Design rationale: Concepts, techniques, and use*. Hillsdale, NJ: Lawrence Erlbaum Associates, 1996. [Chapter 5 in this book.]

MacLean, A., Bellotti, V. M. E., & Shum, S. (1993). Developing the design space with design space analysis. In P. F. Byerley, P. J. Barnard, & J. May (Eds.), *Computers, communication and usability: Design issues, research and methods for integrated services* (pp. 197–219). Amsterdam: Elsevier, North Holland Series in Telecommunication.

MacLean, A., Young, R. M., Bellotti, V. M. E., & Moran, T. P. (1991). Questions, options and criteria: Elements of design space analysis. *Human–Computer Interaction, 6*, 201–250. Also in T. P. Moran & J. M. Carroll (Eds.), *Design rationale: Concepts, techniques, and use*. Hillsdale, NJ: Lawrence Erlbaum Associates, 1996. [Chapter 3 in this book.]

McCall, R. J. (1986). Issue-serve systems: A descriptive theory of design. *Design Methods and Theories, 20*(8), 443–458.

McCall, R. J. (1991). PHI: A conceptual foundation for design hypermedia. *Design Studies, 12*, 30–41.

Minneman, S. (1991). *The social construction of a technical reality: Empirical studies of group engineering design practice* (Tech. Rep. No. SSL-91-22). Palo Alto, CA: Xerox Palo Alto Research Center.

Olson, J. R., & Olson, G. M. (1991). User-centered design of collaboration technology. *Journal of Organizational Computing, 1*, 61–83.

Olson, G. M., Olson, J. S., Storrøsten, M., Carter, M., Herbsleb, J., & Rueter, H. (1995). The structure of activity during design meetings. In T. P. Moran & J. M. Carroll (Eds.), *Design rationale: Concepts, techniques, and use*. Hillsdale, NJ: Lawrence Erlbaum Associates. [Chapter 7 in this book.]

Rein, G. L., & Ellis, C. A. (1991). rIBIS: A real-time group hypertext system. *International Journal of Man–Machine Studies, 24*(3) 349–367. Also in Greenberg S. (Ed.). (1991). *Computer supported cooperative work and groupware* (pp. 223–241). London: Academic Press.

Rittel, H. W. J., & Weber, M. M. (1973). Dilemmas in a general theory of planning. *Policy Sciences*, *4*, 155–169. (Reprinted in *Developments in design methodology*, 1984, pp. 135–144)

Schön, D. A. (1983). *The reflective practitioner: How professionals think in action*. New York: Basic Books.

Sharrock, W., & Anderson, R. (1995). Organizational innovation and the articulation of the design space. In T. P. Moran & J. M. Carroll (Eds.), *Design rationale: Concepts, techniques, and use*. Hillsdale, NJ: Lawrence Erlbaum Associates. [Chapter 15 in this book.]

Shum, S. (1991a). A cognitive analysis of design rationale representation. *Unpublished doctoral dissertation*, Department of Psychology, University of York, England. (Available from Rank Xerox EuroPARC, Cambridge, UK)

Shum, S. (1991b). Cognitive dimensions of design rationale. In D. Diaper & N. V. Hammond (Eds.), *People and computers VI: Proceedings of HCI '91* (pp. 331–344). Cambridge, England: Cambridge University Press.

Singley, M. K., & Carroll, J. M. (1995). Synthesis by analysis: Five modes of reasoning that guide design. In T. P. Moran & J. M. Carroll (Eds.), *Design rationale: Concepts, techniques, and use*. Hillsdale, NJ: Lawrence Erlbaum Associates. [Chapter 8 in this book.]

Tang, J. C. (1991). Findings from observational studies of collaborative work. *International Journal of Man–Machine Studies*, *34*(2), 143–160. Also in Greenberg, S. (Ed.). (1991). *Computer supported cooperative work and groupware* (pp. 11–28). London: Academic Press.

7

The Structure of Activity
During Design Meetings

Gary M. Olson
Judith S. Olson
Marianne Storrøsten
Mark Carter
James Herbsleb
Henry Rueter
University of Michigan

ABSTRACT

The development of schemes to support design, whether behavioral methods or new technologies like groupware, should be based on detailed knowledge about how design occurs. Such data can be used to suggest what kinds of tools people might need as well as to provide a baseline for evaluating the effects of schemes for improvement. We present details of how real groups work in early software design

Gary Olson is a psychologist with an interest in group work and its supporting technologies; he is a Professor of Psychology and Director of the Collaboratory for Research in Electronic Work (CREW) at the University of Michigan. **Judith Olson** is a psychologist with an interest in group work and its support and the design of interfaces to technology for a variety of applications; she is Professor of Computer and Information Systems in the University of Michigan Business School, a Professor of Psychology, and a senior researcher at CREW. **Marianne Storrøsten** was a Ph.D. student at CREW with an interest in technology development for a variety of settings; she is now a researcher at the Norwegian Computing Center in Oslo. **Mark Carter** is systems analyst formerly working at CREW; he is now at the Information Technology Division at the University of Michigan. **Jim Herbsleb** is a psychologist/computer scientist with an interest in software engineering process, who had a postdoctoral fellowship at CREW; he is now at the Software Engineering Institute at Carnegie Mellon University. **Henry Rueter** was a research scientist at CREW with an interest in complex data analysis methods.

CONTENTS

meetings. We studied 10 design meetings from four projects in two organizations. The meetings were videotaped, transcribed, and then analyzed using a coding scheme that looked at both the participants' problem solving and those activities they used to coordinate and manage themselves. We also analyzed the structure of their design arguments. We found, to our surprise, that although the meetings differed in how many issues were covered and the breadth of the discussion of these issues, they were strikingly similar in both how people spent their time and the sequential organization of that activity. Overall, only 40% of the time was spent in direct discussions of design, with many swift transitions between alternative ideas and their evaluation. The groups spent another 30% taking stock of their progress through walkthroughs and summaries. Pure coordination activities consumed about 20%, and clarification of ideas—a cross-cutting classification—took a third of the time, indicating how much time was spent in both orchestrating and sharing expertise among group members. The pattern of transitions revealed these activities were clustered into two general classes, design and management. We focused on the design activities, and described their sequential structure both statistically and grammatically. The structuredness of design as a group activity may lend itself to forms of support such as design rationale, although other findings in the literature suggest that doing so on the fly may be disruptive.

1. INTRODUCTION

The design of complex systems usually involves many people, and during the process of doing their work the designers often get together for face-to-face meetings. Though sometimes such face-to-face meetings can be quite formal, as in doing a presentation of the design to management or a client, or carrying out a formal design review, the actual working sessions of design teams are usually quite informal. A small group will meet in an ordinary meeting room, and in a highly interactive session discuss and debate various design ideas. To the participants in such meetings and to the casual observer of them they appear quite intense and unstructured.

Part of the appeal of schemes for design rationale is that they might provide some order or structure to what seems like a very informal process (e.g., Buckingham Shum, 1995 [chapter 6 in this book]).

We have been studying design meetings in field settings, and in an earlier article (Olson, Olson, Carter, & Storrøsten, 1992) reported on the characteristics of a sample of 10 design meetings taken from four different projects in two different organizations. In that article we focused on an overall characterization of design as an activity, and suggested that despite their informal and highly interactive nature, design meetings contain a fair amount of structure. The earlier article should be consulted for details about the sample of meetings, the coding categories and their reliabilities, and the general characteristics of design activities. In the present chapter we pursue this story further, describing in more detail aspects of the sequential structure of design activity.

Design rationale schemes have at least two kinds of potential functions for designers. Perhaps their primary one is to capture in a systematic way the design decisions that are made, so they are available later as a record of why the design has the characteristics that it does. Our analysis says little about this function. However, design rationale schemes might also be useful as a process aid for designers, helping them structure their discussions, consider a wider range of design alternatives or options, and more systematically map criteria onto alternatives as an aspect of making design decisions. Our work is very relevant to this possibility. It provides information about how variable design behavior is across groups, and the degree of organization or structure already present in current design meetings. It also provides a baseline of current design behavior that could be used to evaluate the effect of providing design teams with design rationale tools. Buckingham Shum (1995 [chapter 6 in this book]) has been studying how QOC (questions, options, and criteria) notation might affect design behavior, and we return to an examination of his findings in our conclusion. As becomes evident through the course of this chapter, we have also found design rationale categories to be very useful in describing the nature of design processes.

2. THE DATABASE

The data reported in this chapter come from naturalistic observations of system design being carried out in field settings. The data were collected at Andersen Consulting (AC) and Microelectronics and Computer Technology Corporation (MCC). We sampled 10 meetings from four different projects for intensive analysis. Figure 1 describes the sample of 10 meetings we analyzed. The meetings ranged in size from three to seven participants, and typically lasted 1 to 2 hours. In all cases, the meeting participants knew each other. Software system design was the principal topic of all of

Figure 1. **The sample of 10 meetings.**

Meeting	Number of Participants	Duration (minutes)	Main Topic
AC-A	7	90	Moving some functions from the mainframe to PCs to make the mainframe's system response time faster.
AC-B1	3	58	Deciding what features to offer the user in printing out several complex diagrams in the reverse engineering process.
AC-B2	5	58	Walk through the existing list of features to offer the user in printing to explicitly evaluate them on whether they can be implemented in the promised time frame.
AC-C1	4	120	Review the current architectural components and the user interface features of the new client-server architecture so that one member can coordinate with their partners in New York in a meeting the following week.
AC-C2	5	123	Use an existing mainframe application to see whether their new architecture can support desired new windowing features.
MCC-1	3	125	How to build advice into the knowledge base and how to make it context sensitive.
MCC-2	4	122	Later the same day, the three designers report their walk in the woods in which they discussed high level issues. They review the design with an expert on knowledge bases, and in particular on how to represent advice.
MCC-3	4	130	How to present aspects of the knowledge base to the user and allow groups to view and edit it.
MCC-4	4	140	Review of the ideas on knowledge editing and the user interface furthering the discussion on several details.
MCC-5	6	52	Later the same day, bringing in two experts in the underlying language of implementation, to review ways to think about knowledge editing and the ideal user interface.

these sessions. The problems were large, requiring many individuals many months to design and code. The systems were both internal products (e.g., an AC systems analysis tool) and prototype ideas for future systems (e.g., a generic architecture for future applications and an exploratory system to edit knowledge bases). In all cases the problem specification was vague at best. The principal focus of the projects during the phase we studied was further development and refinement of the requirements: The major issues involved deciding both what features to offer the client and how to implement those features.

Because the systems being designed were reasonably large and complex, a variety of expertise was required. Thus, the designers usually had different skills, including in many cases expertise on systems that were similar in some respect to the target system, or on the languages or architectures that the new system built upon.

The designers were experienced in working in groups. They were familiar with agendas, assignment of tasks to people (action items), general brainstorming activity, and the concept of loose commitment of alternatives. At AC, all new employees are taught a standard method that they are expected to use in the process of building a system. It includes the kinds of steps one goes through, the coordination required across subgroups, and the documentation of the decisions made. All employees know this method. The designers at MCC were part of a program working on new systems in artificial intelligence. All of them had worked at other companies before coming to MCC. But MCC, unlike AC, did not teach any standard or formal procedures for doing system design. Thus, in some sense the cultures at AC and at MCC were quite different. This makes all the more striking the many similarities we report in later sections of this chapter.

All of the design sessions analyzed here were videotaped and later transcribed. Our analyses focused on the talking by the members of the groups. In many cases we also had artifacts from these design sessions, such as drawings or handouts that were discussed. We have one set of longitudinal data from each organization, and the tapes of face-to-face sessions in these longitudinal projects were supplemented by notes, interviews, and other materials collected during the course of the project. In both the cross-sectional and longitudinal materials, we realize that many important things happened away from the eye of our video cameras. Thus, we have a sample of design activities and materials that are incomplete, but detailed in the most interactive, sustained phases of collaborative design.

3. CODING AND RELIABILITY

The analyses we report in this chapter were based on a coding of the transcripts into 11 major categories that captured the general nature of the design discussion. The basic categories we used are defined in Figure 2.

Figure 2. **Categories of design meeting activity**

Issue	The major questions, problems, or aspects of the designed object itself that need to be addressed. They typically focus on the major topics of "shall we offer this capability to the user" and "how can we implement that." Included in the time is the elaboration of the idea, description not in answer to a group member's question.
	Occasionally, the issues are not stated explicitly but can be inferred (both by the coder and by the members of the design meeting) by the presentation of two alternative solutions. These are counted as short statements of issues, typically marking the first part of the sentence that presents the alternatives.
Alternative	Solutions or proposals about aspects of the designed object. These are typically either features to offer the user or ways to implement the features decided on so far. Included in the time is the elaboration of the idea, description not in answer to a group member's question. Occasionally, there is an elaboration, usually the implications of the idea just presented, that is included in this category as well.
Criterion	The reasons, arguments, or opinions which evaluate an alternative solution or proposal. Occasionally these appear in the form of analogous systems, with the implication that if it worked in this other system, it will be good for this system. Occasionally, an evaluative statement or criterion will be made unconnected to any particular alternative, in background, so that members of the group are reminded to evaluate the upcoming or past proposed solutions in light of it.
Project Management	Statements having to do with activity not directly related to the content of the design, in which people are assigned to perform certain activities, decide when to meet again, report on the activity (free of design content) from previous times, etc.
Meeting Management	Statements having to do with orchestrating the meeting time's activity, indicating that the group members are to brainstorm, decide (and vote), hold off on a discussion, etc.
Summary	Reviews of the state of the design or implementation to date, restating issues, alternatives, and criteria. It is a summary if it is a simple list-like restatement. If it is *ordered* by steps, it is a walkthrough, defined below.
Clarification	Questions and answers where someone either asked or seemed to misunderstand. This includes repetitions for clarification, associations and explanations. Clarifications serve to clear up misunderstandings from other individuals.
Digression	Members joking, discussion side topics (e.g., how to get the computer to make dotted lines), or interruptions having to do with things outside the content of the meeting (e.g., discussion of why the plant was moved over by the window, or that it is beginning to snow and they should leave early today). When the person running the video camera speaks or changes tape, it is considered a digression.

(Continued)

Figure 2. (Continued)

Goal	Statement of the purpose of the group's meeting and some of the constraints they are to work under, such as time to finish or motivating statements about how important this is.
Walkthrough	A gathering of the design so far or the sequence of steps the user will engage in when using the design so far, used to either review or clarify a situation. It usually follows the user's task or the flow of data or messages inside a system architecture.
Other	Time not categorizable in any of the above categories. For example, in one meeting they discussed how their coordination procedures having to do with documentation were insufficient in the past; in another the group members spent a large amount of time in the room hunting within the computer for the answer to a question of whether a language could support a particular function.

These categories were inspired by the extensive discussions of design and design rationale that are emerging in the literature (especially Conklin & Begeman, 1988), and by the categories of group management processes described by Putnam (1981) and Poole and Hirokawa (1986).

Our major focus is on the problem-solving aspects of design. In essence, design discussions are a form of argumentation: Various *issues* are raised either implicitly or explicitly, for each issue various *alternative* design possibilities are presented, and eventually a decision among these possibilities must be made by applying various *criteria* that help select the preferred alternatives. In developing our coding scheme, we expected that large portions of our design meetings would be taken up with discussions of these three things: issues, alternatives, and criteria. Examples of the design categories in the transcripts are given in the Appendix.

We also expected to find a number of other categories of activities that focused more on organizing the work of the group members. These include such coordination activities as discussion of *goals, project management,* and *meeting management* as well as *summaries, walkthroughs,* and *digressions.* The meeting participants engaged in *clarification* of their ideas. This category was defined very strictly as answers to explicit questions. These served not only to answer things unclear, but also to help advance the design itself through refinement of the original ideas. We have not looked at this distinction within this category. Periods of activity that did not fit into any of our categories were put in a catch-all class called *other.* We coded each of the 10 design meetings in Figure 1 using this scheme. We achieved high degrees of reliability in our coding: For the categories in Figure 2 we achieved interobserver correlations for the times in and transitions among these categories that ranged from .83 to .99.

4. GENERAL CHARACTERISTICS OF THE ISSUE DISCUSSIONS

In addition to these general classifications, we derived a complete description of the design discussion in terms of issues, the alternatives raised to that issue, and the criteria used to evaluate those alternatives. The number of issues raised in a meeting varied greatly, from 1 to 44 in our set of 10 meetings. We found that occasionally issues were raised with no alternatives (an "open issue"), but on the whole, 80% of the issues had two or more alternatives considered, with an average of 2.5 per issue. One third of these alternatives were never explicitly evaluated, but when they were, the evaluative criteria were neither heavily positive nor negative. Much more information about these general characteristics of the issue discussions are available in our earlier report (Olson, Olson, Carter, & Storrøsten, 1992).

5. THE USE OF TIME

Virtually all of the time in these face-to-face design sessions was spent talking. The participants in 8 of these 10 meetings spent on average more than 90% of the time talking to each other. The two exceptions were MCC-1 and MCC-2, where the participants spent one long period in each session silently using a workstation in the room to investigate whether the language in which they planned to build the system could do something. Because during this time they were not talking about the design, we excluded these long silent episodes from the MCC-1 and MCC-2 meetings. Indeed, in the analyses about to be reported, we include only time taken up with conversation, because we are concerned with what the groups were talking about.

We first examined the similarity across the 10 meetings in the overall use of time. We obtained the time spent in each of the 11 categories in Figure 2 for each of the 10 design meetings, and then looked at the intercorrelations among the distributions of time use to see how similar the 10 meetings were to each other. The complete table of pairwise correlations among the 10 meetings contains 45 values. All of the correlations were positive, ranging from .17 to .96, with a median of .72. A clustering and scaling analysis of the intercorrelations revealed that the major way in which the meetings varied was the amount of time spent on project management. Subtracting out project management and looking at the distribution of time across the remaining categories increased the median correlations across the meetings from .72 to .81 (range of the 45 correlations, .36 to .99). Thus, in overall use of time, these 10 meetings are quite similar. This is quite remarkable, given that the meetings are from four different projects in two organizations, each involving different sets of people.

The category that consumed the largest amount of time among the 11 was Clarification: more than one third of all the time. Because Clarifications were often about specific aspects of the other categories, we decided to try to associate as much of the Clarification time as we could with one of the other categories. Most of the Clarification time could be handled in this way, but there was still some Clarification time that could not be associated with the other categories. We created two new categories to handle this residual. *Clarification Artifact* refers to those occasions when a meeting participant tried to explain some aspect of a drawing, a list, or other meeting artifact. *Clarification General* refers to those remaining Clarification episodes that could not be associated with any of the other specific Clarification categories.

This revised coding scheme had 22 categories: the 10 original categories (excluding Clarification), an associated clarification for each of these categories, and the two new clarification categories just described. We calculated new coding reliabilities for these 22 categories, using the same procedures and coders as we used for the original 11 categories. The reliabilities were still very acceptable for purposes of establishing the objectivity of these classes.

Figure 3 presents details about the amount of time spent in each of the 22 categories. The direct design activities, namely, the first six categories, account for more than 40% of the time in these meetings. On the other extreme, the last cluster of categories can roughly be thought of as coordination or management activities, and these account for about a fourth of the time. The middle set is more mixed. Summaries are more coordinationlike, whereas Walkthroughs are clearly associated with design.

Alternatives and their Clarification take considerable time. This is hardly surprising, given that one of the core activities in design is to consider options for design issues. Significantly, the largest amount of clarification is associated with Alternatives. People need to communicate fully the ideas they have proposed as solutions to the design issues under discussion. This increases the total time spent on Alternatives to over 20%, making this the single biggest category of activity. Criteria also take up a reasonable amount of time. Interestingly, there is very little direct discussion of issues: only 3.2% of meeting time. This is most likely because to most of the participants in these sessions what issue was being discussed was clear from the context, and thus did not need to be stated explicitly.

6. TRANSITIONS AMONG DESIGN ACTIVITIES

We examined the transitions among the categories. First, we examined how similar the meetings were to each other. For the original 11 categories, the median correlation among the transition matrices for the 10 meetings

Figure 3. Time spent on various categories.

	Percent of Meeting	Range of Values (low)	Range of Values (high)	Combined Categories, Direct + Clarif.	Combined Clusters of Categories	Average Number of Times Visited	Average Length of Stay (seconds)
Issue	3.2	0.3	6.7			16.1	12.3
Issue Clarification	3.2	0.0	10.5	6.4		3.0	57.3
Alternative	11.4	4.1	22.8			41.4	17.1
Alternative Clarification	10.1	3.6	21.8	21.5		12.1	48.3
Criteria	12.9	3.8	22.9			53.1	14.1
Criteria Clarification	2.1	0.0	4.7	15	42.9	5.5	24.6
General Clarification	4.5	0.0	15.7			3.5	75.3
Artifact Clarification	3.5	0.0	15.1	8		2.9	68.0
Summary	13.4	3.8	36.0			12.6	51.9
Summary Clarification	5.0	0.0	14.0	18.5		5.0	53.5
Walkthrough	2.8	0.0	11.7			5.1	37.2
Walkthrough Clarification	0.8	0.0	1.9	3.6	30.1	1.4	34.8
Goal	0.5	0.0	1.5			1.5	18.8
Goal Clarification	0.8	0.0	5.2	1.3		0.9	54.3
Project Management	11.6	0.0	39.7			13.2	44.5
Project Management Clar.	1.5	0.0	6.7	13.1		1.1	54.6
Meeting Management	5.7	0.0	17.5			11.9	25.3
Meeting Management Clar.	0.9	0.0	3.8	6.6		1.1	50.2
Digression	5.2	0.0	11.4			9.7	33.6
Digression Clarification	0.0	0.0	0.5	5.2		0.1	18.0
Other	0.7	0.0	3.2			3.3	14.1
Other Clarification	0.0	0.0	0.3	0.7	26.9	0.4	6.0

Figure 4. **The time spent in various activities and the transitions among them. The size of the circle represents the total time, with direct presentation in white, the clarification in black. The width of the arrow represents the frequency of that transition. Only those transitions above .7% are shown.**

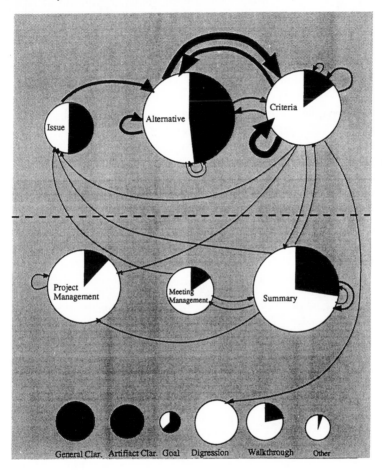

was .66. For the full set of 22 categories the median correlation was .61. Thus, as a first approximation, the patterns of first-order transitions among the meetings were quite similar.

Figure 4 shows in summary form both the total time spent in each category and the frequent transitions among the categories. In this figure the area of the circles corresponds to the amount of time devoted to that activity, whereas the thickness of the arrows corresponds to the frequency of the transitions. Because showing all of the transitions would make for too confusing a diagram, this figure only shows the transitions whose relative frequency was greater than .7%.[1] The black areas within each circle represent clarification time associated with that category, whereas the white areas represent the original presentation time. The exact locations of the heads and tails of the transition arrows show explicitly what coding category is involved in the transition.

There were frequent transitions among the core design categories, especially Criteria. Criteria and Criteria Clarification are involved in 46.1% of all transitions among the 22 categories. Alternatives and their Clarifications account for another large portion of transitions: 40.9% (this includes the transitions with Criteria, so this figure overlaps with the previous one). Together, Alternative and Criteria along with their associated Clarifications accounted for 68.2% of all transitions. In other words, these 4 categories from the total set of 22 categories are involved in over two thirds of the transitions. This is in contrast to time use: These four categories only occupied 36.5% of the total time. Thus, the design meetings we observed had frequent though brief excursions into discussions bearing on design alternatives and criteria for selecting among them.

7. SEQUENTIAL DEPENDENCIES AMONG THE DESIGN CATEGORIES

Figure 4 shows the lag 1 transitions[2] among the coding categories, but it does not capture the full extent of the more far-reaching sequential structure among these categories. We turned to a number of additional analyses to capture the structure of these meetings, and in this section sketch a more global picture of the flow of activities (a more detailed report appears in Olson, Herbsleb, & Rueter, 1994).

In order to identify interesting patterns among the large numbers of transitions that occur in the data, we have used several heuristic techniques. One was to search for patterns from an a priori understanding of design and group interaction. Another was to use sequential statistical analyses to guide the induction of patterns. This search was a mix of top down and bottom up. Each time a pattern was identified using either heuristic, we replaced it with a marker to indicate where the pattern was found. To give a very simple example, when we found that the design team had discussed one or more alternatives in a row, we replaced every occurrence of one or more a with some marker such as A^+. So the sequence

---

1. For this analysis, all of the transitions among the 22 categories summed to 100%.

2. In our use of this term, lag refers to the difference in ordinal position between the occurrence of two categories. Thus, adjacent occurrences have a lag of one, whereas occurrences of two categories that are separated by four intervening occurrences of other categories have a lag of five.

$$mgt \ i \ a \ a \ c \ c \ a \ a \ a \ i \ a \ c \ c \ a$$

would become

$$mgt \ i \ A^+ \ c \ c \ A^+ \ i \ A^+ \ c \ c \ A^+$$

This replacement process generated a new sequence that was examined in order to evaluate the patterns we composed. We used several criteria for this evaluation, including analyzing the rewritten sequence to see if it made theoretical sense, and judging the parsimony of the patterns as descriptions of the data. As new structure was revealed in the rewritten data, the entire process was repeated until we could find no more meaningful patterns.[3]

One of our earliest discoveries was the grouping of the coding categories into two classes: Design and Management. Figure 4 shows a dashed line between the six issue-based categories and all of the others. This division is based on a statistical analysis of the transitions among the 22 categories, looking at lags as long as five. If there were no sequential structure in the data, then the probabilities of going from some category X to any category Y should simply be given by the relative frequency of Y. So if one 10th of all the episodes in the protocol are assigned the category Y, then one would expect that when leaving category X, one would reach category Y 10% of the time. If, on the other hand, there is sequential structure in the data, then the probabilities of going to Y from X may be much higher or lower than 10%. Our statistical analysis looked for transitions that showed statistically significant deviations from chance levels.[4]

We looked at lags from one to five, and found that the six issue-based categories at the top of Figure 4 had 25 transitions among themselves that exceeded chance levels, but only 2 with those categories below the line. Similarly, those below the line had 79 such transitions among themselves, but only 11 that went across the line. Thus, these two sets of categories may constitute functional clusters. It is as though our design meetings consisted of episodes of direct design activity (sequences of the six design categories, which we call Design) intermixed with episodes of management activity (sequences of all the classes below the line, which we call Management). In our subsequent analyses we have treated these as distinct categories, but it will be ultimately important to validate this separation with other measures.

Perhaps the most surprising aspect of this clustering is that Walkthroughs, which certainly seem like a Design activity, cluster with the Management

---

3. We found that placing the strings in a spreadsheet and sorting the data in a variety of ways to be especially useful.

4. See Allison and Liker (1982), Bakeman and Gottman (1986), Gottman and Roy (1990), for the statistical techniques. We used the Z statistic proposed by Allison and Liker.

Figure 5. **The significant conditional transitions (lag one) among categories.**

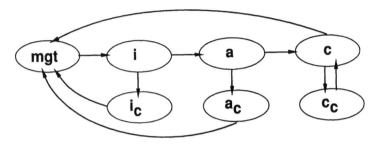

activities. Interestingly, one of the two significant transitions out of the Design category was to Walkthroughs. But this is not very impressive when you think that there were 60 possible transitions to Walkthrough or its associated Clarification from the six Design classes across the five lags. In the other direction, there were no significant transitions from Walkthrough or its Clarification to any of the six Design categories (again, out of 60 opportunities). Thus, despite its surface appearance as a design activity, Walkthrough appears to function in design meetings in a way very analogous to the other Management activities. Our next level of analysis beyond that reported in this chapter, where we track the specific content of the categories as well as who does what, may shed light on this.

Because our primary interest is in design as an activity, we explicated the internal structure of the Design cluster in greater detail. For this purpose, we took the sequences of states for each meeting and replaced any sequence of Management activity as a generic Management category. These new sequences consisted of strings of seven categories: the six Design categories plus a single, undifferentiated Management category. We created one such string for each of the 10 meetings.

We repeated our statistical analyses with the reduced set of seven categories. Figure 5 shows the significant conditional probabilities among the classes at lag one. These categories are arranged spatially to reflect the fact that Management discussions seem to play an anchoring role in the Design discussions, often leading initially to an Issue discussion. For example, all 10 meetings began with a Management discussion immediately followed by an Issue discussion. Overall, 54% of issue discussions were preceded by Management (whereas only 12% were followed by Management). So the impression created by Figure 5 that issue discussions flow primarily from Management discussions appears to be correct.

As we continued the process of formulating meaningful patterns and searching for occurrences of them in the data, one thing that became obvious was that there were extended discussions of alternatives and criteria

Figure 6. Diagrammatic representation of the various types of patterns found in our grammar.

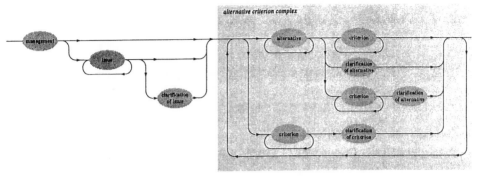

that were punctuated by discussions of issues and management. So we first described the structure of these alternative and criteria discussions. There were a number of ways in which these could arise, and we used a series of rewrite rules to capture these regularities. The shaded area on the right side of Figure 6 shows four main paths that summarize in diagrammatic form the structure of these various categories of alternative-criteria strings. Further analysis using these four categories resulted in a general pattern we called Alt_Crit that contained many interestingly structured strings. The overall structure of the gray area shows this. The general characteristic these strings have is that discussions of alternatives tend to precede discussions of criteria, though there is often a complex alternation involving this general pattern.

Next, we developed patterns that used these Alt_Crit complexes as building blocks. Issue-based design sequences consisted of an issue complex followed by an Alt_Crit. We called these IAC episodes. If these were preceded by a Management discussion, we called these MIAC episodes. Finally, we also defined patterns consisting of Alt_Crit complexes, without issues, that were preceded by Management. We called these MAC episodes, reflecting the fact that there are often design discussions in which the issue is implicit. The left-hand side of Figure 6 shows these various paths.

The Appendix shows examples of actual transcript segments that correspond to the most common kinds of sequences. Here it is possible to see what kind of talk is actually included in these categories. One thing this makes apparent is that there are some additional important analyses to do, such as coding the specific content of the issue, alternative, and criteria categories, or coding by speaker. These analyses are in progress.

To evaluate the extent to which these patterns were structure per se and not what you might expect from just the base frequencies of the categories, we compared the fit of our patterns in real meetings to their fit to 10 random "meetings" whose overall frequency of the seven classes of activities matched that of the 10 real meetings. We then performed the

Figure 7. Comparison of pattern frequencies for real and randomly generated
meetings.

Rewrite Rule	Real Meetings	Randomly Generated Meetings
IAC Episodes		
(not involved in MIAC episodes)	37	25
MIAC Episodes	45	6
MAC Episodes	56	32

same rewriting exercise with each of these 10 random meetings and
calculated comparable statistics about the fit.

Figure 7 shows the comparisons for the three patterns described in
Figure 6. Overall, there are 82 IAC episodes in the real meetings, but only
31 in the random ones. The majority of the IAC episodes in the real data
are in turn included in MIAC episodes, whereas only a small number of
those in the random data fit into this more complex structure. This means
that in the real design meetings there is a very strong trend to have design
sequences that have an issue stated at their beginning in turn preceded
by a management discussion. Design sequences that do not contain an
issue complex at their beginning are also more often preceded by Management in the real data than the random (the MAC episodes in Figure
7), though this trend is not as strong as the MIAC episodes.

Another way of looking at this is in terms of how many elements were
replaced by the search pattern, given the opportunity. For IAC episodes, 47%
of the available appropriate tokens (the intermediate issue complex and
Alt_Crit categories) were replaced for the real meetings, but only 20% of the
same tokens for the random meetings. For the MIAC episodes, 28% of
available tokens were replaced for the real meetings, but only 3% for the
random ones. Thus, the real and random meetings were quite different in
their fit to the higher level patterns that describe overall design sequences.

The fitting of patterns to data like these involves significant judgments
about how far to elaborate them. Clearly, with more and more patterns,
especially ones that are driven bottom up rather than top down by theory,
the fit can be made arbitrarily good. The question was when to stop. We
settled on a set of patterns that gave us a sense of the sequential structure
of these meetings that could be understood in terms of meaningful sequences of discussion and argumentation.

We have also looked at the similarities among the 10 meetings. When
assessed by the fit of the patterns in Figure 6 they tend to be quite similar.
Thus, at the level of sequential structure as captured by patterns, the

participants in the 10 meetings organized their activity in very similar ways. However, there was at least one striking difference between the meetings drawn from the two organizations: The KBE meetings at MCC have much longer sequences of Design discussions than the AC meetings. The average number of Design categories between Management discussions was 6.2 for the KBE meetings, and 3.2 for the AC meetings. Another way to state this is that the AC meetings have many more Management discussions than the KBE ones, and these tend to break up the design discussions into smaller episodes. Another symptom of the greater role of Management discussions in the AC meetings is that the tendency for issue discussions to be preceded by Management discussions was somewhat stronger: AC = 58%, MCC = 42%. This difference in the pattern of Management discussion may reflect the differences in the work cultures of the two organizations that we described earlier. However, these differences should be thought of as parametric differences in the average length of strings generated but not in their structure. The pattern of the Design discussions in the meetings of the two organizations is quite similar. Of course, there may be other things that could be measured in these meetings that would show striking cultural differences. We are only claiming that the pattern of activity as captured by our coding scheme made the two sets of meetings look similar.

Thus, the picture that emerges is that at this higher level of abstraction, design as an activity in our 10 meetings is quite structured. It consists of episodes of Design and Management that are distinct from each other, and within the Design episodes there is an orderly flow of argumentation and discussion.

8. SUMMARY

These were all very informal, intensely interactive meetings. On the surface they seem to be quite chaotic in organization. But our analyses tell us a different story: Design discussions at the level we described them are quite structured and orderly. There are several reasons why this may be so. First, it may be that design as a task is relatively structured, and that in order to do design at all one must proceed with some degree of sequential orderliness. Second, our designers were quite experienced. The AC designers had been trained on how to do design and meetings, and the MCC designers had considerable experience in industry prior to coming to MCC. Third, the orderliness may have been due to the high level of our analyses, without concern for who said what or how the issues interplay.

The fact that the 10 design meetings are more similar than different is perhaps our most surprising finding. We expected much less uniformity in the groups' behavior, due to differences in organizations, projects, and participants. Because this is the largest sample of field design meetings analyzed in this level of detail that we know of, we are inclined to conclude that there is substantial regularity to this behavior in general.

What are the implications of our analyses for design rationale? Clearly we have shown that descriptive categories based on general design rationale schemes are an interesting and useful way to analyze the activities in design meetings. But we have also shown that design as an activity is already quite structured, at least within the limits of these analyses. Even more astounding is that this structure is common across four design groups in two corporate cultures. This raises interesting questions as to the usefulness of using design rationale schemes as a way of organizing design as an activity. It may be that because design as an activity is relatively well structured, it would be straightforward to capture rationale information on the fly. However, several lines of work suggest that this may not be so straightforward. Buckingham Shum (1994 [chapter 6 in this book]) has reported that asking designers to use QOC notation during design sessions can be quite disruptive. Conklin and Burgess-Yakemovic (1991 [chapter 14 in this book]) reported that gIBIS only worked when there was a dedicated scribe who was willing to devote enormous amounts of time to the capture and analysis of rationale information. The difficulty of doing this is underscored in our own work, where trying to capture the design rationales of our meeting discussions takes an enormous amount of coder time off line. In the final analysis, of course, we need to study this directly. It would be useful to carry out the kind of detailed analysis of time use, depth and breadth of discussion, and sequential behavior of the sort reported here for design sessions where rationale notation was being used.

Overall, the data we have reported here give us a more detailed picture of group design in its natural field settings than we have had available before. Having such a picture available is crucial both to the development of tools for design tasks as well as for the evaluation of the effects of these tools (see Olson, Olson, Storrøsten, & Carter, 1992, for a study of simple workspace tools).

NOTES

Acknowledgments. Many people have participated in the collection and analysis of the data reported here. Libby Mack played a key role in arranging for the data collection at Andersen Consulting, and helped us with their interpretation. Nancy Pennington collected the data at MCC, Barbara Smith transcribed them, and Bill Curtis arranged for access to them. Kevin Biolsi assisted in data analysis. We are grateful to Tom Moran and Allan MacLean for discussions about this work. This work was begun during sabbatical leaves by the Olsons in Cambridge, England, and we are grateful to Rank Xerox EuroPARC and the Applied Psychology Unit for providing supportive environments for thinking about design and collaboration.

Support. This work has been supported by the National Science Foundation (Grant IRI-8902930) and by the Center for Strategic Technology Research (CSTaR) at Andersen Consulting.

Authors' Present Addresses. Gary Olson and Judith Olson, CREW, 701 Tappan Street, University of Michigan, Ann Arbor, MI 48109-1234; James Herbsleb, Software Engineering Institute, Carnegie Mellon University, Pittsburgh, PA 15213; Marianne Storrøsten, Norwegian Computing Center, Gaustadalleen 23, P.O. Box 114 Blindern, N-0314 Oslo, Norway; Mark Carter, Pencom Systems Incorporated, 1801 Alexander Bell Drive, Suite #210, Reston, VA 22091; Email: gmo@umich.edu, jsolson@umich.edu, jherbsle@sei.cmu.edu, marianne.storrosten@nr.no, markc@pencom.com

---

REFERENCES

Allison, P. D., & Liker, J. K. (1982). Analyzing sequential categorical data on dyadic interaction: A comment on Gottman. *Psychological Bulletin, 91,* 393–403.

Bakeman, R., & Gottman, J. M. (1986). *Observing interaction: An introduction to sequential analysis.* Cambridge, England: Cambridge University Press.

Buckingham Shum, S. (1995). Analyzing the usability of a design rationale notation. In T. P. Moran & J. M. Carroll (Eds.), *Design rationale: Concepts, techniques, and use.* Hillsdale, NJ: Lawrence Erlbaum Associates. [Chapter 6 in this book.]

Conklin, E. J., & Begeman, M. L. (1988). gIBIS: A hypertext tool for exploratory policy discussion. *ACM Transactions on Office Information Systems, 6,* 303–331.

Conklin, E. J., & Burgess-Yakemovic, KC. (1991). A process-oriented approach to design rationale. *Human–Computer Interaction, 6,* 357–391. Also in T. P. Moran & J. M. Carroll (Eds.), *Design rationale: Concepts, techniques, and use.* Hillsdale, NJ: Lawrence Erlbaum Associates, 1996. [Chapter 14 in this book.]

Gottman, J. M., & Roy, A. K. (1990). *Sequential analysis: A guide for behavioral researchers.* Cambridge, England: Cambridge University Press.

Olson, G. M., Herbsleb, J., & Rueter, H. (1994). Characterizing the sequential structure of interactive behaviors through statistical and grammatical techniques. *Human–Computer Interaction, 9,* 427–472.

Olson, G. M., Olson, J. S., Carter, M. R., & Storrøsten, M. (1992). Small group design meetings: An analysis of collaboration. *Human–Computer Interaction, 7,* 347–374.

Olson, J. S., Olson, G. M., Storrøsten, M., & Carter, M. (1992). How a group editor changes the character of a design meeting as well as its outcome. *Proceedings of CSCW '92,* 91–98. New York: ACM.

Poole, M. S., & Hirokawa, R. Y. (1986). *Communication and group decision making.* New York: Sage.

Putnam, L. L. (1981). Procedural messages and small group work climates: A lag sequential analysis. *Communication Yearbook 5* (pp. 331–350). New Brunswick, NJ: Transaction Books.

APPENDIX

Examples of Coding Categories and Common Patterns in Complete Episodes (Nothing Deleted)

1a. *IAC episode* (issue-alternative-criterion-alternative-criterion)

John:	Are we accounting for printing multiple levels of the control flow diagram?	issue
Tim:	Well, I think we need to decide that.	
	Certainly we want to print the top level. The control flow diagram . . . George, Greg, and I thought through this a little bit, and	alternative
	there are some easy things to implement from a user interface standpoint for printing out selected parts of it.	criterion
	Uh, specifying the starting block and the ending block on the printing is probably	alternative
	the easiest way to print out sequential block numbers, if you only want to see first half of it, you can get a pretty good shot at printing knowing the first half of it.	criterion

1b. *IAC episode* (issue-alternative-alternative-criterion-criterion)

Don:	If knowledge editing was fundamentally collaborative then how would it be different than it is in GKR lab, that's the question.	issue
Bill:	Well, there's really two different answers to that. One is you want to be able to do group work which is sort of the Colab thing where you actually have people at the same time working in the same space and getting a lot of negotiation going but that isn't the one that I've concentrated most on	alternative
	but rather the one of how you can mediate people working together through time where basically somebody . . . I mean if you look at advice, the packaging of advice in computations is really sort of a method for being able to help someone at a different time in the future. You sort of set things up in a little mouse trap sort of thing and when they, at some future date, stick their finger in the mouse trap instead of getting snapped, the cheese is over here or something, they get help.	alternative

Don:	We want something that says, I knew you'd look here for the cheese but the reason you looked here is this and this and you were wrong so look over there.	
Bill:	So that's the one that I'm probably more interested in	criterion
	although I think the other one is very valuable.	criterion

2. *MIAC episode* (Management-issue-alternative-criterion-alternative-clar-alternative-alternative-criterion-digression-criterion)

Don:	I was just going to change the topic a little bit because I keep . . .	meeting management → mgt
	The thing you said after we started about how knowledge editing is if you do it alone, you haven't done . . . I mean, it's kind of fundamentally collaborative, right, because it's not just a process of building an artifact, it's a process of a group of people buying into a particular axiomatization of the world.	issue
Bill:	Yeah. that's what's important, co-representation.	
Tom:	Yeah. One of the things I really liked about the Colab thing, I don't know if this is off the track, and what I always wanted to do when we were talking about putting this collaborations and this interaction thing is to have somebody else's frame up in the corner of my window so I could see what they were doing, kind of tiny, as I was doing whatever I was doing. That's what they did in the Colab thing. In a sense they had one little window up in the corner that was common to everybody and that you interacted with but you could still do your own interaction locally, and then maybe propagate it (unclear).	alternative
Stu:	So that might be as easy as just like having a shared inspector or something, I mean, having one of the inspector designs that everybody could write into it.	criterion
Bill:	It would be nice if you could have a window to another guy's interface to see what he's doing and somehow the thing would shrink in a cognitively appropriate way so that the meaningful, sort of the highlights would stand out and you could tell what he was doing, but a lot of the detail would fade.	alternative

Don:	Say it again.	clarification of
Bill:	It's the idea that you could say I want a	alternative

window on Don's activity so it makes a little
window of your whole screen but it's smart enough
so that it doesn't just shrink pixel by pixel and
then do the hypercard thing where you really can't
see what your old frames are exactly but it somehow
shrinks smartly so that maybe you just get the names
of the units that are . . . and only slots that have
been touched with the mouse and all other slots
never get seen or something. That would be a way
to sort of keep up with you but it would also be, I
guess, according to your vision, I could actually go
there and just do work. I could make things happen
in your interface. If it's the Colab idea I have to be
able to go up there, well, hot it's either a spy on
your personal work area or it really is a group work
area where I have license to go make modifications.

Tom: Yeah, that's what it seems like to me. It seems like   alternative
you want the both of those things. You want a group
area where everybody sees it and sort of everybody
is interacting at the same time in that window. And
then you want your own work area and then maybe
you want to be able to see what somebody else is
doing in their work area.

I don't know how you handle these on limited   criterion
screens.

Don: A million pixels is not enough.

Bill: Well, you know though, we should be thinking . . .   digression
I hate to think for the long term.

Don: Golly, that would be bad.

Bill: But we will have the work surface someday and I   criterion
don't think we should really try and do it but we
should think with incompatible ways.

3. *MAC episode* (Management-alternative-criterion)

Walter: Each diagram essentially has an ID,   summary → mgt:
through the simplification, it has those number
dash number, right? You always know what it is.
But we can always give him some other attributes.
That's simple enough, and that's the least
constrained at this point.

John: It'd be nice if you filled in the dialog boxes   alternative
         because you know what diagram I called you from.
Tim: That's easy.   criterion

8

Synthesis by Analysis: Five Modes of Reasoning That Guide Design

Mark K. Singley
Educational Testing Service

John M. Carroll
Virginia Polytechnic Institute and State University

ABSTRACT

This chapter presents a psychological analysis of the ways in which designers bridge the gap between an understanding of existing artifacts and the synthesis of new artifacts. We propose a taxonomy of design reasoning, five distinct ways of bringing psychological constraints to bear on the design process: assessing an artifact genre, hillclimbing from a predecessor artifact, process modeling, scenario envisionment, and formative evaluation. We explore how these modes of reasoning complement one another and how their unique contributions can be integrated through design rationale. In a case study exploring the interaction of the modes, we construct a psychological design rationale for a key component of an intelligent tutoring system for Smalltalk, a system-generated display of user goals called the Goalposter.

Kevin Singley is a cognitive psychologist with an interest in theories of learning and transfer and the psychology of instructional design; he is a research scientist in the Division of Cognitive and Instructional Science at the Educational Testing Service. **John Carroll** is a cognitive psychologist interested in the analysis of human learning and problem solving in human–computer interaction contexts and in the design of methods and tools for instruction and design; he is Professor of Computer Science and Psychology and Head of the Computer Science Department at Virginia Polytechnic Institute and State University.

CONTENTS

1. SYNTHESIS BY ANALYSIS

We are developing a new representation for psychological knowledge that reifies the causal connections between features of software artifacts and the psychology of users (Carroll, Kellogg, & Rosson, 1991; Carroll & Rosson, 1991 [chapter 4 in this book]). These causal connections, called *claims*, can serve as building blocks for an applied psychological science of design. Because the language of design is the language of artifacts, psychological knowledge can be brought to bear more directly on the design process if it is indexed by particular features of artifacts. This eliminates some of the inferencing and indirection required in deriving prescriptions from a descriptive psychology. Also, as new artifacts are built, a claims-style theory, being cast at a certain level of abstraction, makes predictions regarding the psychology that is operative in the new artifact, assuming the new artifact shares features with older, analyzed artifacts. Thus, a claims-style theory provides a vehicle for generalization and cumulation of the science base as new artifacts are brought under the umbrella of analysis.

Claims are not only used to represent how the design of a particular software artifact affects the psychology of the user; they can also be used to drive the design process. According to a conceptualization of the design process known as the task-artifact cycle (Carroll & Rosson, 1991 [chapter 4 in this book]), design takes place in a rich historical context. No serious design work goes on without rich influences from previous, similar designs. This means that the process of design can be conceptualized as essentially the redesign (e.g., specialization, extension, simplification) of existing artifacts. Given a thorough claims analysis, redesign is driven by the following simple heuristic: Within the constraints of the design task, try to eliminate or mitigate the claims involving negative psychological consequences while preserving or enhancing the claims involving positive psychological consequences. Claims-driven design is a systemization and reification of a natural process that has the added value of reserving placeholders for principled scientific psychological knowledge to ground rationale for the design.

To what extent can psychology guide the design of software artifacts? Certainly now, and probably always, psychological principles will underdetermine design. In the task-artifact framework, claims do not determine the outcome of design but rather act as a kind of heuristic control on the process: Claims direct the designer's attention to problematic aspects and also highlight features worth preserving. By characterizing important features at a possibly higher level of abstraction than the physical device space, they also provide a new "parse" of the artifact and in a sense help redefine the design problem space. Claims essentially elaborate the representation of the design problem and add constraints on the solution.

The open-ended, nondeductive nature of design is captured in the task-artifact framework by a kind of analytic-synthetic gap: The design of a new artifact is driven largely by analyses of existing artifacts that support existing tasks. However, the new artifact, having new features and new functionality, could radically change the kinds of tasks performed, and this could radically change the operative psychology. So, the psychological analysis that has driven the redesign may no longer be valid in the new artifact. This situation potentially blunts the power of principled psychological analyses and could be especially problematic if we are to take to heart the contextualist critique of Whiteside and others (e.g., Whiteside & Wixon, 1987), who say that principled interface design is fundamentally intractable, and can only be done through an iterative design process driven by early and continued feedback from users. Others claim that psychological theory has very little to say to designers and can provide very few constraints on the design process (e.g., Landauer, 1991).

However, we feel this view is too gloomy, and in fact does not characterize what designers actually do. It is certainly not right to characterize the design process as blind search waiting to be informed by formative evaluation. It is true that bridging the gap between an analysis of existing artifacts and the design of a new artifact is essentially a creative act. But this creative act can be constrained by bringing to bear psychological knowledge from a variety of sources. Whether these constraints have a positive or negative effect on the resulting design depends on the comprehensiveness and correctness of the constraints themselves. But in good design constraints are continually being applied. In our work, we cast these constraints in the form of psychological claims whose upsides are to be preserved and whose downsides are to be mitigated.

2. THE FIVE MODES

In this chapter we characterize some ways in which designers bridge the gap between analysis of existing artifacts and synthesis of new artifacts. We are designing an intelligent tutoring system for Smalltalk, and reflecting on our own design processes to explore how we ourselves bring psychological constraints to bear. In our introspections, recollections, and records of our work on the tutor, we see evidence for five modes of reasoning that drive design through the task-artifact cycle. Our interest in this chapter is not to view them as common design activities but rather as complementary modes of reasoning with characteristic strengths and weaknesses. Furthermore, we explore how the unique contributions of these modes can be integrated through design rationale.

The modes are primarily scenario based, in that they have to do with analyzing the interactions of users with artifacts in task settings. The artifacts and tasks may be more or less abstract; for example, the analysis may be of a concrete artifact supporting a real task, a genre of artifact supporting a generic task, or it may be of an envisioned artifact. The modes differ primarily in terms of the precise nature of the artifacts and tasks under analysis and therefore ultimately in terms of the source of the constraints. The five modes are as follows.

Hillclimbing From Predecessor Artifact. Often design decisions are motivated by an analysis of the strengths and weaknesses of an existing artifact that is closely related to the envisioned artifact in either functionality, interface techniques employed, or tasks supported. A common strategy is simply to itemize the shortcomings of the existing artifact and remedy those shortcomings in the envisioned artifact. This may amount to fixing a perceived problem or perhaps adding a feature. Hillclimbing expresses itself most clearly in those artifacts whose development can be characterized by incremental improvements. For example, Carroll (1991) characterized 10 years of spreadsheet development as an evolutionary process driven largely by hillclimbing from one spreadsheet program to the next.

Process Modeling. Sometimes an interactive process exists in the world that can serve as an inspirational model or guiding metaphor for an artifact. Perhaps a designer uses the observed or idealized interaction between two people to serve as a model for the interaction between the envisioned artifact and its user. In these cases through independent means the designer has become convinced of the efficacy of the model and, rather than subjecting the model to a thorough evaluation of strengths and weaknesses as in hillclimbing, the task is simply to extract the principles of the model so that they may be implemented in the target situation. Whereas hillclimbing implies reworking and improving, process modeling implies capturing and mimicking. Of course, to be effective, a lot of critical thought should go into the selection of a model.

An example of process modeling is the classic automation situation, where some process exists in the world and the goal is to embody that process in a computer program, for example, modeling the operations of a bank teller in an automated teller machine (ATM). Sometimes, the model is well known to the designer, in which case the modeling can proceed analytically, but occasionally, the model is obscure, in which case the modeling must have an empirical discovery component. An example of the latter is the capture of human expertise for use in an expert system. Although the term modeling implies a fairly straightforward mapping from process to target artifact, in fact the process can require a fair amount of interpretation on the part of the designer.

Assessing an Artifact Genre. Hillclimbing and modeling are characterized by fairly concrete mappings from individual predecessor to target artifacts. However, the analysis may also take place at a higher level of abstraction if the artifacts can be viewed as instances of a genre. This amounts to an analysis of the fundamental strengths and weaknesses of an entire class of artifacts, a kind of state-of-the-art critique.

The existence of a genre implies a fairly mature technology with a variety of instances, each of which exemplifies some defining set of principles. It is this set of principles that is being evaluated when assessing the state of the art. In other work, we have explored the use of second-order artifacts as repositories of abstract design principles (Carroll, Singley, & Rosson, 1992). In this view, concrete artifacts inherit claims from one or possibly several second-order artifacts (or genres).

Envisioning Scenarios. Hillclimbing, modeling, and assessing the genre can all be used to help define goals and provide constraints on the design. Ultimately, however, these constraints are intensional and do not specify in concrete detail what form the new design will take. So we find ourselves at the precipice of the analytic-synthetic gap, with no recourse but to do some constructive problem solving and propose some unique organization of features and functions in an attempt to satisfy the many constraints. One method for concretizing a design is scenario envisionment. In this mode of design reasoning, one or more user scenarios is conjectured and can then be developed in arbitrary detail to initially explore the efficacy of various features of the design for users.

In essence, any version of scenario envisionment is simulation. The medium of the simulation can be as informal as paper and pencil, or even mental images. It can also be very refined: Scenarios can be mocked up with video and programmed with special effects. Scenarios can even be represented as symbolic structures.

We regard scenario envisionment as the linchpin for psychological design rationale. We are not interested merely in understanding designs that have already been implemented. We wish to use design rationale to assess designs that have not yet been implemented. We do this by envisioning systems and analyzing the design rationale for the "system" as envisioned. What we envision are sets of user scenarios, selected to cover our best guess as to the critical and typical types of things users will do and experience. We develop these scenarios as textual narratives and then analyze what seem to be the important causal relations in them (that is, the salient psychological user consequences of the various features in our envisioned designs).

So, in our methodology, psychological design rationale is brought to bear on both sides of the analytic-synthetic gap. In a particular design situation, claims drawn from hillclimbing, modeling, and assessing the genre steer us to a well-defined ledge of the precipice and influence our initial trajectory as we leap. Once we cross into the realm of the new artifact, claims drawn from envisioned scenarios soften our landing and reduce the likelihood of crashing and burning.

Formative Evaluation. Once scenarios have been envisioned in rich enough detail, interface designers begin to commit their designs to code. As the system begins to take shape, it can be subjected to formative evaluation. Although this implies a clean transition between the two modes, the transition becomes somewhat blurred with the use of rapid prototyping tools like Smalltalk: Often scenario envisionment proceeds hand in hand with code development in these kinds of environments. We can define the beginning of formative evaluation as the endpoint of scenario envisionment where the scenarios have been rendered in the medium of executable code. This allows for the evaluation of richly interactive scenarios, which is quite useful when the behavior of a complex system is difficult to envision. The ultimate step is to evaluate the interaction of the fledgling system with a user. There will always be more to observe than could ever be envisioned with scenarios.

Of course, this does not mean that the entire system has to be implemented before formative evaluation begins. It is often useful to evaluate pieces of the design before committing to the design as a whole. Missing pieces may be simulated to approximate the target usage situation, as is done in the classic Wizard-of-Oz studies (Kelley, 1983). For speculative design exercises way beyond the state of the art, it is sometimes useful to simulate missing functionality that in actuality cannot be supplied. For example, Gould's listening typewriter studies (Gould, Conti, & Hovanyecz,

1983) or Carroll and Aaronson's (1988) studies of intelligent help can be conceived essentially as formative evaluation with very little of the functionality actually implemented and with little hope of full implementation in the near future.

In the remainder of the chapter, we explore a case study of design and try to show how these five modes interact, support, and supplement each other to guide the design of an artifact. We discuss how we designed a component of an intelligent tutoring system for Smalltalk called the Goalposter, and how these five modes of reasoning played a role in providing the rationale.

3. PSYCHOLOGICAL DESIGN RATIONALE FOR THE GOALPOSTER

We organize the discussion of the design of the Goalposter into sections corresponding to the five modes outlined previously. We do this for expository purposes primarily, although the modes do have a certain natural chronological ordering: Assessing the genre, hillclimbing, and modeling tend to occur before scenario envisionment, which in turn tends to occur before formative evaluation. Of course this ordering is only partial and even then not without exceptions: It is possible to assess the genre while doing formative evaluation; it is usually inefficient to do so. Although we discuss each of the modes separately here, this is not meant to imply that the modes operate independently. We discuss issues of interaction and convergence later in the chapter. Also, our description of the design process is not exhaustive; we have selected aspects of the process that exemplify the five modes and their interaction.

Throughout this case study, we explore the use of psychological claims to represent the design considerations generated by each of the modes. We do this for two reasons. First, we would like to assess the use of claims as a unified representation for design reasoning. Rationale may converge from multiple sources and strengthen a design argument, and this should be captured in a unified representation. Alternatively, rationale may conflict and a unified representation may ease the management of the problem and point the way to a resolution. Second, although we hope that the claims representation can accommodate each of the modes, we also want to investigate whether the modes typically generate claims of a certain kind, or build on existing claims in a certain way. By viewing the modes through the claims they generate, it may be possible to further distinguish the modes and better understand the distinct roles they play in design.

In what follows we present a reconstruction of our design reasoning. Our design discussions were peppered with claims terminology: We stated claims, discussed upsides and downsides, and reasoned about their interactions.

However, the claims we present here are not literally drawn from our design protocol. They are crafted with the benefit of hindsight, and are therefore clearer and more succinct than our real-time appeals to them. The tutor project is small (the authors plus Sherman Alpert) and much of our design discussion relied on shared context, memory, and vision. However, we feel that the following fairly well captures what actually happened.

3.1. Assessing the Genre

The design of the Goalposter began with a concrete problem. Smalltalk, although recognized as a good platform for rapid prototyping and software reuse, is widely regarded as difficult to learn. We decided to build an intelligent tutoring system for Smalltalk to address this learning problem. This decision was influenced by our experience with tutoring systems in general and programming tutors in particular. Tutoring systems as a genre have a psychological rationale (e.g., see Anderson, Boyle, Farrell, & Reiser, 1989), and to the extent our design inherits properties from tutoring systems it inherits some of the rationale as well. Our analysis of the tutoring system genre revealed that a great deal of instructional leverage comes from an accurate and complete task analysis of the domain and the successful communication of that analysis to the learner. This is especially true for the more strategic aspects of a skill. The higher level cognitive structures, such as goals and plans, are what drive problem solving, yet their explication is typically absent from traditional pedagogy.

We may express this generalization about the usefulness of communicating goals and plans as a psychological claim associated with the tutoring systems genre. Again, a claim posits a causal connection between an artifact feature and its psychological consequences for the user. Positive consequences are conveyed in the upside clause of the claim, and negative consequences in the downside clause:

Reifying Goals claim:
Reifying a learner's goals and plans promotes an abstract understanding of task structure, which leads to efficient action and broad transfer.
(But the goal and plan language may be unfamiliar, abstract, and therefore difficult to comprehend.)

The upside of this claim is drawn from Anderson's cognitive principles of intelligent tutoring (Anderson et al., 1989), and is backed by myriad studies of problem solving and instruction. For example, Singley (1990) showed that having learners post goals to a goal blackboard improved measures of both learning and transfer. However, the usefulness of this principle may be limited by the fact that it may be difficult to find a transparent representation for goals and plans in some domains. The problem is that the planning space is typically positioned at a level of abstraction far above what Newell and Simon (1972) called the physical problem space, the space where actual operations are carried out. This danger, that learners will be confused by abstract goal and plan terminology, constitutes the downside of the claim.

The principle of goal reification is present in tutoring systems generally. As stated previously, it is not tied to or qualified by any particular artifact or instructional setting. As a result, the principle is quite abstract. The claim says nothing about how goal reification is to be realized, or how the particulars of an instructional setting might influence that realization. Instantiating the claim sensibly in a particular design context (e.g., a tutoring system for Smalltalk) requires significant design activity. However, by considering what rationale we wish to inherit from the tutoring systems genre up front, we define our theoretical commitments and cast the design problem we wish to address at a higher level of abstraction.

3.2. Hillclimbing From Predecessor Artifacts

Hillclimbing From Related Tutoring Systems. In our attempt to promote the tutoring genre claim in our design, we found it useful to consider some concrete artifacts that instantiate the claim in instructional settings that are similar to our own. There are examples of systems in which the physical and planning problem spaces are distinct and yet learners are asked to generate plan structures prior to solving the problem in the physical problem space. The most closely related predecessor artifact of this type is Bridge, a tutoring system for Pascal programming (Bonar, 1988). In Bridge, learners are first asked to code an algorithm in terms of planning icons that are themselves a fully executable programming language. It is our suspicion, however, that this brute force approach to conveying abstract knowledge probably overwhelms the learner with additional complexity. It is quite possible that learning to use these formalisms is at least if not more difficult than mastering the domain itself. We decided to look for a more indirect approach.

We can summarize our analysis of Bridge by the following claim:

Full-Blown Planning Language claim:
Having learners first generate a solution in terms of an abstract planning language and then translate it into code promotes the learning of the plans and provides a holistic view of how plans organize code.
(But the planning language is an additional language to learn.)
(But learners only make mappings from plans to code and never from code to plans.)

The two aforementioned downside clauses highlight some of the contrasts between Bridge and its own predecessor artifact, Proust (Johnson, 1985). In Proust, planning structures are not manipulated directly by learners but instead are used "under the covers" by the system to analyze learner code. Error messages generated by the system make use of the plan analysis by referring to the underlying plans when describing learner errors and prescribing fixes. The Proust position on goal reification can be summarized by the following claim:

Planful Error Messages claim:
Having system-generated error messages couched in terms of goals and plans brings the proper abstractions to bear when they are critically needed and spares the learner the burden of generating goal and plan descriptions themselves.
(But learners may develop superficial understandings when errors are not made.)
(But goal-plan dialogues are restricted to errors and are not under learner control.)

Proust minimizes the burden on the learner by keeping its goal and plan structures for the most part to itself. However, the cost is that these abstractions are not readily available and as a result have no influence on the majority of learner problem solving. The only way a learner can get goal and plan information is to make an error, and then the information is restricted to the goals and plans pertinent to that error. The learner never receives a holistic analysis of the abstract structure of a problem. From our analysis of Bridge and Proust, we decided to try to provide learners with a holistic analysis of abstract structure without requiring them to program in an abstract language.

Hillclimbing From Smalltalk. Every domain is unique, with its own earmarks and peculiarities. We wanted to make sure our design solution was custom-fit to the demands of our particular domain. To this end, we decided to do some empirical work on problem solving in the standard Smalltalk environment. Our goal was to better understand the processes of planning and goal management in Smalltalk so that we could support them without sacrificing what we felt were the numerous strengths of the Smalltalk environment. Again, this goal can simply be restated as preserving the upsides of Smalltalk claims while mitigating the downsides.

We observed four persons of varying levels of expertise undertake an introductory yet representative project in Smalltalk: to create a window that transforms and displays all of its keyboard input in upper case. Although this project involves writing only several lines of code, it typically takes an intermediate-level Smalltalk programmer about 30 minutes to find the appropriate reusable components and test the solution. All subjects performed the task on their own with no help from the experimenters. They were asked to give verbal protocols as they worked, and all were videotaped. From these observations, we tried to understand how features of the Smalltalk environment were influencing the behavior of our subjects, and further, how the problematic aspects of those features might be mitigated in the Goalposter. (For a fuller description of these protocols, see Singley, Carroll, & Alpert, 1991.)

Many of our observations concerned problems associated with goal management. Smalltalk provides a rich set of tools to help manage the complexity of the language, but these tools initially represent additional overhead for learners, making the Smalltalk mountain that much harder to climb. Anderson and Jeffries (1988) documented the problems plaguing beginning LISP programmers, and claimed many of their errors could be attributed to losses of information from working memory. One might expect that these problems would be even worse in Smalltalk. Indeed, our protocols contained striking examples of programmers getting confused, losing their place, and forgetting to perform stated objectives.

Helping our learners manage goal structures is complicated by the fact that Smalltalk's rich environment encourages incremental development and task switching. This kind of flexibility is good in terms of supporting rapid prototyping, but can only worsen task management problems for learners. For example, early in the project, one subject switched freely between searching for the character input functionality, searching for the uppercase transformation functionality, and generating test code. As she switched between tasks, she spent time reestablishing the local goal context and reinstating her understanding of the state of her progress.

Our analysis of goal management in Smalltalk can be represented by the following claim:

Rich System Tools claim:
Unconstrained, modeless access to Smalltalk's rich system tools supports task switching, opportunistic problem solving, and an empirical approach to coding.
(But programmers may lose their place or waste time reestablishing context.)
(But programmers may adopt an overly reckless attitude to code development.)

We decided the Goalposter should support task switching by providing some kind of external record of tasks attempted, tasks completed, and their relation to the current task.

Learning Smalltalk involves becoming familiar with a large set of existing object classes, methods, and their inheritance relationships. Much of the time in initial projects is spent conducting extended searches for the right components to reuse. Of course, given the complexity and sheer magnitude of the hierarchy, learners will often have trouble finding the right class or method. Often, the search is biased by the alphabetical listings in the browser: Because there are often dozens of classes or methods to consider, those at the top of the list tend to get more consideration than those at the bottom. For example, one of our subjects began her consideration of the more than 100 TextPane methods at the very top of the list. She spent much time opportunistically modifying, testing, and ultimately rejecting the first few methods on the list that looked potentially relevant. Judgments of relevance were based totally on the names of the methods, which in most cases provided little guidance. Finally, she found the method she wanted buried in the middle of the list, but only after following many false leads.

Extended searches can waste the learner's time and energy and lead to working memory problems. Our analysis of browsing in the Smalltalk environment can be summarized by this claim:

> *Extensive Class Library claim:*
> Smalltalk's extensive library of classes and methods encourages a code reuse strategy.
> (But learners must familiarize themselves with the existence and functional roles of components in the hierarchy.)
> (But learners are given little information to make choices among candidate classes and methods.)

Our challenge is to provide an enhanced browsing environment through the Goalposter so that learners can evaluate what they find and find what they want.

3.3. Process Modeling

In order to further inform the design of our tool, we had two additional subjects perform our uppercase window project with the aid of local Smalltalk guru acting in the role of a coach. The coach was instructed to intervene whenever it was felt to be appropriate, but to not be too directive or reveal too much of the solution in her comments. Our motivation in doing this was to observe the coach's behavior and thereby supply ourselves with a possible model for the Goalposter. This can be viewed as a fairly standard knowledge engineering situation, where the task is to extract the principles of operation from a highly regarded yet poorly understood model (an experienced, articulate Smalltalk coach).

One thing we observed was that the coach reinstated learner goals when learners began to flounder. One subject began the project by spending a protracted period in the Class Hierarchy Browser viewing numerous Window classes. He finally chanced upon the characterInput: method in the TextPane class, which was the method he needed to modify. By this time, however, he was so completely engrossed in the vagaries of the Window subhierarchy that he had completely forgotten that the project specification had included the transformation of keyboard input to upper case, and in fact seemed to have forgotten that the project had specified any transformation whatsoever. He began to invent subgoals for himself, saying "Now what can I do with this method? Well, I could encode characters, or count them . . ." At this point, the coach intervened, "What do you have to do for this project?" The subject was surprised. "Oh, I guess I forgot." The coach reinstated the goal of the uppercase transformation, and the subject was on his way again.

This episode is consistent with other observations of human tutoring (e.g., McKendree, 1990) and provided convergent rationale for the idea of the posting goals:

Reinstating Goals claim:
Reinstating goals for learners facilitates the resolution of problem-solving impasses.
(But this may rob learners of the opportunity of managing their own goals.)
(But the tutor's intervention may be disruptive to ongoing problem-solving efforts.)

Although our coach reinstated goals when learners got into trouble, most interesting from the standpoint of modeling was what she did when learners were ostensibly on track. What we observed was that the coach delivered feedback on progress even when the learner was proceeding without difficulty. For example, our earlier subject often looked to his coach for confirmatory feedback for hypotheses and goals he considered pursuing. When searching for the class that handles character input, he asked, "Will it be [a class] under Application Window?" He continued to seek confirmation as he selected and ultimately rejected candidate classes. The coach spent a significant amount of her time reassuring the subject by saying, "Yes, that's right."

This protocol shows that even when learners are correct feedback can be useful. Such feedback encourages learners to pursue appropriate courses of action and builds feelings of confidence. This analysis leads to the following claim:

Confirmatory Feedback claim:
Providing confirmation that learners are pursuing legitimate goals fosters confidence and allows learners to focus attention on the chosen course of action.
(But this may rob learners of the opportunity of evaluating their own goals.)

The downside of this claim is similar to the first downside of the Reinstating Goals claim. Whenever a system provides support there is the danger that that support will become a crutch.

3.4. Scenario Envisionment

What we have presented to this point is much of the rationale we generated from our consideration of predecessor artifacts that had a bearing on the eventual design of the Goalposter. The tutoring genre claim presented us with a high-level vision of goal reification and its associated costs and benefits. We then considered two tutoring systems in which this claim was realized, and considered the various trade-offs associated with these concrete instantiations. Finally, we extracted claims associated with the Smalltalk environment itself, in order to fully situate the design reasoning in our target domain. Some of these claims characterized the interactions of a human coach with Smalltalk learners. We now provide a structural account of how these various claims played a role as we envisioned scenarios for the Goalposter.

Figure 1 provides an integrative preview of how these claims relate to and influence the Goalposter claims about to be discussed. This claims map not only provides rationale for the current state of the design, but also summarizes the design history of the Goalposter (the claims are roughly chronological from top to bottom). The links in the figure simply specify "responds to" relations among the claims; that is, a design decision may embody a claim that responds to a claim in a prior design decision or analysis.

System-Generated Goals.   Perhaps the defining feature of the Goalposter is that the goals are system generated, based on an ongoing analysis of learner behavior. As learners browse the class hierarchy or write code, a plan parser automatically updates the contents of a goal blackboard to reflect the system's current interpretation of the learner's actions. The Goalposter is useful after a project is finished in that goal traces are saved for later inspection. It is our hope that these goal traces, if properly annotated, will support analogical problem solving outside of the context of the tutor. Thus, the Goalposter makes the following claim:

System-Generated Goals claim:
Having the system generate goal structures throughout the course of problem solving promotes abstract understandings without requiring learners to generate solutions in an abstract language.
(But learners must draw correspondences between their own behavior and the display to get help.)
(But learners who recognize may have less retention than those who generate goals.)

This claim directly addresses the downsides of the Bridge and Proust claims while preserving the upsides. A problem with Bridge is that learners are

Figure 1. A claims map for the Goalposter. Claims are shown in rough chronological order, as links denote a "responds to" relation from top to bottom. Claim texts are abridged versions of claims in the body of the chapter; names are the same.

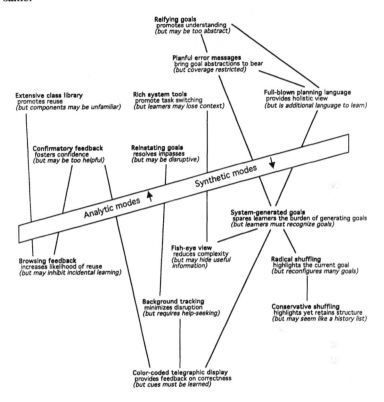

forced to generate solutions in an abstract language. A problem with Proust is that information about goals was provided only intermittently.

Figure 2 shows the Goalposter as we envisioned it in the uppercase window scenario. Here, the learner is looking for character input functionality and has just selected the characterInput: method defined in the Window class. The contents of the Goalposter reflect the system's interpretation of this activity.

Presently, only the Goalposter is permitted to post goals to the display. In order to get support from the tool, learners are required to recognize the posted goals and make the correspondence between the posted structures and their own behavior. Of course, a well-known psychological finding is that generation of information leads to better memory than recognition, and this constitutes a downside for the claim. Also, recognition carries with it the burden of mapping one's own behavior onto the structure of

Figure 2. A screen image from the intelligent tutoring system for Smalltalk showing the Goalposter tool in the lower-left pane. Other panes show the Smalltalk V/PM class hierarchy browser and other tutor components. At this point, the learner has just selected the characterInput: method defined in the Window class.

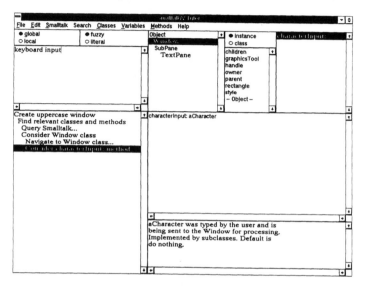

the display. This opens up a host of usability issues, and much of the remaining design work we report here deals with these issues.

Positive and Negative Feedback.   Informing a learner about the correctness of actions is perhaps the most basic sort of feedback a computer-based instructional system might provide. In the Smalltalk domain, this is especially true in browsing situations where untold numbers of unfamiliar classes and methods may be considered in service of a particular goal. Our subject's time-consuming search episodes reported earlier could have been curtailed by informing her that some of her choices should not have been pursued. However, if feedback is too intrusive it can undermine planfulness (Carroll & Aaronson, 1988), and can thwart opportunities for incidental learning. The Goalposter takes a compromise position: It provides feedback on the correctness of goals and actions but it remains in the background of the learner's work and does not prescribe action:

Browsing Feedback claim:
Providing feedback on browsing increases the likelihood of finding an appropriate component to reuse and reduces frustration.
(But it may inhibit incidental learning.)

Background Tracking claim:
Tracking the learner's behavior automatically in a separate window mini-

mizes the disruption of the learner's work.
(But learners must manage their own help seeking.)

Providing feedback on browsing responds directly to the downsides of the Extensive Class Library claim. The threat to incidental learning that our proposal entails is mitigated somewhat by the fact that learners can in fact browse wherever they like. The Goalposter simply comments on the relevance of that browsing to the current task.

Once we had committed to the idea of system-generated goals, we could have proposed a heavy-handed implementation in which learners were asked to respond in some way to each new posting. By allowing learners to consult the display at their own discretion, however, we respond to the second downside of the Reinstating Goals claim. Indeed, one might expect that a computer tutor would be less artful at managing interventions than a human tutor. We avoid this downside (but introduce another) by allowing learners to manage their own help seeking. In the Goalposter, we are deliberately blurring the distinction between a tutor and a task-oriented help system.

Because learners must understand the display in order to know when and how to get help, we must strive for simplicity in its structure and presentation. In dealing with these usability issues, we are essentially grappling with the downside of the Reifying Goals claim characteristic of the genre in the context of our particular design. The Goalposter displays the text of an inferred goal colored either red or green. Green signifies an appropriate goal for the learner to attempt. (Thus, we have chosen to model to some extent the behavior of our human coach, who provided confirmatory feedback. We import the Confirmatory Feedback claim wholecloth.) Red indicates a "buggy," or incorrect or irrelevant, goal. The aim of this coloring scheme is to make salient whether the goal should be pursued. Although textual, the presentation of individual goals is telegraphic: several words at most. At any point, the learner can expand any of the telegraphic nodes in the goal blackboard (by making a menu selection) to display a fuller explanation of why the goal is a worthwhile one to pursue or not. Thus, the system provides feedback not only on the correctness of an action but also an interpretation of why it is correct and where it fits into the overall solution. We hope this will help communicate the conceptual structure of the solution as it develops. This leads to the following claim:

Color-Coded Telegraphic Display claim:
A color-coded telegraphic display of goals provides quick feedback on the correctness of actions as well as hooks for further information.
(But learners must learn the display's cues.)

This claim is somewhat special in that it is the first we have presented that responds primarily to the downsides of Goalposter claims, that is,

claims drawn from the realm of the envisioned artifact rather than predecessor artifacts (although, as we have mentioned, in some sense these
downsides are simply variants on the downside of the Reifying Goals claim).
The fact that the goals are system generated (System-Generated Goals
claim) means that learners must draw mappings from their own behavior
to the goal structures to understand the display. The fact that the display
is updated in the background of the learner's work (Background Tracking
claim) means that learners must easily perceive when and how to get help.
This is our first example of how extracting and responding to claims from
envisioned scenarios can provide additional design constraints on the far
side of the analytic-synthetic gap.

Goal Management.   In addition to thinking about the static properties of
the goal display, we considered how the display would be updated dynamically in response to learner action. Earlier in our discussion of Smalltalk
learner protocols, we saw how one subject switched freely between tasks but
had to spend time reestablishing the local goal context and reinstating her
understanding of the state of her progress (our analysis was captured in the
Smalltalk Rich System Tools claim). That scenario made us think carefully
about how to update the goal display in such situations.

The Goalposter employs a variety of goal management techniques. As
goals become active, they are posted to the display in green. They remain
in green until they are satisfied, at which time their color changes to black.
Hierarchical goal-subgoal relationships are shown using indentation.
Whenever a goal is suspended in favor of some other goal, its structure
collapses; that is, only the top-level goal is shown in the display. The
retention of the top-level goal in green serves as a reminder that the goal
had been pursued but is now suspended. Once the goal is resumed, the
structure expands and new subgoals are added to the existing structure.
Thus, the Goalposter provides a kind of fish-eye view (Furnas, 1986) of
the current state of the project, with higher resolution shown for the
currently active goal:

> *Fish-Eye View claim:*
> Providing dynamic fish-eye views of the display reduces its complexity and
> increases the probability that learners will be able to draw correspondences
> between the display and their own behavior.
> (But this may hide parts of the goal tree that are of interest.)
> (But this may complicate the learner's mental model of the display.)

This claim responds to the downsides of both the Rich System Tools
claim and System-Generated Goals claim, and therefore forges considerations drawn from two modes of reasoning. Task switching is supported by
maintaining a historical record of suspended tasks. Mappings between the

display and behavior are supported by focusing learner attention on the fine structure of the current task.

Although the display will track the learner's behavior and expand and contract in accordance with its interpretation of the learner's current goal, it may be that the learner would like to explore parts of the display that are not currently shown in fine detail. To support this scenario, we made the fish-eye view a browsable structure: Learners can expand or contract any of the goals themselves by simply double-clicking on its text. This design move responds to the second downside of the previous claim as well as the second downside of the Proust claim. The Goalposter embodies the notion that goal dialogues should span the entire task and be under the control of the learner. This, then, is another case in which rationales drawn from hillclimbing and scenario envisionment have combined to influence a design decision.

3.5. Formative Evaluation

The design of the Goalposter is ongoing, and we have not yet subjected the system to a thorough empirical evaluation. However, we have used formative evaluation analytically. At a certain point it became too difficult to envision the operations of the Goalposter with regard to goal management and task switching. It became time to at least partially commit our design to code and interact with it. One feature of the system we were evaluating was the way in which the goal structures reformatted when learners switched tasks. We had tentatively decided to have a global reordering of the tree in order to make those goals that represented the current task appear at the bottom of the list of goals. This would happen as that part of the tree that represented the suspended task would be collapsed to its parent node. Our thinking on this issue could be represented by the following claim:

Radical Shuffling claim:
Shuffling the tree so that the current subgoal structure appears at the bottom of the display allows learners to easily find the current goal.
(But this changes the position of many goals in the tree and may disrupt understanding of the display.)

This is in essence a further response to the downside of the System-Generated Goals claim. Once we had this feature implemented, however, and could evaluate its dynamics, we concluded that it was too jarring visually and would require learners to completely reparse the tree. We proposed a more local reordering, keeping the major subgoals of the solution spatially fixed with respect to each other but shuffling the leaves of the tree when appropriate. In this scheme, other cues, such as highlighting the current goal and expanding the current subgoal structure, bear more of the burden of signaling the current focus of attention. Here is the claim the Goalposter now makes:

Conservative Shuffling claim:
Shuffling only the leaves of the tree, along with highlighting the current goal and expanding its subgoal structure, allows learners to find the current goal and retain their grasp of previously posted goals.
(But learners may become confused about whether the Goalposter is a history list or a conceptual structure.)

4. ASSESSING THE MODES

We have proposed a taxonomy of design reasoning, specifically, five distinct ways of bringing psychological constraints to bear on the design process. We have also proposed psychological claims as a design representation that succinctly captures and integrates the reasoning and findings across these design modes. We have shown through a case study how claims can play an integrative role throughout design: from a designer's earliest commitments to a particular genre of artifact to the final stages of formative evaluation. We have tried to demonstrate how the resulting map of claims (shown in Figure 2) is a relatively succinct summary of the design history as well as the current state of the design. In our view, a design may be well described in terms of the positive psychological effects it promotes, as well as the negative effects it mitigates or finally ignores. A design history represented in terms of claims acknowledges that positive and negative effects often inhere in the same design decision, and that often design decisions are made to mitigate the negative consequences of earlier decisions.

An important question is whether it is useful to draw distinctions between modes of reasoning in the design history. Of course, there is the usual case for descriptive accuracy. Beyond this, there are issues of convergence and coverage. To the extent that different modes of reasoning generate similar or consistent constraints, then we might be entitled to greater confidence in the quality of those constraints. For example, we emerged from our assessment of the tutoring genre with a firm belief in the value of goal- and plan-based instruction, and in particular of reifying these structures for learners. This belief was founded in part on how difficult it is for learners to formulate and manage goals in complex problem-solving situations. Then, when we did our own empirical evaluation of a Smalltalk, we observed myriad specific problems associated with goal management. Thus, the modes of assessing the genre and hillclimbing provided convergent rationale. Also, each of the modes has its own characteristic strengths and weaknesses. Perhaps it would be beneficial to make a conscious effort to employ several modes of reasoning in design, in order to increase the chances of uncovering useful constraints and compensate for weaknesses or blind spots in individual modes (the metaphor here is the well-diversified investment portfolio).

The strength of assessing the genre is that it gives the designer a broad view of the artifact space and thus is an antidote to local maximization. Its weakness is that the constraints that are derived tend to be quite abstract and it may be difficult or inappropriate to apply the abstractions to the current design. (This is a version of the contextualist critique of the role of theory in design.)

A variety of factors influence the utility and validity of hillclimbing. Of course, one is on firmer footing as one's experience with and understanding of the predecessor artifact becomes more extensive. Of critical importance is the amount of overlap or "distance" between the predecessor artifact and the target. As alluded to earlier, if the distance is too great, the analysis will not have much coverage and is in danger of being completely dominated or washed out by unaccounted-for features of the target artifact. Unfortunately, because the target is derived, one cannot anticipate this distance a priori. Furthermore, artifact similarity is not a well-understood concept. Perhaps overlap in claims is promising as a qualitative metric.

Hillclimbing is more situated than assessing the genre in that one is mapping from one concrete situation to another rather than from an abstraction. This makes it easier to reason and, if the overlap is great between predecessor and target, tends to enhance the validity of the result. A weakness is that one is essentially reasoning analogically from a single example, and this may encourage local maximization. One may defeat this somewhat by considering multiple predecessor artifacts. For example, in our work we have considered other programming tutors as well as the unadorned Smalltalk environment as multiple predecessor artifacts.

Finally, hillclimbing carries with it a danger of thrashing. Sometimes, in the zealous pursuit of eradicating flaws or supplying missing functionality, designers can unwittingly disrupt the overall integrity of an artifact and hurt overall usability. Perceived weaknesses may be attacked with little awareness of the possible damage done to associated strengths. Claims, by bringing into focus the linkage between downsides and upsides, may help overcome this bias. For example, in our consideration of Bridge as a predecessor artifact, we rejected the notion of programming in an abstract planning language. However, instead of rejecting the Bridge approach entirely, we examined the upside of the Bridge claim and decided to retain the associated feature of providing a holistic view.

When modeling, it may not always be possible to find an appropriate model. Also, although the model may be universally perceived as good and worth emulating, it may not be well understood what in the model is responsible for its valued effect. For example, in the case of human tutors, there are classic studies by Bloom and his students (1984) documenting the value of one-on-one human tutoring. However, there has never been any controlled experimentation on the source of the effect. Those interested in motivating the design of intelligent tutoring systems have generally attributed the effect to the ability of the human tutor to offer immediate, personalized feedback based on a well-articulated student model. But the source of the effect could lie elsewhere. For example, one explanation is that the student is simply more attentive with a figure of authority working in such close proximity. If this is the source of the effect, then the method we have espoused of using human tutors as models for computer tutors is completely wrong-headed. Rather than modeling the cognitive aspects of human tutors, we should be modeling the motivational, interpersonal aspects. Of course, this is probably much harder if not impossible to model in a computer program.

Whereas hillclimbing can be biased toward mitigating downsides, modeling can be biased toward preserving upsides. Modeling generally encourages the designer to be in a rather uncritical state of mind, which may result in an inadvertent importation of downsides. For example, MacLean, Young, Bellotti, and Moran (1991 [chapter 3 in this book]) describe how their subjects rarely considered the downsides of their design proposals. Again, the claims representation may help in this regard. Rather than mimic the delivery of feedback of our Smalltalk coach exactly, we realized by examining the downsides of the coach claims that her style was probably too instrusive for our instructional situation.

Scenario envisionment is perhaps the most critical mode in that it is to our knowledge the single way to proactively bridge the analytic-synthetic gap. Unfortunately, the selection of scenarios is subject to biases because of incomplete knowledge of users and tasks. In general, the problem seems to be one of undergeneration. For example, designers sometimes fixate on single scenarios (Bellamy & Carroll, 1992) and find it particularly difficult to generate plausible error scenarios (Payne & Singley, 1990). In other work (Carroll & Rosson, 1991 [chapter 4 in this book]), we are experimenting with structured approaches to scenario generation that may help to overcome some of these biases.

The strength of formative evaluation is that the data are very pertinent; the designer is not mapping from one artifact space to another but rather is working directly in the artifact space of interest. Also, many of the details that are so important in determining the design outcome have been specified; thus the overlap between test and actual usage situations is maximized. The weakness is that at a late stage it is very difficult to undo major commitments in the design and the result tends to favor local fixes rather than a global maximum.

Although we have proposed a taxonomy of five modes of reasoning, the taxonomy can alternatively be viewed as a five-stage process. Of course, in actual design work the stages are strongly interleaved (Goel & Pirolli, 1989; MacLean et al., 1991 [chapter 3 in this book]). Scenario envisionment is the crux of this process, the place where real design work is done. In order to provide an integrative framework for the modes, the design process can be thought of as a classic generate-and-test problem-solving situation (Newell & Simon, 1972). In this view, constraints generated by other modes are brought to bear on either the generation or the testing of design features. There are modes typically associated with applying constraints to the generation of features (assessing the genre, hillclimbing, modeling) and there are modes typically associated with the testing of features (later stages of scenario envisionment, formative evaluation). One might regard these as upstream and downstream modes, respectively. (Casaday, 1995 [chapter 12 in this book], makes a similar distinction between construction-driven and evaluation-driven design.) Upstream modes tend to provide broad constraints that define one's neighborhood in the design space, whereas downstream modes tend to provide finer constraints that are geared toward local maximization within the neighborhood.

Upstream and downstream modes obey a cost-correctness trade-off: Generally, the earlier one can apply constraints the better in terms of cost and efficiency. It is never too late to retract a decision; it is just more costly to do so. However, it is also more difficult to judge the correctness of a constraint early in the process.

Our basic objective here is to enhance the rigor of what is already common practice, and one we believe is important and appropriate in human–computer interaction (HCI): justifying design by appealing to user psychology. Ultimately, we just cannot afford to see interface design as little more than a shot in the dark waiting to be informed by user testing. Formative evaluation is a well-understood, sanctioned method in HCI, a popular and surefire way to bring psychological constraints to bear on design. But we suggest here that time is well spent upstream in the design process thinking about user psychology. By strengthening the other modes of design reasoning we will increase our chances of more broadly optimizing designs with respect to user requirements.

NOTES

Acknowledgments. Many thanks to Sherman Alpert who participated in early discussions of this work. We are grateful to Tom Moran and two anonymous reviewers for helpful comments on earlier versions of this chapter.

Authors' Present Addresses. Mark K. Singley, Division of Cognitive and Instructional Science, Educational Testing Service, Princeton, NJ 08541. Email: ksingley@rosedale.org; John M. Carroll, Department of Computer Science, 562 McBryde Hall, Virginia Tech (VPI&SU), Blacksburg, VA 24061. Email: carroll@cs.vt.edu

REFERENCES

Anderson, J. R., Boyle, C. F., Farrell, R., & Reiser, B. J. (1989). Cognitive principles in the design of computer tutors. In P. Morris (Ed.), *Modelling cognition* (pp. 93–134). New York: Wiley.

Anderson, J. R., & Jeffries, R. (1988). Novice LISP errors: Undetected losses of information from working memory. *Human–Computer Interaction, 1,* 107–131.

Bellamy, R. K. E., & Carroll, J. M. (1992). Structuring the programmer's task. *International Journal of Man–Machine Studies, 35,* 503–527.

Bloom, B. S. (1984). The 2 sigma problem: The search for methods of group instruction as effective as one-to-one tutoring. *Educational Researcher, 13,* 4–16.

Bonar, J. (1988). Intelligent tutoring with intermediate representations. *Proceedings of ITS '88 Intelligent Tutoring Systems Conference,* 25–32. New York: ACM.

Carroll, J. M. (1991). History and hysteresis in theories and frameworks for HCI. In D. Diaper & N. V. Hammond (Eds.), *People and computers VI* (pp. 47–55). Cambridge, England: Cambridge University Press.

Carroll, J. M., & Aaronson, A. (1988). Learning by doing with simulated intelligent help. *Communications of the ACM, 31,* 1064–1079.

Carroll, J. M., Kellogg, W. A., & Rosson, M. B. (1991). The task-artifact cycle. In J. M. Carroll (Ed.), *Designing interaction: Psychology at the human–computer interface* (pp. 74–102). New York: Cambridge University Press.

Carroll, J. M., & Rosson, M. B. (1991). Deliberated evolution: Stalking the View Matcher in design space. *Human–Computer Interaction, 6,* 281–318. Also in T. P. Moran & J. M. Carroll (Eds.), *Design rationale: Concepts, techniques, and use.* Hillsdale, NJ: Lawrence Erlbaum Associates, 1996. [Chapter 4 in this book.]

Carroll, J. M., Singley, M. K., & Rosson, M. B. (1991). Toward an architecture for instructional evaluation. In *Proceedings of the International Conference on the Learning Sciences* (pp. 85–90). Charlottesville, VA: AACE.

Casaday, G. (1995). Rationale in practice: Templates for capturing and applying design experience. In T. P. Moran & J. M. Carroll (Eds.), *Design rationale: Concepts, techniques, and use.* Hillsdale, NJ: Lawrence Erlbaum Associates. [Chapter 12 in this book.]

Furnas, G. W. (1986). Generalized fisheye views. In M. Mantei & P. Orbeton (Eds.), *Proceedings of CHI '86: Human Factors in Computing Systems* (pp. 13–23). New York: ACM.

Goel, V., & Pirolli, P. (1989, Spring). Design within information processing theory: The design problem space. *AI Magazine,* pp. 18–36.

Gould, J. D., Conti, J., & Hovanyecz, T. (1983). Composing letters with a simulated listening typewriter. *Communications of the ACM, 26,* 295–308.

Johnson, W. L. (1985). *Intention-based diagnosis of errors in novice programs.* Unpublished doctoral dissertation, Yale University, New Haven, CT.

Kelley, J. (1983). An empirical methodology for writing user-friendly natural language computer applications. *Proceedings of the CHI '83 Human Factors in Computing Systems.* New York: ACM.

Landauer, T. K. (1991). Let's get real: A position paper on the role of cognitive psychology in the design of humanly useful and usable systems. In J. M. Carroll (Ed.), *Designing interaction: Psychology at the human–computer interface* (pp. 74–102). New York: Cambridge University Press.

MacLean, A., Young, R. M., Bellotti, V. M. E., & Moran, T. P. (1991). Questions, options and criteria: Elements of design space analysis. *Human–Computer Interaction, 6,* 201–250. Also in T. P. Moran & J. M. Carroll (Eds.), *Design rationale: Concepts, techniques, and use.* Hillsdale, NJ: Lawrence Erlbaum Associates, 1996. [Chapter 3 in this book.]

McKendree, J. (1990). Effective feedback content for tutoring complex skills. *Human–Computer Interaction, 5,* 381–413.

Newell, A., & Simon, H. A. (1972). *Human problem solving.* Englewood Cliffs, NJ: Prentice-Hall.

Payne, S. J., & Singley, M. K. (1990). *Imagination bias: Cognitive origins of interface design flaws* (IBM Research Report). Yorktown Heights, NY: IBM T. J. Watson Research Center.

Singley, M. K. (1990). The reification of goal structures in a calculus tutor: Effects on problem solving performance. *Interactive Learning Environments, 1,* 102–123.

Singley, M. K., Carroll, J. M., & Alpert, S. R. (1991). Psychological design rationale for an intelligent tutoring system for Smalltalk. In J. Koenemann-Belliveau, T. Moher, & S. Robertson (Eds.), *Empirical studies of programmers: Fourth workshop* (pp. 196–209). Norwood, NJ: Ablex.

Whiteside, J., & Wixon, D. (1987). Improving human–computer interaction: A quest for cognitive science. In J. M. Carroll (Ed.), *Interfacing thought: Cognitive aspects of human–computer interaction* (pp. 353–365). Cambridge, MA: MIT Press.

DESIGN RATIONALE TOOLS
IN DESIGN PRACTICE

9

Making Argumentation
Serve Design

Gerhard Fischer
University of Colorado

Andreas C. Lemke
ALCATEL-SEL

Raymond McCall
University of Colorado

Anders I. Morch
University of Oslo

Gerhard Fischer is a computer scientist interested in design and design support systems, particularly in domain-oriented design environments and how they make argumentation serve design by supporting reflection-in-action; he is the director of the Center for Lifelong Learning and Design, a professor of Computer Science, and a member of the Institute of Cognitive Science at the University of Colorado. **Andreas Lemke** is a computer scientist with interest in tools to support knowledge workers; he is a researcher at Alcatel, a major telecommunications company, working on multimedia and mobile communication. **Ray McCall** is a design theorist interested in issue-based argumentation and hypermedia-based CAD systems that enable argumentation to serve design; he is an Associate Professor in the Division of Environmental Design at the University of Colorado. **Anders Morch** is a computer scientist with interests in human-computer interaction, object-oriented programming, and design rationale to bridge between the two; he was a member of technical staff at the NYNEX Science and Technology Center and is now pursuing a Ph.D. in Informatics at the University of Oslo.

CONTENTS

ABSTRACT

Documenting argumentation (i.e., design rationale) has great potential for serving design. Despite this potential benefit, our analysis of Horst Rittel's and Donald Schön's design theories and our own experience has shown that there are the following fundamental obstacles to the effective documentation and use of design rationale: (a) A rationale representation scheme must be found that organizes information according to its relevance to the task at hand; (b) computer support is needed to reduce the burden of recording and using rationale; (c) argumentative and constructive design activities must be explicitly linked by integrated design environments; and (d) design rationale must be reusable. In this chapter, we present the evolution of our conceptual frameworks and systems toward integrated design environments, describe a prototype of an integrated design environment including its underlying architecture, and discuss some current and future work on extending it.

1. INTRODUCTION

Documenting argumentation (i.e., design rationale) has great potential for improving design. In addition to being invaluable for maintenance, redesign, and reuse, it promotes critical reflection during design. Despite such potential benefits, our experience has shown that there are fundamental obstacles to the effective documentation and use of design rationale. Argumentation does not naturally serve design; it must be made to do so.

The structure of this chapter follows the history of our work, driven by the development and evaluation of conceptual frameworks and prototype systems. In Section 2, the term *design rationale* is characterized. In Section 3, we discuss *Issue-Based Information Systems* (IBIS) and *Procedural Hierarchy of Issues* (PHI), two frameworks for representing argumentation. We show that IBIS has fundamental problems. IBIS represents neither dependency relationships between issues nor nondeliberated issues. PHI is a variant of IBIS that remedies these problems. In the past, argumentation has been considered in isolation from the activity of solution construction. The major breakthrough in our thinking, based on observing the shortcomings of the two isolated approaches, was the realization that argumentation must be integrated into the context of construction. In Section 4, we describe approaches to devising tools for construction to reduce the transformation distance from application domain to implementation domain by supporting human problem-domain communication. In Section 5, we discuss *integrated design environments* that unify construction and argumentation. The theoretical basis for this integration is Schön's theory of reflection-in-action. In Section 6, we describe current and future work on adaptive and reusable domain-oriented issue bases, enriched catalogs, and improved representations of the task at hand.

Throughout the chapter, we use the JANUS system (Fischer, McCall, & Morch, 1989a, 1989b) as an "object-to-think-with." The chapter discusses aspects of the JANUS system only as they are relevant to our theme; details about JANUS can be found in the given references.

2. DESIGN RATIONALE

Design. In order to define design rationale, we must first define the term *design*. Like design theorists Cross (1984), Rittel (1984), Schön (1983), and Simon (1981), we see design not only as problem solving but also as continual problem finding. It is a process of dealing with the kind of "messy situations" that are characterized by uncertainty, conflict, and uniqueness. It is an evolutionary process in which "understanding the problem is identical with solving it" (Rittel, 1972, p. 392), and it can best be characterized by creativity, judgment, and dilemma handling, rather than by objective scientific methods.

We agree with Donald Schön's view of design. For Schön (1983) designing is not primarily a form of problem solving, information processing, or search, but a kind of making: "I shall consider designing as a conversation with the materials of the situation" (p. 78). This definition covers a wide range of fields, including architectural (building) design, urban design, software design, hardware design, and various types of engineering design. We call the transactions of designers with materials and artifacts *construction,* which is the activity of creating the actual form of the solution. Construction cannot always be physical, but may have to be carried out in the abstract (e.g., on the drafting board). Physical interactions with materials may be too expensive or too dangerous.

Design Rationale. In our approach, design rationale means statements of reasoning underlying the design process that explain, derive, and justify design decisions. A truly complete account of the reasoning relevant to design decisions is neither possible nor desirable. It is not possible because some design decisions and the associated reasoning are made implicitly by construction and are not available to conscious thinking. Some of the rationale must be reconstructed after design decisions have been made. Many design issues are trivial; their resolution is obvious to the competent designer, or the design issue is not very relevant to the overall quality of the designed artifact. Accounting for all reasoning is not desirable because it would divert too many resources from designing itself.

Design rationale in our approach is a synonym for argumentation. Rittel (1972) was the first to advocate systematic documentation of design rationale as part of design. He saw design problems as fundamentally open-ended and controversial in the sense that there are no objective criteria for closing problem definitions and settling disagreements. Such closing and settling are necessary for design, but for the designer the decisions on closing and settling are judgmental and political in nature. The design rationale takes the form of a network of issues (design questions), selected and rejected answers, and arguments for and against these answers (see Section 3).

The Promise of Design Rationale. Design rationale serves design if it helps designers (a) to improve their own work, (b) to cooperate with other people holding stakes in the design, and (c) to understand existing artifacts (i.e., communicate with past designers). Design rationale can trigger critical thought in the individual designer. Writing an idea down allows the designer to make the transition from simply creating that idea to thinking about it.

Design rationale can serve as a memory aid not only to individuals but also to groups (Conklin & Begeman, 1988) by providing a forum for airing issues crucial for coordinating group activities. It is useful for triggering and focusing discussion among members of a project team. By making the processes of reasoning public, it extends the number of people who can participate in the critical reflection on decisions. This reduces the chances of missing some important consideration and it rationalizes discussion.

To alter a design sensibly—adding, fixing, or modifying features—it is crucial to have an understanding of why it has been designed the way it has. Without knowing the rationale, a designer is apt to violate constraints and to repeat errors by ignoring what previous designers have learned. That is, the rationale created in one design project may be a resource for future, related design projects. Even if the difficulties encountered in a project are not overcome, they might still be informative for future designers. The mere existence of unforeseen problems is itself valuable information. Often, design is based on mistaken predictions of how the artifact will perform in use. If these predictions are documented, they can be compared to actual use. This would allow the development of better theories for predicting performance.

3. SUPPORT FOR ARGUMENTATION

On the basis of his theory of Wicked Problems, Rittel (1984) rejected the efforts by the majority of design methodologists to automate design reasoning. The argumentative approach tries to enhance design by improving the reasoning underlying it and is aimed at supporting the reasoning of human designers rather than replacing it with automated reasoning processes (Fischer, 1990; Stefik, 1986).

3.1. IBIS

IBIS (Kunz & Rittel, 1970) is a method (not a computer system) for structuring and documenting design rationale. The central activity of IBIS is deliberation, that is, considering the pros and cons of alternative answers to questions. The questions deliberated are called *issues*. Proposed answers—including ones that are mutually exclusive—are called *answers* or *positions*. Statements of the pros and cons of answers are called *arguments*. The decision as to which answers to accept and reject is called the *resolution of the issue*.

The various issue deliberations are connected by a variety of interissue relationships. The original IBIS included "more general than," "similar to," "replaces," "temporal successor of," "logical successor of," and others. Graph diagrams with labeled nodes and links representing issues and their relationships were used for visualization. Such diagrams, called *issue maps*, were meant to facilitate navigation through the IBIS "problemscape."

From 1970 to 1980, a variety of projects were undertaken that attempted to use IBIS in real-world settings. These projects included IBIS systems for the United Nations, the Commission of European Communities, the (West) German Parliament, the German Federal Office of the Environment, and the German Office of Health (Reuter & Werner, 1983). None of these systems got past the pilot project stage. At the end of this stage each was judged as somehow failing to adequately serve the design tasks for which it had been created.

After a decade of intensive and generally well-funded efforts to implement IBIS, it became difficult to believe that the failures to do so were coincidental. Clearly, there were fundamental problems with the IBIS method or the approach to implementing it.

The identification and solution of fundamental problems in the creation and use of issue-based design rationale has been a central concern of our research. The first step in this research was a critique of IBIS and an improved issue-based method called PHI (McCall, 1979). The next step was the proposal of a new sort of software technology, hypertext, to handle issue-based rationale (McCall, Mistrik, & Schuler, 1981). We next look at these suggested improvements and the results of their implementation and use.

3.2. PHI and the Critique of IBIS

McCall (1979) suggested that there are two related types of information that are omitted from IBIS but that are required for an issue-based approach to serve design effectively. The first and most basic is dependency relationships between issue resolutions, that is, relationships representing the fact that the answering of issues often depends on how other issues are answered. IBIS has no way of representing such dependencies. Instead it treats issue-resolution processes as if they were separable.

The second type of information omitted from IBIS is questions that are not deliberated—that is, questions for which pros and cons of alternative answers are not considered. IBIS ignores these in favor of those questions with which debate and controversy are most likely to be associated. Yet nondeliberated questions occur frequently in design and can influence the resolutions of issues. Furthermore, many such questions themselves have answers that depend on the resolutions of issues.

In an effort to overcome these limitations of IBIS, McCall (1991) developed the PHI approach to documenting design rationale. PHI, like IBIS, is a design method rather than a piece of software. It differs from IBIS in two crucial respects: It uses a broader definition of the concept issue, and it uses a new principle for linking issues together.

In IBIS, the term issue denotes a design question that is deliberated; in PHI, however, every design question counts as an issue, whether deliberated or not. PHI also abandons the interissue relationships proposed by Rittel (1980)—"temporal successor of," "similar to," "replaces," and so forth. Instead it uses *serve* relationships. We say Issue A serves Issue B if and only if the resolution of A influences the resolution of B. The dominant type of serve relationship used in PHI is the "subissue of" relationship, which indicates that resolving one issue is a subtask of the task of resolving another. More formally, we say Issue A is a subissue of Issue B if and only if A serves B and B is raised before A. Note that this means that A's being

Figure 1. An IBIS issue map.

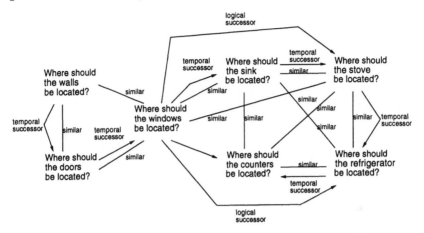

Figure 2. A PHI issue map. The starred issues, which are not deliberated, are dealt with by PHI but not by IBIS.

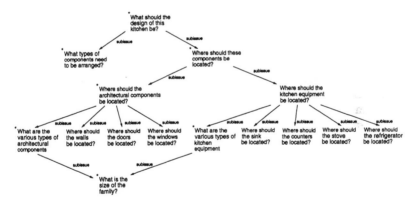

a subissue of B implies that A serves B, but A's serving B does not in itself imply that A is a subissue of B.

In Rittel's IBIS, as evidenced by the many years of real-world and student projects, an issue map is characteristically a dense and tangled network of issues connected by a half dozen different relationships (see Figure 1). In PHI, however, an issue map is a simple quasi-hierarchical structure connected only by serve relationships and having a single root issue (see Figure 2). This structure is treelike but is seldom a pure tree, because issues can share subissues (see Figure 2). The root of a PHI issue map is an issue that represents the project as a whole. For example, if one is designing a kitchen, the root issue might be "What should be the design of this kitchen?"

PHI has been in nearly continual test use both with and without computer support since 1977. This testing has been informal rather than in the framework of a formal, experimental setting. Furthermore, the testing has emphasized intensive use by a relatively few users at a time rather than extensive use by many people—a style we have found to be especially informative for system-building efforts. Testing began in 1977 and 1978 with students at the University of California, Berkeley (McCall, 1979). It continued in Heidelberg, Germany, from 1979 to 1984. Since 1984, continual test use has been made of PHI at the University of Colorado, Boulder.

Testing in Berkeley used eight undergraduates and was spread out over a 2-year period. The most important results of this were the generation of issue bases (i.e, networks of issues) that showed the applicability of the PHI to student design projects.

3.3. PHI Hypertext

In Heidelberg, testing of PHI began with the attempt in 1979 to use PHI with only typewriters and word processors. By 1980, these efforts ran into severe difficulties in managing the issue-based information. In particular, the information management tasks were so labor intensive and error prone that the decision was made to attempt to develop computer support for PHI. The system developed, called MIKROPLIS (McCall et al., 1981), became the first issue-based hypertext system.

The defining characteristics of hypertext are nonlinear structure and navigation. The need for the former was understood at the beginning of the MIKROPLIS project (McCall, 1979). The need for the latter emerged in 1982 from working with early users of MIKROPLIS who repeatedly pointed to displayed nodes and asked how to retrieve the nodes linked to them.

Since the beginning of the MIKROPLIS project, a number of other issue-based hypertext systems have been developed. These include Rittel's own system (Conklin, 1987), gIBIS (Conklin & Begeman, 1988), JANUS-ARGUMENTATION (Fischer et al., 1989b), and PHIDIAS (McCall et al., 1990). MIKROPLIS, PHIDIAS, and JANUS-ARGUMENTATION differ from the others by using PHI rather than IBIS.

To further test PHI and the computer support being developed for it, the MIKROPLIS team kept a PHI issue base for the design of the system. As soon as MIKROPLIS became usable, this issue base was maintained using the MIKROPLIS system itself. This "self-referential use" encouraged a certain level of awareness and honesty about the performance of PHI and MIKROPLIS.

Additional testing of PHI and MIKROPLIS involved the development of issue bases with MIKROPLIS by a dozen users of various kinds over a period of 3 years. These users included MIKROPLIS project members, people from other project groups within the organization in Heidelberg, and several "knowledge workers" from other organizations. In 1984, an American physician was hired to test the system on a full-time basis for 3 months by attempting to develop an issue base on health care policy. In 11 weeks he developed a tightly structured issue base equivalent to exactly 500 single-spaced pages in length. This was taken by the physician and others as evidence of the usefulness and usability of both PHI and MI-KROPLIS. In particular, the physician felt that he could not have achieved these results with alternative methods or technologies.

Despite this success, there were still problems with using both PHI and MIKROPLIS. The artifact the physician was trying to produce was the issue base itself. To those for whom the issue bases were only means for designing other kinds of artifacts, the use of PHI involved a great deal of work over and above the ordinary work of design. MIKROPLIS substantially reduced the errors and secretarial work of creating an issue base, but there remained a large amount of conceptual and editorial work. Many people were therefore disinclined to use PHI because the costs of invested effort exceeded the immediate payoff. For them, even with MIKROPLIS support, PHI still did not serve sufficiently the design task at hand.

3.4. Grounding Argumentation in Construction

PHI hypertext with domain-oriented issue bases reduced the cost and increased the benefits of design rationale. But as our systems dealt successfully with this aspect of design rationale, another, more fundamental obstacle was revealed. There is a crucial design activity not supported by argumentative hypertext: *construction*. In fields such as architectural design, construction is a graphic activity traditionally done by drawing. Construction is the *sine qua non* of design, for no design project can be completed until the construction is done. Argumentation gets its usefulness in design only by influencing construction. For argumentation to serve design it must serve construction.

Test use of PHI at the University of Colorado, Boulder, provided evidence for the need to integrate argumentation with construction. The test use began with two junior-level undergraduate environmental design studios, each with about 20 students—taught by Raymond McCall in 1985. Each studio involved the same semester project: designing a neighborhood shopping center at a particular location in Boulder. Students were asked to record their rationale in PHI form during the project. In both studios, this worked well until students began working out the details of the solution form, that is, actual drawings of buildings. At this point, it became effectively impossible to get students to document their rationale.

To see if these difficulties were independent of instructor and project, two independent study students were asked to document a studio on housing design taught by a nationally known architect. In an effort to keep this inquiry unbiased, the students who did the documentation were not told anything about the hypothesis being investigated and were given only minimal supervision by McCall. The students produced a 175-page document in PHI form, representing the work of a project group of five students in the studio. Again, the documentation of rationale ceased shortly after the construction of solution form began. According to the students who did the documentation, the project group members became unable or unwilling to talk to the documenters as form generation began.

The difficulties encountered in attempting to document the studio projects suggested that there was a fundamental incompatibility between form construction and PHI. To understand what this incompatibility might be, McCall made a series of three videotaped think-aloud protocols of student designers from the College of Environmental Design.

The first protocol involved two juniors who worked for 6 weeks on the design of a store. The second involved a senior working for 10 weeks on the design of a house. The third involved a single senior working for 3 weeks on the design of a kitchen. All were analyzed informally and the second was selected for intensive formal analysis. In particular, representative sections of the form construction process were transcribed and compared to the structures of PHI on a sentence-by-sentence and drawing-by-drawing basis.

These results suggested some possible additions to PHI. One would be to enable explicit representation of criteria on occasion—as in the "goals" of DRL (see Lee & Lai, 1991 [chapter 2 in this book])—though not to always require this—as in QOC (see MacLean, Young, Bellotti, & Moran, 1991 [chapter 3 in this book]) (also see Buckingham Schum, 1995 [chapter 6 in this book] on use of criteria in QOC). Another addition would be to enable better representation of hypothetical reasoning—something not provided by any of the rationale representation schemes presented in this book. On the whole, however, our protocol studies showed a clear match between the processes the student used in form generation and the processes represented in PHI.

The student who created the 10-week protocol was asked whether he felt the conclusions of this analysis were accurate. Before being shown the actual videotapes of his protocol he claimed that he would not be able to think in PHI form while he was designing. When shown the videotapes and their analysis he agreed that the analysis was correct, but he professed great surprise at this fact.

At first, these results were quite puzzling. It seemed that students claimed not to be able to use exactly the kind of thinking that they in fact used. Eventually, we found a solution to this puzzle in Schön's (1983) theory of reflection-in-action, which is explained later. This theory suggests that the problem was not that students could not think in a PHI-type manner while they devised a solution form, but rather that they could not be self-consciously aware of doing so. The principle is the same as that which makes it impossible to watch one's own fingers while playing the piano and which incapacitated the fabled centipede who attempted to think about his feet while running.

In the past 3 years, additional informal testing of PHI has gone on at Boulder, Colorado, within the framework of an undergraduate course on design theory and methods. Each of the three times this course has been offered a consistent pattern has emerged that confirms the earlier results: Students do not deal issues of form construction until given a project that requires them to do so. To do this project, students rely heavily on taped protocols.

One reason for the need to support construction is that design argumentation is densely populated with deictic references to parts of the partially constructed solution. Without the ability to relate construction and argumentation to each other, it is impossible to discuss the solution. Without construction situations, design rationale cannot be contextualized. Students using our systems to generate issue-based design rationale invariably left out all the issues dealing with construction. They instead concentrated on philosophical discussion, requirements, programmatic analysis, and other preparatory issues rather than actually "getting into the design."

Another problem was that serve relationships were often not effective in helping the designer to generate the important rationale. Designers tended to waste time on issues with little impact on the outcome of the project. This too resulted from lack of support for construction. Designers were often unable to judge the relative merits of issues because they could not see their influence on construction. It is only by being relevant to construction that issues serve the project. The serve relationships of PHI showed that resolving one issue was valuable for resolving another. They could not, however, guarantee that any issue served the project as a whole, for this depended on its influencing construction. This lack of relevance to construction promoted what architects call "talkitecture" (i.e., extended discussion having little impact on the solution).

In a good design project, construction generates and regulates argumentation. Argumentation arises out of construction, and is often tested by construction. Creating good design rationale requires support for construction.

4. SUPPORT FOR CONSTRUCTION

Construction, a subactivity of design, is the composition of elementary building blocks or materials to form an artifact. Sometimes the designer constructs the artifact directly, but in many domains the designer constructs it by making a model or plan of the artifact to be realized by others. The elementary building blocks and materials available for construction activities form the *design substrate*.

Construction is a crucial aspect of design because it creates situations that can "talk back" to the designer:

> Typically [the designer's] making process is complex. There are more variables—kinds of possible moves, norms, and interrelationships of these— than can be represented in a finite model. Because of this complexity, the designer's moves tend, happily or unhappily, to produce consequences other than those intended. When this happens, the designer may take account of the unintended changes he has made in the situation by forming new appreciations and understandings and by making new moves. He shapes the situation, in accordance with his initial appreciation of it, the situation "talks back," and he responds to the situation's backtalk. (Schön, 1983, p. 79)

Human Problem-Domain Communication. The substrate used to design computer-based artifacts typically consists of low-level abstractions (e.g., statements and data structures in programming languages, and primitive geometric objects in engineering computer-aided design). Abstractions at that level are far removed from the concepts that form the basis of thinking in the application domains in which these artifacts are to operate. The great transformation distance between the design substrate and the application domain (Hutchins, Hollan, & Norman, 1986) is a reason for the high cost and the great effort necessary to construct artifacts using computers. To reduce this transformation distance, high-level, domain-oriented substrates are required. Akin (1978) and others have shown that designers design with meaningful abstractions at different levels. For example, architects use domain-related chunks or parts of buildings such as clusters of rooms, individual rooms, areas, and furniture when they design.

Rather than communicating with computers, designers should perceive design as communication with an application domain; the computer should become effectively invisible. Human problem-domain communication (Fischer & Lemke, 1988) tries to achieve this goal. It provides a new level of quality in human–computer communication because the important abstract operations and objects in a given area are built directly into the computing environment. In an environment supporting human problem-domain communication, designers build artifacts from application-oriented building blocks according to the principles of that domain—not the principles of software or geometry.

Construction Kits. Construction kits (Fischer & Lemke, 1988) support human problem-domain communication by offering domain-oriented building blocks presented in a palette and a work area for construction by direct manipulation. Interacting with a computer-based construction kit does not provide the same "back-talk" afforded by designing with real objects.

Figure 3. JANUS-CONSTRUCTION: The work triangle critic. JANUS-CON-STRUCTION is the construction part of JANUS. Building blocks (design units) are selected from the *Palette* and moved to desired locations inside the *Work Area.* Designers can reuse and redesign complete floor plans from the *Catalog.* The *Messages* pane displays critic messages automatically after each design change that triggers a critic. Clicking with the mouse on a message activates JANUS-ARGU-MENTATION and displays the argumentation related to that message (see Figure 5).

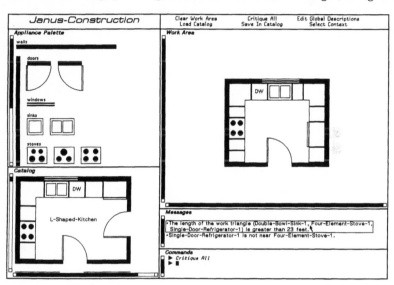

However, construction kits are an active medium that can react to the designer's actions in ways that are different from real objects. To illustrate the concept of a construction kit, we describe JANUS-CONSTRUCTION, a part of the JANUS system for the domain of residential kitchen design.

JANUS-CONSTRUCTION is a construction kit for the domain of kitchen design. The palette of the construction kit contains domain-oriented building blocks called design units, such as sink, stove, and refrigerator (Figure 3). Designers construct by obtaining design units from the palette and placing them into the work area. They can thus see how different configurations fit the floor plan and how requirements about storage space, work flow, and other considerations can be satisfied. A situation is constructed that can "talk back" to a skilled designer.

In addition to design by composition (using the palette and constructing an artifact from scratch), JANUS-CONSTRUCTION also supports design by modification. Existing designs can be modified by retrieving them from the catalog and manipulating them in the work area. The catalog can also serve as a learning tool. The user can copy both good and bad examples into the work area. The system can critique such designs to show how they can be improved, thus allowing users to learn from negative examples. Designers can learn about the good features of prestored designs as well.

Designers using JANUS-CONSTRUCTION expressed that they experienced a sense of accomplishment in using the system because it enabled them to construct something quickly without having detailed knowledge about computers. But construction kits do not in themselves lead to the production of interesting artifacts (Fischer & Lemke, 1988; Norman, 1986). Construction kits do not help designers perceive the shortcomings of an artifact they are constructing. In that they are passive representations, constructions in the work area do not talk back unless the designer has the skill and experience to form new appreciations and understandings when constructing. Designers often do not see characteristics that lead to breakdowns in later use situations. As Rittel put it: "Buildings do not speak for themselves."[1] Designers who are unaware of the work triangle rule do not perceive a breakdown if that rule is violated (i.e., if the total distance between stove, sink, and refrigerator is greater than about 23 feet).

Critics.   Critics operationalize Schön's (1983) concept of a situation that talks back. They use knowledge of design principles to detect and critique suboptimal solutions constructed by the designer.

The critics in JANUS-CONSTRUCTION identify potential problems in the artifact being designed. Their knowledge about kitchen design includes design principles based on building codes, safety standards, and functional preferences. An example of a building code is "the window area shall be at least 10% of the floor area"; an example of a safety standard is "the stove should be at least 12 inches away from a door"; and an example of a functional preference is the work triangle rule (Jones & Kapple, 1984; Paradies, 1973). Functional preferences may vary from designer to designer, whereas building codes and safety standards should be violated only in exceptional cases.

Critics detect and critique partial solutions constructed by the designer based on knowledge of design principles. Critics' knowledge is represented as relationships between design units. The stove design unit, for example, has critics with the following relations: `away-from stove door`, `away-from stove window`, `near stove sink`, `near stove refrigerator`, and `not-immediately-next-to stove refrigerator`. These critics are implemented as condition-action rules, which are tested whenever the design is changed. The changes that trigger a critic are operations that modify the design in the work area. When a design principle is violated,

---

1. One of Rittel's favorite sayings, it was a standard part of the lectures for his Architecture 130 course at Berkeley.

a critic will fire and display a critique in the messages pane of Figure 3. In the figure, the work triangle critic fired telling the designer that the "work triangle is greater than 23 feet." This identifies a possibly problematic situation (a breakdown), and prompts the designer to reflect on it. The designer has broken a rule of functional preference, perhaps out of ignorance or by a temporary oversight.

Users can modify and extend JANUS-CONSTRUCTION by modifying or adding design units, critic rules, and relationships (Fischer & Girgensohn, 1990). This end-user modifiability allows for evolution of the environment as design practice and requirements change. Designers can also modify critic rules when they disagree with the critique given. Standard building codes (hard rules) should not be changed, but functional preferences (soft rules) vary from designer to designer and thus can and should be adapted. Designers have the capability to express their preferences. For example, if designers disagree with the design principle that the stove should be away from a door, they can edit the stove-door rule by replacing the away-from relation between stove and door with another relation (selected from a menu) such as near. After this modification, they will not be critiqued when a stove is not away from a door.

Lack of Argumentative Support.   The advantage of constructing something is that the constructed artifacts and situations can talk back to the designer. The back-talk of the situation is enriched in our framework with the critics, but the short messages the critics present to designers cannot reflect the complex reasoning behind the corresponding design issues. To overcome this shortcoming, we initially developed a static explanation component for the critic messages (Lemke & Fischer, 1990; Neches, Swartout, & Moore, 1985). The design of this component was based on the assumption that there is a "right" answer to a problem. But the explanation component proved to be unable to account for the deliberative nature of design problems. Therefore, argumentation about issues raised by critics must be supported, and argumentation must be integrated into the context of construction.

5. INTEGRATED DESIGN ENVIRONMENTS

Separate systems for construction and argumentation have major deficiencies (as articulated in the previous sections and by Fischer et al., 1989b). If argumentation is to serve design, it must do so by informing construction. If construction acknowledges the nature of design processes (messy situations that are characterized by uncertainty, conflict, and uniqueness), it must have access to the argumentative component. This can happen only if construction and argumentation are explicitly linked in an integrated design environment.

5.1. Reflection-in-Action

Our original attempt at integrating construction and argumentation was to have construction take place within the framework of argumentation—in other words, to raise an issue for each construction step ("What should the next step be?"), deliberate it, and turn the resolution into a constructive action. Unfortunately, trials of this approach with design students showed that it did not work (Section 3). A reason for this failure can be found in Schön's (1983) theory of design. Schön portrayed design as a continual alternation between two radically different and mutually exclusive types of design processes: "knowing-in-action" and "reflection-in-action." As he explained, "In a good process of design, this conversation with the situation is reflective. In answer to the situation's back-talk, the designer reflects-in-action on the construction of the problem, the strategies of action, or the model of the phenomena, which have been implicit in his moves" (p. 79).

Knowing-in-action is the unself-conscious, nonreflective doing that controls the situated action of constructing the actual artifact. *Reflection-in-action* is the self-conscious, rational process of reflecting about this action within the "action-present," that is, the time period during which reflection can still make a difference to what action is taken. Reflection is required when there is a breakdown in knowing-in-action. Such a breakdown typically occurs when action produces unforeseen consequences, either good or bad. When a breakdown occurs, reflection can be used to repair the breakdown situation, and then action can continue.

Schön's (1983) concepts do not in themselves tell us what the architecture of design support environments should be. His concepts must be further operationalized and substantially augmented if they are to provide a basis for computerbased systems. In our work, we interpret *action* as "construction" and *reflection* as "argumentation." For argumentation to get used, it must be part of reflection-in-action, implying that it should be brought to the designer's attention only in breakdown situations. Construction cannot be done within an argumentative framework because the former implies unself-conscious, nonreflective engagement in creating the solution whereas the latter implies self-conscious, reflective thinking about the solution. Argumentation must take place within the "action present." If the time required to read and/or record the argumentation is greater than the action present, design is disrupted and the required context is lost. Design rationale can aid reflection by informing it with design knowledge, principles, and ideas, and by triggering critical thought in the designer. Schön's theory, when operationalized, can then be used as the basis for a system architecture.

Figure 4. A multifaceted architecture. The links between the components are crucial for exploiting the synergy of the integration.

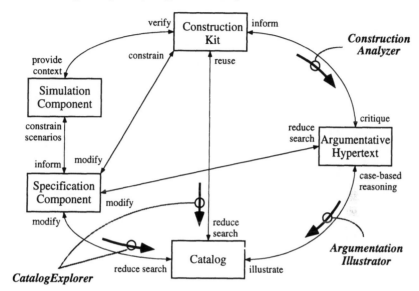

5.2. An Architecture for Integrated Design Environments

Over the last few years, we have developed an integrated, multifaceted architecture for design environments (see Figure 4). The multifaceted architecture consists of the following five components:

1. A construction kit is the principal medium for implementing design. It provides a palette of domain-specific building blocks and supports the construction of artifacts using direct manipulation and form filling.
2. An argumentative hypertext system contains issues, answers, and arguments about the design domain. Users can annotate and add argumentation as it emerges during design processes.
3. A catalog provides a collection of prestored design examples illustrating the space of possible designs in the domain and supporting reuse and case-based reasoning.
4. A specification component allows designers to describe some characteristics of the design they have in mind. The specifications are expected to be modified and augmented during the design process, rather than to be fully articulated at the beginning. They are used to retrieve design objects from the catalog and to filter information in the hypertext.

5. A simulation component allows designers to carry out "what-if" games simulating usage scenarios with the artifact being designed.

Integration. The multifaceted architecture derives its essential value from the integration of its components and links between the components. Used individually, the components are unable to achieve their full potential. Used in combination, however, each component augments the value of the others, forming a synergistic whole. At each stage in the design process, the partial design embedded in the design environment serves as a stimulus to users for suggesting what they should attend to next.

Links among the components of the architecture are supported by various mechanisms (see Figure 4):

1. *CONSTRUCTION ANALYZER:* Users need support for construction, argumentation, and perceiving breakdowns. Experience with our early systems has shown that users too often fail to hear the situation talk back; breakdowns do not occur that trigger reflection-in-action. Additional system components are needed to signal breakdowns. This is the role of the CONSTRUCTION ANALYZER in the multifaceted architecture. The CON-STRUCTION ANALYZER is a version of the critics described in Section 4 enhanced with pointers into the argumentation issue base. The firing of a critic signals a breakdown to users and provides them with entry into the exact place in the argumentative hypertext system at which the corresponding argumentation is located.

2. *ARGUMENTATION ILLUSTRATOR:* The explanation given in argumentation is often highly abstract and very conceptual. Concrete design examples that match the explanation help users to understand the concept. The ARGUMENTATION ILLUSTRATOR helps users to understand the information given in the argumentative hypertext by finding a catalog example that realizes the concept (Fischer, 1990).

3. *CATALOG EXPLORER:* This helps users to search the catalog space according to the task at hand (Fischer & Nakakoji, 1991). It retrieves design examples similar to the current construction situation, and orders a set of examples by their appropriateness to the current specification.

A typical cycle of events supported by the multifaceted architecture is: (a) Users create and refine a partial specification or construction, (b) breakdowns occur, (c) users switch and consult other components in the system made relevant by the system to the partially articulated task at hand, and (d) users refine their understanding based on the "back talk of the situation." As users go back and forth among these components, the problem space is narrowed, a shared understanding between users and the system evolves, and the artifact is incrementally refined. This chapter

Figure 5. JANUS-ARGUMENTATION: Rationale for the work triangle rule. JANUS-ARGUMENTATION is an argumentative hypertext system based on the PHI method. The *Viewer* pane shows a diagram illustrating this answer generated by the ARGUMENTATION ILLUSTRATOR. The *Visited Nodes* pane lists in sequential order the previously visited argumentation topics. By clicking with the mouse on one of these items, or on any bold or italicized item in the argumentation text itself, the user can navigate to related issues, answers, and arguments. Hypertext access and navigation is made possible using this feature, inherited from the SYMBOLICS DOCUMENT EXAMINER.

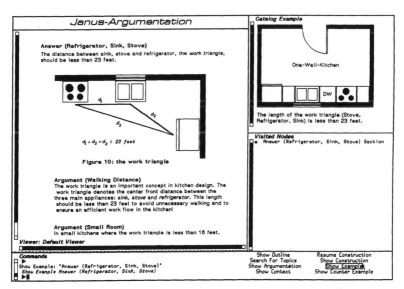

focuses on the integration of construction and argumentation. Other components of the multifaceted architecture are described elsewhere (Fischer, 1990; Fischer & Nakakoji, 1991).

JANUS-ARGUMENTATION: The Argumentation Component of JANUS. JANUS-ARGUMENTATION is the argumentation component of JANUS (Figure 5). It is an argumentative hypertext system based on the PHI method and implemented using the SYMBOLICS DOCUMENT EXAMINER (Walker, 1987). JANUS-ARGUMENTATION offers a domain-oriented, generic issue base about how to construct residential kitchens. This design knowledge has been acquired from protocol studies (Fischer et al., 1989a) and from kitchen design books (Jones & Kapple, 1984). In JANUS-ARGU-MENTATION, designers explore issues, answers, and arguments by navigating through the issue base. The starting point for the navigation is the argumentative context triggered by a critic message in JANUS-CONSTRUC-

TION. Clicking with the mouse on a critique in JANUS-CONSTRUCTION (see Figure 3) activates JANUS-ARGUMENTATION and accesses the issue and answer corresponding to the critique. At any place in the issue base, designers can invoke the ARGUMENTATION ILLUSTRATOR to obtain an example from the catalog that implements the current issue answer.

5.3. Evaluation, Shortcomings, and Limitations of JANUS

Evaluation. We have informally evaluated JANUS with subjects ranging from neophyte to expert designers and from neophyte to expert computer users (Fischer et al., 1989b). The subjects were tested in an experiment consisting of two tasks: a learning task and a design task. The learning task consisted of improving a "bad" kitchen design from the catalog (see Figure 3), and the design task consisted of designing a "good" kitchen given a set of constraints. The constraints were imposed to test the various operations of the system. Users unfamiliar with the computer system were given help by the experimenter during an initial learning task. A final questionnaire was given to the subjects after the experiment.

Designers with limited domain knowledge were able to understand the critics and learn from them to create reasonable kitchen designs. For example, several students did not know that building codes require that at least one of the entrances into a kitchen should be at least 36 inches wide. One user also learned that the stove should be away from a door, based on safety requirements with respect to fire and burn hazard. He found this to be especially relevant to his own home where small children are constantly running in and out of the kitchen.

The critics were appreciated, but were often ignored when they actively critiqued the user during construction. One user replied to this by saying that too much information was presented and that she could give attention to only one thing at a time. She preferred to complete some part of the design and then ask the system for a critique by using the Critique All command. Other users explained that they ignored critics because they had already been aware of them, either by a previous critique or by the fact that they already knew about them (such as that the sink should be in front of a window). In the questionnaire, all users found that critiquing was helpful in reminding them about design rules they did not think about while they were designing.

Users uncertain about a critique from the system or interested in more background information about design principles entered the hypertext system by clicking on the critique message. No users got "lost" in the hyperdocument, but one found that some of the arguments were not justified from his point of view. He would have liked to add his own counterarguments to it. Currently end-user modifications of the issue base are not supported. Another user found that some arguments did not go into enough depth in order to be persuasive. For example, he would have liked to know why a building code requires that a kitchen entrance should be greater than 36 inches wide.

Shortcomings and Limitations. Our integrated design environments in their current form still suffer from a number of major limitations:

1. Design environments need to evolve for the following reasons: (a) The world modeled in these design environments changes (Curtis, Krasner, & Iscoe, 1988), and (b) the background knowledge for a design domain cannot be fully articulated—it is tacit and requires breakdown situations to be activated (Ehn, 1988; Winograd & Flores, 1986). End-user modifiability is a prerequisite for evolution because the breakdown situations are experienced by the domain experts using these systems, not by the knowledge engineers who built them originally. Fischer and Girgensohn (1990) described a mechanism to make JANUS-CONSTRUCTION end-user modifiable. RE-FLACT (described in the next section) is an effort to make the argumentative component adaptive. One reason that JANUS-ARGUMENTATION failed to achieve this goal is that the DOCUMENT EXAMINER (Walker, 1987) is only a reader's interface to the hypertext system and requires a different writer's interface (Walker, 1988). Therefore, JANUS-ARGUMENTATION primarily serves as a design information system and does not allow the addition of new design rationale in a contextualized manner.

2. The back-talk of the situation must be enhanced further with a simulation component providing us with insights that argumentation does not capture. This requirement became obvious in our experiments with professional kitchen designers who tested their design by running mental simulations of specific situations (e.g., preparing a fancy dinner, imagining work-flow patterns with more than one person working in the kitchen).

3. The issue base of JANUS-ARGUMENTATION is generic; that is, it is used for any kitchen design project. The issue base is also static in that it does not adapt to the individual design projects. Some issues are only relevant to some of the design projects addressed by the issue base. For example, if the kitchen has no eating area, then issues relating to the eating area in the kitchen are irrelevant. Structures in the issue base irrelevant to the task at hand make the issue base unwieldy and make it difficult to find the relevant information. To filter out irrelevant information from a generic issue base, the serves relationship must be dynamically computed from the task at hand. The exploratory nature of design makes any static argumentative hypertext system, such as JANUS-ARGUMENTA-TION, inadequate. A dynamic hypertext system adapts to design decisions such as adding or removing an eating area.

6. CURRENT AND FUTURE WORK

Our current and future work is focused on four ways to make argumentation better serve design: (a) Static issue bases are made extensible and dynamic, (b) the reusability of issue bases is being improved, (c) design rationale is being added to the examples in the catalog, and (d) a system component for articulating and representing the task at hand is being developed. Some of these extensions are being carried out as separate efforts later to be integrated into the overall environment.

Adaptive Issue Bases. In response to the problems caused by the static nature of JANUS-ARGUMENTATION, we are exploring ways to make reusable issue bases more active and responsive to the situation, thus increasing the immediate benefit of issue bases. We have implemented these methods in REFLACT, a PHI-based hypertext system (Lemke, 1990). In REFLACT, the designer not only consults the issue base but also indicates design decisions—whether deliberated or not—by selecting one or more answers. The selected answers determine which issues the system raises from its issue base. This is done with the help of the PHI subissue relationship. In REFLACT, issue bases are fully modifiable and extensible by end users. Designers can add, modify, or delete issues, answers, and arguments without leaving REFLACT.

Reusable Domain-Oriented Issue Bases. A design rationale is a large additional product of the design process. Creating and representing a design rationale is a great effort. Reuse of existing issue bases has the potential to dramatically reduce this effort. Every project is unique in some respects; few if any projects are unique in all respects. Therefore, the contents of a project issue base are not entirely unique to that project. Similar projects overlap substantially in issues, answers, and arguments. This is not to say that the issues are resolved in the same way, but merely that a great deal of the reasoning is shared by projects.

Reusable issue bases can serve as seeds that grow with each new design project. Each project extends and enhances the reusable issue base. The issue base being reused provides information about how to decompose the task, possible answers to issues, and principles of design. The issue base also warns designers of potential dead ends and unproductive solution directions. This is important because designers need better access to domain-oriented information (Curtis, et al., 1988). Even expert designers can no longer master all the relevant knowledge, especially in technologically oriented design, where growth and change of the knowledge base are incessant (Draper, 1984; Norman, 1988).

Domain-oriented issue bases also amplify the designer's ability to reflect on issues. Recurring design issues could be researched intensively and the results of this could then be stored at the appropriate location in the issue base for use by future designers encountering similar decisions in the future. This would, for example, allow the "folk theories" of designers to be subjected to rigorous scientific scrutiny. Cumulative domain-oriented issue bases could also foster communication among designers, researchers, and users about recurring matters of design.

The PHI subissue relationship is crucial to making issue bases reusable. The hierarchical grouping of issues allows argumentation systems to be built that filter issue bases according to the specifics of the new task. REFLACT filters issue bases using its mechanism of issue conditions. The system provides a common issue base for all projects in a domain such as kitchen design. This issue base includes issues, answers, and arguments at all levels of generality. As pointed out earlier, not every issue applies in each design project, even if it falls into one general domain.

Enriched Catalogs. The JANUS catalog does not currently contain the design rationale for the designs it contains. By adding the rationale to each catalog example, designers can better understand the examples, can more easily find examples that are similar to the kitchens they are designing, and can reuse the rationale.

Representation of the Task at Hand. More support to incrementally capture the task at hand is needed. Beyond the information contained in the construction situation, our specification component needs to be further developed to let designers articulate the specifics of their design effort. This knowledge can be used by REFLACT to filter out irrelevant information from a reusable issue base. An initial effort in this direction is described in Fischer and Nakakoji (1991).

7. CONCLUSIONS

Approximately 25 years ago, Horst Rittel began the first work on design argumentation (i.e., rationale). The central idea of his approach was to represent rationale as the argumentative evaluation of alternative answers to questions. Half the chapters in this book deal with some variant of that idea. As our experience and the experiences of others (see Buckingham Shum, 1995 [chapter 6 in this book]; Conklin & Burgess-Yakemovic, 1991 [chapter 14 in this book]) make clear, this approach has run into fundamental difficulties. Argumentation does not naturally and easily serve design; it must be made to do so. The central challenge facing advocates of design rationale capture is to understand and overcome these difficulties. This chapter has confronted this challenge.

We began by identifying fundamental difficulties in making argumentation serve design. Creating and using design rationale is a time-consuming process that must be carried out in addition to standard design activities, and there is little immediate reward. Recording and accessing design rationale can disrupt design and interfere with reflection-in-action. Argumentation that is removed from construction loses relevance to the task at hand. Without tight integration of argumentation and construction, designers fail to apply argumentation in the construction activity.

To understand how to overcome these difficulties, we analyzed them within the design theories of Schön (1983) and Rittel (1984). This gave us a constructive understanding that suggested the following solution approaches. First, the IBIS method should be modified to emphasize relevance to (i.e., serving) the task at hand. This notion resulted in the development of the PHI method. Refinement of the rationale representation scheme is not enough; substantial computer support is also needed. Hypermedia systems can be created to reduce the secretarial work of managing issue bases. Support is also needed to reduce the conceptual work of creating project rationale whose content and form correspond to the designer's changing understanding of the problem. This can be done through support for reuse of issue bases. Argumentation must be made to serve the construction of solution form. To do this support for construction must be integrated with support for argumentation. We developed tools for construction that support human problem-domain communication, and integrated them with tools for argumentation via critics. These integrated design environments form a synergistic whole by causing the construction situation to talk back to the designer.

A final word on the generality of our approach. JANUS was used as an "object-to-think-with" in this chapter. We used the same basic approach for user interface design (Lemke & Fischer, 1990), development and maintenance of Cobol programs (Atwood et al., 1991), river basin planning and operations (Lemke & Gance, 1990), computer network design (Fischer et al., 1991), knowledge editing, and design and planning of lunar habitation. As these systems get used in realistic work environments, we will get valuable feedback about the viability, the strengths, and the weaknesses of this approach.

NOTES

Background. This chapter first appeared in the *Human–Computer Interaction* Special Issue on Design Rationale in 1991.

Acknowledgments. The authors thank the members of the Human–Computer Communication group at the University of Colorado, who contributed to the conceptual framework and the systems discussed in this chapter.

Support. The research was supported by the National Science Foundation under Grants IRI-9015441, CDA-8420944, IRI-8722792, and MDR-9253425; by the Army Research Institute under Grant MDA-903-86-C0143; and by grants from the Intelligent Interfaces Group at NYNEX, and from Software Research Associates (SRA) in Tokyo.

Authors' Present Addresses. Gerhard Fischer, Department of Computer Science and Institute of Cognitive Science, University of Colorado, Boulder, CO 80309-0430; Andreas C. Lemke, ALCATEL-SEL, ZFZ/SW1, Stuttgart, Germany; Raymond McCall, Division of Environmental Design and Institute of Cognitive Science, University of Colorado, Boulder, CO 80309-0314; Anders Morch, Department of Informatics, University of Oslo, P.O. Box 1080, Blindern, 0316, Oslo, Norway. Email: gerhard@cs.colorado.edu, alemke@rcs.sel.de, mccall_r@cubldr.colorado.edu, and andersm@ifi.uio.no

REFERENCES

Akin, O. (1978). How do architects design? In J. Latombe (Ed.), *Artificial intelligence and pattern recognition in computer aided design* (pp. 65–104). New York: North-Holland.

Atwood, M. E., Burns, B., Gray, W. D., Morch, A. I., Radlinski, E. R., & Turner, A. (1991). The Grace integrated learning environment—A progress report. *Proceedings of the Fourth International Conference on Industrial & Engineering Applications of Artificial Intelligence & Expert Systems (IEA/AIE 91)*, 741–745. New York: ACM Press.

Buckingham Shum, S. (1995). Analyzing the usability of a design rationale notation. In T. P. Moran & J. M.Carroll (Eds.), *Design rationale: Concepts, techniques, and use.* Hillsdale, NJ: Lawrence Erlbaum Associates. [Chapter 6 in this book.]

Conklin, E. J. (1987). Hypertext: An introduction and survey. *IEEE Computer, 20*(9), 17–41.

Conklin, E. J., & Begeman, M. (1988). gIBIS: A hypertext tool for exploratory policy discussion. *Proceedings of the Conference on Computer Supported Cooperative Work*, 140–152. New York: ACM.

Conklin, E. J., & Burgess-Yakemovic, KC. (1991). A process-oriented approach to design rationale. In T. P. Moran & J. M. Carroll (Eds.), *Design rationale: Concepts, techniques, and use.* Hillsdale, NJ: Lawrence Erlbaum Associates, 1996. [Chapter 14 in this book.]

Cross, N. (1984). *Developments in design methodology.* New York: Wiley.

Curtis, B., Krasner, H., & Iscoe, N. (1988). A field study of the software design process for large systems. *Communications of the ACM, 31*(11), 1268–1287.

Draper, S. W. (1984). The nature of expertise in UNIX. *Proceedings of INTERACT '84, IFIP Conference on Human–Computer Interaction*, 182–186. Amsterdam: Elsevier Science Publishers.

Ehn, P. (1988). *Work-oriented design of computer artifacts*. Stockholm, Sweden: Almquist & Wiksell International.

Fischer, G. (1990). Cooperative knowledge-based design environments for the design, use, and maintenance of software. *Software Symposium '90* , 2–22. Tokyo, Japan: Software Engineering Association.

Fischer, G., & Girgensohn, A. (1990). End-user modifiability in design environments. *Human Factors in Computing Systems, CHI '90 Conference Proceedings*, 183–191. New York: ACM.

Fischer, G., Grudin, J., Lemke, A. C., McCall, R., Ostwald, J., & Shipman, F. (1991). Supporting asynchronous collaborative design with integrated knowledge-based design environments (Tech. Rep.). Boulder: University of Colorado, Department of Computer Science.

Fischer, G., & Lemke, A. C. (1988). Construction kits and design environments: Steps toward human problem-domain communication. *Human–Computer Interaction*, 3(3), 179–222.

Fischer, G., McCall, R., & Morch, A. (1989a). Design environments for constructive and argumentative design. *Proceedings of the CHI '89 Conference on Human Factors in Computing Systems*, 269–276. New York: ACM.

Fischer, G., McCall, R., & Morch, A. (1989b). JANUS: Integrating hypertext with a knowledge-based design. *Proceedings of Hypertext '89*, 105–117. New York: ACM.

Fischer, G., & Nakakoji, K. (1991). Empowering designers with integrated design environments. In J. Gero (Ed.), *Proceedings of the First International Conference on Artificial Intelligence in Design* (Edinburgh, Scotland) (pp. 191–209). Cambridge, England: Butterworth-Heinemann, Ltd.

Hutchins, E. L., Hollan, J. D., & Norman, D. A. (1986). Direct manipulation interfaces. In D. A. Norman & S. W. Draper (Eds.), *User-centered system design. New perspectives on human–computer interaction* (pp. 87–124). Hillsdale, NJ: Lawrence Erlbaum Associates.

Jones, R. J., & Kapple, W. H. (1984). *Kitchen planning principles—Equipment-appliances*. Urbana-Champaign: University of Illinois, Small Homes Council—Building Research Council.

Kunz, W., & Rittel, H. W. J. (1970). *Issues as elements of information systems (Working Paper No. 131)*. Berkeley: University of California, Berkeley, Center for Planning and Development Research.

Lee, J. & Lai, K.-Y. (1991). What's in design rationale? *Human–Computer Interaction*, 6, 251–280. Also in T. P. Moran & J. M. Carroll (Eds.), *Design rationale: Concepts, techniques, and use*. Hillsdale, NJ: Lawrence Erlbaum Associates, 1996. [Chapter 2 in this book.]

Lemke, A. C. (1990). Framer-hypertext: An active issue-based hypertext system. *Proceedings of the Workshop on Intelligent Access to Information Systems*, 34–38. Darmstadt, Germany: GMD-IPSI.

Lemke, A. C., & Fischer, G. (1990). A cooperative problem solving system for user interface design. *Proceedings of AAAI-90, Eighth National Conference on Artificial Intelligence*, 479–484. Cambridge, MA: AAAI Press/MIT Press.

Lemke, A. C., & Gance, S. (1990). End-user modifiability in a water management application (Tech. Rep.). Boulder: University of Colorado, Department of Computer Science.

MacLean, A., Young, R. M., Bellotti, V. M. E., & Moran, T. P. (1991). Questions, options, and criteria: Elements of design space analysis. *Human–Computer*

Interaction, 6, 201–250. Also in T. P. Moran & J. M. Carroll (Eds.), *Design rationale: Concepts, techniques, and use.* Hilldale, NJ: Lawrence Erlbaum Associates, 1996. [Chapter 3 in this book.]

McCall, R. (1979). *On the structure and use of issue systems in design.* Unpublished doctoral dissertation, University of California, Berkeley.

McCall, R. (1991). PHI: A conceptual foundation for design hypermedia. *Design Studies, 12,* 30–41.

McCall, R., Bennett, P., d'Oronzio, P., Ostwald, J., Shipman, F., & Wallace, N. (1990). PHIDIAS: A PHI-based design environment integrating CAD graphics into dynamic hypertext. In A. Rizk, N. Streitz, & J. Andre (Eds.), *Hypertext: Concepts, systems and applications* (pp. 152–165). Cambridge, England: Cambridge University Press.

McCall, R., Mistrik, I., & Schuler, W. (1981). An integrated information and communication system for problem solving. *Proceedings of the Seventh International CODATA Conference,* 107–115. London: Pergamon.

Neches, R., Swartout, W. R., & Moore, J. D. (1985). Enhanced maintenance and explanation of expert systems through explicit models of their development. *IEEE Transactions on Software Engineering, SE-11*(11), 1337–1351.

Norman, D. A. (1986). Cognitive engineering. In D. A. Norman & S. W. Draper (Eds.), *User-centered system design* (pp. 31–62). Hillsdale, NJ: Lawrence Erlbaum Associates.

Norman, D. A. (1988). *The psychology of everyday things.* New York: Basic Books.

Paradies, K. (1973). *The kitchen book.* New York: Wyden.

Reuter, W., & Werner, H. (1983). Thesen und empfehlungen zur anwendung von argumentativen informationssystemen [Theses and recommendations about the use of argumentative information systems] (Working Paper). Institut fuer Grundlagen der Planung, University of Stuttgart, Germany.

Rittel, H. W. J. (1972). On the planning crisis: Systems analysis of the first and second generations. *Bedriftsokonomen, 8,* 390–396.

Rittel, H. W. J. (1980). APIS: *A concept for argumentative planning information systems* (Working Paper 324). University of California, Berkeley: Institute of Urban and Regional Development.

Rittel, H. W. J. (1984). Second-generation design methods. In N. Cross (Ed.), *Developments in design methodology* (pp. 317–327). New York: Wiley.

Schön, D. A. (1983). *The reflective practitioner: How professionals think in action.* New York: Basic Books.

Simon, H. A. (1981). *The sciences of the artificial.* (2nd ed.) Cambridge, MA: MIT Press.

Stefik, M. J. (1986). The next knowledge medium. *AI Magazine, 7*(1), 34–46.

Walker, J. H. (1987). Document examiner: Delivery interface for hypertext documents. *Hypertext '87 Papers,* 307–323. Chapel Hill: University of North Carolina Press.

Walker, J. H. (1988). Supporting document development with concordia. *IEEE Computer, 21*(1), 48–59.

Winograd, T., & Flores, F. (1986). *Understanding computers and cognition: A new foundation for design.* Norwood, NJ: Ablex.

10

Supporting Software Design: Integrating Design Methods and Design Rationale

Colin Potts
Georgia Institute of Technology

ABSTRACT

Many software development organizations are now using standard design methods. Because a method concentrates the designer's attention on different issues at different times, it is fruitful to analyze, compare, and support different methods in an issue-based framework. In our work, we have extended the IBIS model in several ways to accommodate design methodological concepts. In this chapter three such extensions are reviewed, illustrated, and compared: (a) the embellishment of IBIS by method-specific constructs, (b) the specialization of IBIS in method-specific ways, and (c) the incorporation into IBIS of the results of objective analysis and simulation. We conclude that the second approach is the most extensible, but that relaxation of the IBIS model (as is possible in the third approach) is necessary for practical use on design projects.

Colin Potts, formerly a cognitive psychologist, is now a computer scientist interested in collaborative design practices and software requirements determination; he is Associate Professor of Computer Science in the College of Computing at Georgia Institute of Technology.

CONTENTS

1. SOFTWARE DESIGN METHODS

In recent years, the software industry has started to adopt a number of design methods to improve the productivity and quality of software. These include structured methods, object-oriented methods, and a variety of special-purpose and hybrid methods (see Birrell & Ould, 1985, for an excellent, albeit now dated summary). Because a method concentrates the designer's attention on different issues at different times and provides the rationale for many decisions, it is fruitful to analyze, compare, and support different methods in an issue-based framework. Yet despite this obvious relationship between software development methods and design rationale research, few researchers have attempted to analyze, compare, or support methods or develop new methods in an issue-based framework. That is what this chapter sets out to do.

There are several reasons for investigating the relationship between design methods and design rationale. One is that design methods are more specific than generic models of design rationale such as IBIS (Rittel, 1984). By incorporating domain knowledge—in this case, knowledge about how to design software—the design rationale representation can be made more powerful. Another reason is that an explicit representation of design decision making—as provided by design rationale research—allows us to compare methods on a rational basis by examining what questions they address.

Despite their many differences, software development methods have three things in common:

1. Every method has a design philosophy or strategy with which the practitioner is supposed to attack the problem. This is nothing other than a view, usually an implicit view, of what software design is. Structured analysis, for example, is predicated on the simple philosophy that design is the recursive process by which a designer refines a functional description of the problem into smaller subproblems. Object-oriented design requires the adoption of the belief that software design is the process by which a system is divided into a number of "objects," each of which stands for some concrete thing in the problem domain with its own responsibilities and interfaces to other objects.

2. In addition to the grand strategy provided by a method's design philosophy, methods also have "tactics" or heuristics that the method's authors recommend to practitioners in different situations. Methods vary widely in the quantity and prescriptiveness of the tactics they contain.

3. Methods differ in the kinds of artifacts they assume will be produced by applying the method's strategy and tactics. Eventually, of course, all software development will yield executable program code. There are big differences, however, in the intermediate, nonexecutable representations, models, descriptions, or specifications of the system that are produced before the executable code. We refer to these products as *artifacts*. Because different artifacts contain different types of information at different levels of detail, artifacts should be produced in a rational order. Thus the artifacts are related to the strategy and tactics of the method.

Rather than provide an introduction to a wide range of design methods in this chapter, I describe three methods in enough detail for the reader who is not a software engineer or computer scientist to understand the principles on which the method is based. The methods are Liskov and Guttag's (1986) abstraction-based design approach, JSD (Jackson, 1983), and a method for designing and simulating distributed systems using VERDI (Shen, Richter, Graf, & Brumfield, 1990). Having described each method, I illustrate its application in a realistic design situation, a temporal fragment of designing. And I show how the evolution of the design can be characterized as a history of episodes in which design commitments are made. I use Rittel's (1984) IBIS as the main descriptive apparatus for the commitments themselves, but to represent the interaction between those commitments and the artifacts that describe the evolving system I introduce some extensions to the model. IBIS also requires extension to show how the issues, positions, and arguments that are the atoms of deliberation are specialized into method-specific and recurrent types of issues, positions, and arguments encapsulating the method's strategy and tactics.

2. SOFTWARE DESIGN EVOLUTION AND CHOICE

2.1. Liskov and Guttag's Abstraction-Based Method

In their book *Abstraction and Specification in Program Development*, Liskov and Guttag (1986) presented a method for developing medium-size sequential programs. Liskov and Guttag based their method on the concept of *abstractions*, which are pieces of software that have hidden internal details that implement a simple, public interface. Liskov and Guttag considered three kinds of abstraction: data abstractions, procedures, and iterators. In this chapter, I ignore iterators and address the other two kinds of abstraction: the data and procedural types.

A data abstraction is a way of clumping together all the information needed about a piece of data and the operations that can affect it. For example, in a chess-playing program, there might be a data abstraction to represent a position on the board. How the board is implemented inside the computer is irrelevant at the application level, and other parts of the program that interrogate the board (e.g., to see whether a piece is on a particular square) or change it (e.g., by moving or taking a piece) should not need to know the internal details of the board data structure. The board data abstraction would have a collection of operations that changed the board or returned the positions of specified pieces.

Procedures, or procedural abstractions, are possibly more familiar to casual programmers. These are also known variously as subprograms and subroutines. A procedure, like a data abstraction, has an externally visible interface, and a hidden implementation.

Data abstractions and procedures are by no means unique to Liskov and Guttag's (1986) method. They are central to many other design methods (e.g., Berzins & Luqi, 1991) and are supported in many high-level programming languages (e.g., Ada and C++).

Liskov and Guttag (1986) were among the first authors to provide a coherent account of how a designer should approach a problem when using abstractions. For example, they introduce the "task," an informal concept that captures the intuition of some responsibility the system must fulfill. To illustrate, in the example of a chess-playing program, one task the program must do is to choose a move when it is its turn. Another of its many tasks is to reset the board for a new game. Procedures and data abstractions can both be used to implement tasks: A procedure can implement a task directly (e.g., there could be a single, albeit complex, procedure that selects a move when provided with a position), and a data abstraction can implement a task through one of its operations (e.g., the board abstraction might have operations that clear or reset its constituent squares). Liskov and Guttag provided a number of heuristics for helping the designer to choose which type of abstraction is the most suitable for a given task.

2.2. An Example of Abstractions in Use

To illustrate the role of tasks in Liskov and Guttag's (1986) method, let us consider a simple design episode from an extended example they present in their book. The application is a simple text formatter similar to the Unix formatters *nroff* and *troff*. It reads and formats documents that contain embedded command lines. A command line starts with a period followed by a command name. In Liskov and Guttag's example, there are only a few formatting commands, and their syntax is very simple. The formatter's output is a formatted text file that has been formatted in accordance with the embedded commands. Default parameters also affect output properties like the page length, margin width, and so on.

Liskov and Guttag (1986) pursued the example through to implementation, and presented a 300-line CLU program that implements the formatter. Thus the example is small but nontrivial.

Consider some of the design decisions that must be made by the designer of this formatter. First of all, is the formatter as a whole a procedural abstraction? Because it takes well-defined inputs (text files with embedded formatting commands) and transforms them into well-defined outputs (formatted documents), it appears to be a single procedure.

But having identified a single procedure as the first level of abstraction for the problem, where do we go from there? Some methods would encourage the designer to think hard about formatting and break it down into about seven (plus or minus two) smaller procedures. Liskov and Guttag (1986) had more explicit advice. They recommended that the designer identify the tasks that the formatter procedure performs and then decide whether these tasks correspond to smaller procedures or whether they can be operationalized in another way. For example, the formatter is responsible for reading input, formatting the input, and writing output. Each of these could be encapsulated in various ways. For example, the task of reading input could be performed by operations belonging to a text file abstraction, by a procedural abstraction (effectively a subroutine), or it could be performed by the formatter itself by inline code.

This is a standard, recurring type of issue in abstraction-based design, an *encapsulation issue.* Many of the decisions documented by Liskov and Guttag (1986) are resolutions of issues of this type. This particular issue they recommended resolving by having a line-by-line input procedure. They justified the decision to read text line by line by pointing out that the problem is inherently line oriented: A line is either text to be formatted or it is a formatting command.

Other decisions in the evolution of the text formatter design are more complicated and do not lead to simple one-to-one correspondences between tasks and abstractions. For example, the interpretation of input lines and output of formatted text cannot easily be done by single procedures. (It is here that Liskov and Guttag's [1986] advice scores heavily over simple top-down decomposition, which, if adopted without too much thought or inadequate application knowledge, tends to lead a designer to structure programs into an input-process-output triad, whether or not that is appropriate to the problem.) The problem with an `interpret_line` procedure is that how a line is interpreted depends not just on the command in the line, if it is a command, but on a lot of state information set by previous commands or by default settings. For example, how a text line is processed depends on how justification has been set and what the page margins are. If all this were done by a procedure, it would have to have a large number of parameters or would have to access globally visible variables, neither of which are satisfactory because they increase the risk that hidden dependencies will be forgotten when the program is modified in the future. A better solution, in a situation in which there is a great deal of relevant state information, is to hide the information in a data abstraction. In this case, the data abstraction represents the entire document, and it encapsulates operations that turn justification on and off, alter the margins, and so on.

2.3. An Extension to IBIS to Capture Design Evolution

Design decisions such as the resolution of the aforementioned encapsulation issues have an obvious similarity to the issue-based reasoning of IBIS. However, whereas IBIS supports problem formulation and exploration in general, a design method supports a more specific design process. The differences between IBIS and Liskov and Guttag's (1986) design method are several:

1. IBIS has an ontology that captures the speech acts of a general design discussion. Raising issues, responding with positions, and arguing are the primitive speech acts of IBIS theory. In abstraction-based design, the ontology seems to need to be supplemented by method-specific categories, in the case of Liskov and Guttag's method by *tasks.*

Figure 1. **Ontology for design rationale in Liskov and Guttag (1986) method. To the left of the dashed line is IBIS. To the right is the Liskov and Guttag ontology. In the center is the bridge, consisting of method-specific IBIS constructs.**

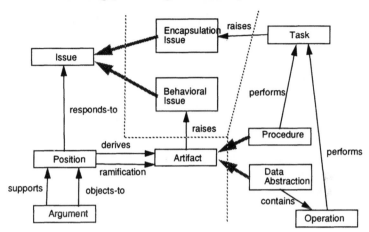

2. IBIS involves a detached discussion about a problem space. The artifact being discussed does not appear in the discussion itself and is not linked to it in a well-defined way. In the case of the design episode mentioned earlier, the discussion is about the further refinement of an abstraction, the formatter procedure, and it results in the identification of the need for new abstractions, a procedure that reads lines and a document data abstraction. Thus three design artifacts, the specifications or code skeletons of the formatter and do_line procedures and the doc data abstraction are connected by an episode of IBIS-like reasoning. To support the evolution of a software design, the IBIS fragments must be linked to the space of artifacts in a well-defined way that is easily managed.

3. In IBIS, issues are not divided into domain-specific classes. Rittel (1984) did talk about classes of issues that affect the ways they are resolved: Thus deontic issues are resolved by discussions about values, whereas factual issues are resolved by appeals to evidence. It would be better to classify issues in terms that correspond directly to the method's strategy and tactics. Encapsulation issues are a particular kind of issue that can be resolved in a particular way: by reasoning about the tasks performed by the to-be-refined design component.

To address these limitations of the unadorned IBIS model, Glenn Bruns and I developed a simple model of abstraction-based design decision making that extended IBIS with method-specific categories (Potts & Bruns, 1988). (In fact, we used a different terminology, referring to "alternatives" and

Figure 2. Evolution of text formatter design. The do_line and doc abstractions result from a discussion about the formatter abstraction and the tasks it performs.

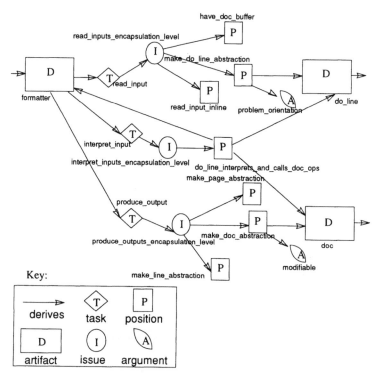

"justifications," instead of "positions" and "arguments," but for present purposes the underlying speech act model can be considered to be IBIS, and in this chapter I revert to IBIS terminology.) The ontology is shown in Figure 1. On the left is IBIS; on the right is the Liskov and Guttag (1986) design ontology; and in the center is a bridge consisting of specialized speech acts—specialized according to Liskov and Guttag's heuristics.

2.4. A Liskov and Guttag Design Episode Analyzed

When the ontology is instantiated on real problem, the issue-based reasoning becomes the "glue" in the design's evolution. Figure 2 illustrates the point. The diagram shows the evolution of a design, with artifacts giving rise to still more refined artifacts, just as the formatter gave rise to the do_line and doc abstractions. The rationale for the derivations is represented by IBIS substructures that mediate between the artifacts. A node's

Figure 3. **Example nodes from Liskov and Guttag (1986) formatter example.**

formatter = PROC (ins, outs, errs: stream)
SIGNALS (badarg(string))
MODIFIES:
 ins, outs, errs
EFFECTS:
 If ins is not open for reading, or outs or errs is not
 open for writing, badargs(s) is signaled, where s
 identifies an argument that was opened
 improperly (e.g. badarg "input stream")). Other
 wise format proceeds as described in the text
 [i.e. Liskov and Guttag(1986) pp. 271–273],
 ....
TASKS:
 read_input
 interpret_input
 produce_output
EFFICIENCY:
 Where n is the number of characters in ins, time
 is O(n), space added (the storage for outs)
 is O(n) and temporary space<< n.

read_input = TASK
DONE–BY:
 format
SUMMARY:
 formatter reads input

read_inputs_encapsulati
on_level = ISSUE
CONCERNING:
 read_input
SUMMARY:
 How should the format
task
 read_input be encap-
sulated?
ALTERNATIVES:
 read_input_inline
 have_doc_buffer
 make_do_line_ab-
straction

make_do_line_abstraction =
POSITION
SELECTED
SUMMARY:
 Make a do_line procedural
 abstraction to process each line.
RESPONSE–TO:
 read_inputs_encapsulation_level
ARTIFACT:
 do_line

problem_orientation = ARGUMENT
CONCERNING:
 make_do_line_abstraction
SUMMARY:
 1. The problem is line–oriented (lines
are
 either text or commands)
 2. Reading the entire document into a
 buffer contravenes one of formatter's
 efficiency constraints.

content depends on the type of node it is. Some examples are given in Figure 3. More are given in Potts and Bruns (1988).

We have implemented a hypertext of the Liskov and Guttag formatter design history in Planetext (see Conklin, 1987).

2.5. Discussion of First Extension to IBIS

The ontology of Figure 1 is a good start, because it meets several of the needs set out previously: The issue-based model is used at a level of detail that captures the method's design strategy and tactics; and the choices made by the designer are clearly connected to the evolving set of artifacts.

However, there are some shortcomings in the ontology and its application to the design episode described earlier. First, what are tasks doing in the ontology? Liskov and Guttag (1986) emphasized the role played by "helper abstractions" (e.g., do_line and doc) in elaborating a design component (viz., formatter), and to capture their heuristics, it is necessary to be able to talk about tasks. But is it necessary to reify them? A more thorough analysis of Liskov and Guttag's method reveals several other significant concepts (e.g., exceptions) that also appear in their heuristics. Are these to be reified too? In that case, the ontology would become rather crowded and too method-specific. It would be better to represent tasks and these other concepts as types of issues that are raised or triggered as the result of representative design steps or analyses.

Nevertheless, the ontology is still fairly simple. The same cannot be said of its instantiation in a design history (Figure 2). The connection between the deliberation process and the resulting artifacts is present but camouflaged by a tangle of details. What the current ontology lacks is a way of encapsulating recurrent patterns of deliberation in the same way that artifacts encapsulate the results of many decisions.

The need for encapsulated or summarized records of rationale was made clear when Jeff Conklin and I analyzed a much larger design, an elevator control system, in the design method JSD (see next section for an improved treatment). We omitted tasks from the ontology, because they are specific to Liskov and Guttag's (1986) approach; instead we linked issues directly with the artifacts whose analysis or scrutiny triggered them. Instead of using Planetext, we covered a wall (literally) with typewritten nodes and hand-drawn links. With this volume of data, it becomes difficult to find information, change anything, or to see the forest for the trees.

3. SOFTWARE DESIGN METHODS AND RECURRENT DECISIONS

3.1. The JSD Method

JSD (Jackson, 1983) is a systems design method that has been very influential in the United Kingdom and Scandinavia, especially for conventional data-processing applications, but also for embedded real-time systems. It is not possible in this chapter to give a detailed description of JSD, but the following summary imparts enough information about JSD for the reader to follow the principles being demonstrated by the example. I describe only the early, preimplementation phases of JSD.

A JSD project consists of a sequence of steps, integrated by the use of a common set of representations and informal transformations. In the early steps of JSD great significance is placed on modeling the system's environment. Jackson (1983) argued that the nature of the environment is more fundamental to the structure of the system than is a functional architecture. It changes more slowly and less radically during the lifetime of the system, and therefore forms a more stable basis. Much of the early decision making in a JSD project concerns whether entities or actions that occur in the real world of the system need to be modeled by the system. JSD provides a set of criteria and heuristics to aid in this conceptual modeling process.

By the end of the *entity-action step*, the designer has identified the principal entities of the system's environment, their attributes, and the actions in which they are involved. In the next step, *entity-structuring*, the life cycle of each entity is described as a regular expression of action occurrences. Structure diagrams are used to represent an entity's life cycle. At the *initial modeling step*, attention shifts away from the environment toward the system. Each entity in the environment is assumed to communicate with a monitoring process in the system. Monitoring processes have an algorithmic structure that is derived from the structure of the entity they monitor, but with communication operations added. Thus the real entity and the monitoring process can be thought of as communicating sequential processes. They are defined using conventional imperative control constructs and communicate by means of one of two mechanisms: buffered data streams, or state vector inspection.

In keeping with the downplaying of the significance of functional architecture, JSD includes no formal requirements document. Instead, system outputs are listed and briefly described in a *command list*. The functions responsible for generating these outputs (e.g., actuating a motor or generating a summary report) are of two basic types: those that can be embedded inside existing monitoring processes (these tend to be simple online status reporting functions) or those that require additional processes to merge the outputs of other processes and perform computations. JSD provides some analytic techniques for assessing whether a function can be embedded in a given process or whether a new communicating process is necessary.

As the *function step* continues, the initial model becomes overlaid by additional function processes, their outputs and interactions. The network of resulting processes and communication channels is represented in a system specification diagram (SSD).

3.2. An Example of JSD in Use

As with the Liskov and Guttag (1986) method, I illustrate JSD with an example provided by the author of the method. In his book, Jackson (1983) gave three detailed, worked examples, one of which is an elevator control system. One design episode from the evolution of the elevator system design suffices to convey the spirit of JSD.

Figure 4. **A collection of artifacts from the elevator system design.**

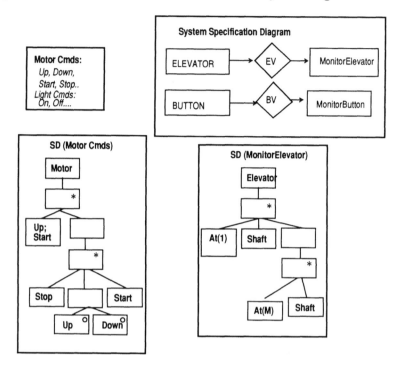

Figure 4 shows a set of JSD artifacts from the elevator system: a command list, which lists and describes system outputs (the descriptions have been omitted from the figure); a structure diagram for the motor commands that specifies the order in which they can occur, and one for the behavior of the MonitorElevator process; and an SSD, which shows how the monitoring processes are connected to the real-world entities they track.

None of these artifacts explains how or when the system outputs (the motor commands) are generated. This issue is similar to the question asked of the tasks performed by the text formatter: Namely, which abstraction should be responsible for each task? In JSD, the analogous question is: Which process is responsible for producing each system output? There are only three possible answers: Either the MonitorElevator process is responsible, or the MonitorButton process, or a new process that does not yet appear in the design. That a new process is necessary is apparent by a process of elimination: Controlling the elevator involves knowing when and where to send it, and this requires in turn knowing where the elevator is at any time and what requests have been made by passengers pushing buttons. Neither the MonitorButton nor the MonitorElevator process knows enough: A third process that collates their information is required. (Note, however,

Figure 5. Ontology for recurrent methodological issues. IBIS is on the left; extensions on the right. Method-specific argumentation is implemented as subtypes of issues, positions, and arguments.

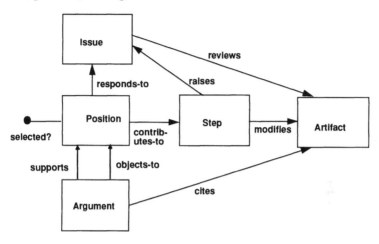

that additional processes are not always necessary: E.g., to illuminate the lights associated with buttons it is necessary only to know that they have been pressed, and this may be done directly by the `MonitorButton` process.) A more formal way of describing the same chain of reasoning is to analyze the temporal structure of the motor commands and the process structures of the candidate processes to find out whether they are compatible (i.e., whether one structure can be superimposed on the other).

As a result of this method-specific reasoning, the SSD is revised to include a new `ElevatorControl` process, which merges the elevator and button data. So far, we do not yet know what structure this process will have, only that it should be in the design. For further elaboration of this episode, see Jackson (1983) and Potts (1989).

3.3. Extensions to IBIS to Capture Recurrent Methodological Issues

The ontology to capture recurrent methodological issues is somewhat different from the first extension to IBIS. As can be seen from Figure 5, there is no method-specific submodel. At the bottom of the diagram are the IBIS objects and relations, and at the top are artifacts and steps (elements of design process that create, revise, derive, or analyze artifacts). In particular, two of the problems with the earlier ontology have been addressed. First, there are no method-specific objects in the model: These are accommodated by defining subtypes of the objects in the model. And deliberation can now be encapsulated and viewed independently of the design history: The design

Figure 6. Part of the artifact, step, and IBIS taxonomy for JSD.

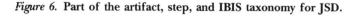

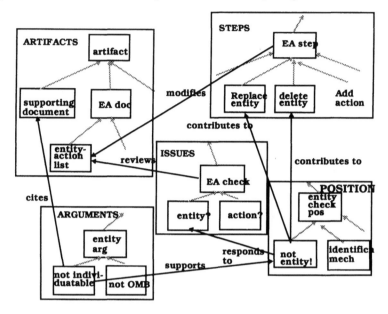

history is a network of evolving artifacts and the steps that mediate changes; rationale "hangs off" of the steps and artifacts.

Method-specific artifacts are simply subtypes of artifact. Method-specific stages, refinements, and analyses are subtypes of step. Method-specific heuristics are captured by subtypes of issues, positions, and arguments. Figure 6 shows a small portion of the taxonomy that relates to entity identification. Because JSD is an "outside-in" method, the ultimate quality of the design rests heavily on the appropriateness of the analysis of the system's environment and subject matter (see Potts, 1989, for an illustration regarding the identification of entities in the elevator control system). This is why the issues include the type Entity?, which queries whether a given candidate entity should be modeled by the designer and play a part in the design. (In the elevator system, e.g., are passengers entities? How about motors?) Jackson (1983) provided many criteria for arguing in favor of or against including a candidate entity in the model, among which, in our terminology, are NotIndividuatable (i.e., individual instances of the entity cannot be identified by the system; this argument contributes to ruling passenger as an entity in most elevator systems), and NotOMB (i.e., not outside the model boundary, an argument that defends inclusion of entities that, according to the designer's understanding of the problem, are within the narrow subject matter of the system; this would defend including

Figure 7. **Analysis of JSD design episode.**

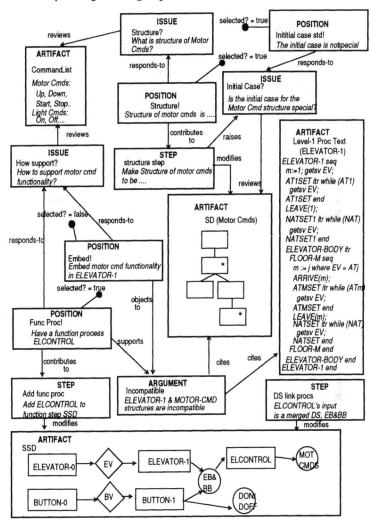

passengers in the elevator system, but rule out the maintenance engineer if the system were not planned to support maintenance functions).

3.4. A JSD Design Episode Analyzed

Let us now go through the example again, this time paying attention to the *types* of artifacts, design steps and issues, positions, and arguments, and how these are connected in a rational design episode (see Figure 7).

In JSD, whenever some temporally related commands or data items are identified the designer should ascertain the temporal structure of the items. For example, we know that there are a number of motor commands that the system can issue to the motor, but are there any constraints on the ordering of these events? For example, is it possible to go down and then go up, without first stopping? These kinds of temporal constraints are at the heart of JSD, because the ordering of inputs and outputs is reflected directly in the structure, or pseudocode, of the processes that produce and consume them. Structuring issues, and the contexts in which they are raised, are therefore an important component of the JSD design strategy. The issue of type Structure? that reviews the command list (see top of Figure 7) represents this question. Its resolution leads to the derivation of the SD(MotorCmds) structure diagram artifact (in the center of Figure 7).

Another kind of structuring issue, but this time of more tactical significance, is to ask whether a proposed temporal structure (of a process or set of data items) is a "white lie" that is true of the general case, but does not account for start-up conditions (see top right of Figure 7). Exploring this kind of InitialCase? issue results in a more detailed specification of the initialization of a process. It is a type of issue is automatically triggered by the step that creates the structure diagram: Whenever a structure diagram is created it should (not necessarily immediately) be reviewed to see whether it correctly describes the entire life history of the entity or data stream it represents, or whether it erroneously describes only the steady state once the initial conditions have passed. In the example, no refinement of the SD(MotorCmds) structure diagram is necessary.

Returning to the command list, another issue that is addressed is how the motor command output functionality is to be supported by the system. This is a standard issue of type HowSupport? (see top left of diagram). There are two standard alternatives: Embed the functionality in a monitoring process (position type Embed!, the best candidate process being MonitorElevator) or add a function process (position type FuncProc!, with the process called, in this case, ElevatorControl). An argument in favor of creating the function process ElevatorControl and against embedding the motor control inside MonitorElevator is that the structure of the monitoring process, and the output data stream structure, MotorCmds are incompatible. That is, one structure cannot be superimposed on the other. Given that ElevatorControl is necessary, the SSD is updated to include it.

3.5. Discussion of Second Extension to IBIS

The second refinement to IBIS has better claim to being a generic model for representing design methods, the title of the article in which it was described, because method-specific details are treated by specialization and instantiation, rather than being embedded in the ontology itself. However, it has practical problems.

First, although the ontology lends itself to automated support, and we have implemented GERM (Bruns, 1988) networks capturing its use, there is no built-in connection between the treatment of artifacts or steps and any existing design tool (e.g., a JSD support tool). Thus, although it is possible to discuss and analyze an existing design history *in vitro*, the ontology itself does not support design decision making by a designer *in vivo*. For that to happen, it would be required to integrate the technology that supported the ontology with design automation tools (cf. Fischer, Lemke, McCall, & Morch, 1991 [chapter 9 in this book]). Given the current lack of integration standards in software technology, it is not feasible to do this in a generic way. However, the next section describes an experimental system that does this for distributed systems designs. First, though, it is necessary to say something about the role of simulation and analysis in design and why these must be tightly integrated with design decision making and design synthesis.

4. INTERTWINING OF SYNTHESIS AND ANALYSIS

The behavior of distributed systems, which are notoriously difficult to reason about informally, often requires formal analysis or simulation. The most common and problematic design flaw in a distributed system is deadlock. In a deadlock, two or more processes reach states where each is waiting for the others to complete some action. Because of the deadlock, none of the effected processes can proceed, and the system will "hang," or become inexplicably dormant. Deadlocks are obvious when they occur, but their presence is not easily detected in a design.

Deadlock is a qualitative property. Other important distributed systems properties are quantitative; for example, average response time, rate of transaction throughput, and the size of any information bottlenecks. These, too, are difficult to estimate: Analysis or simulation is usually essential.

Thus what constitutes "good" or "better" in the design space of distributed systems can require a careful analysis or simulation of the design in addition to the application of informal heuristics. No popular design methods for distributed systems have yet appeared. There is, of course, a large research and tutorial literature on distributed systems, but it focuses primarily on system architectures, implementation techniques, and languages. The strategies and tactics of possible distributed systems design methods are there, but they lie buried in this literature.

During the 1980s at MCC, we explored a design method that more closely intertwines analysis and synthesis in distributed system design. One of the objectives of this work was to develop a well-founded approach to refinement of distributed systems designs and methodological guidelines for making

Figure 8. A VERDI design refinement from a three-party to a two-party interaction.

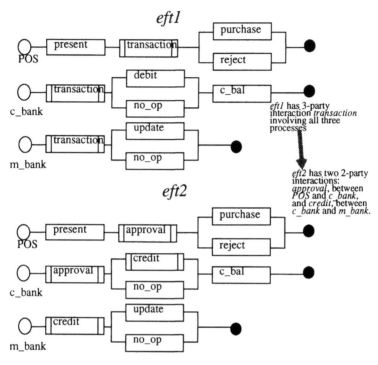

refinement choices. Most of this work used the visual design language and environment VERDI (Shen et al., 1990). This section addresses the relationship of design rationale, and methodological guidelines for refining and evaluating distributed system design using VERDI. No knowledge of VERDI is assumed.

4.1. The Visual Design Language VERDI and Its Method

VERDI (Shen et al., 1990) is the name of both an executable visual language for designing distributed systems and a simulation tool for executing those designs. VERDI can be used to validate qualitative behavior (e.g., absence of deadlock) and quantitative properties (i.e., performance).

A VERDI design consists of teams of interacting processes or procedures. For example, the two simplified designs of an electronic funds transfer (EFT) system shown in Figure 8 comprise a point of sales terminal (POS) process, a customer bank (c_bank) process, and a merchant bank (m_bank) process. These three processes have their own control structures; for example, in design eft2, the c_bank repeatedly cycles through the following behavior: It interacts with POS to approve or reject the proposed transaction (approval); it then either does nothing (no_op) or credits the merchant's account (credit), depending on whether the transaction was approved; and finally archives the transaction and customer's balance to an audit trail (c_bal). Note that interactions are shown by boxes with double bars in more than one process. (The approval interaction occurs jointly in POS and c_bank.) Local operations are shown as simple rectangles. Choices, such as that between no_op and debit, are shown as forks in the control flow. Without going into the semantics of VERDI interactions, that is all the reader needs to know at this point to read VERDI diagrams.

In VERDI, it is possible to describe a complex system at many levels of abstraction. The model of the EFT system in Figure 8 is very simple and easy to understand. However, it does not describe how EFT systems really work. The simplified description requires the POS and c_bank (and later the POS and m_bank) to synchronize unnecessarily. In a real system, that would lead to degraded performance. To evolve an abstract description of a system into a more efficiently implementable, albeit more complex design, the VERDI designer can use a number of well-defined refinement rules. For example:

1. Multiparty interactions can be thinned into sequences of two-party interactions. For example, eft2 is derived by thinning a three-party interaction (transaction) in eft1. A similar refinement replaces symmetrical interactions, in which several processes send information to each other in a single interaction by several asymmetrical send/reply pairs of interactions. These types of refinement reduce unnecessary waiting in some processes.
2. Processes may be introduced as intermediaries between others. For example, in a real EFT system, the POS does not interact directly with banks but with a switch process, similar to a telephone switch, and it is the switch that interacts with the banks.
3. Synchronous interactions, which are easy to understand and reason about, must be replaced eventually by asynchronous message passing, which is more easily implemented in modern networks but more difficult to reason about.

Each time such a refinement is made, the resulting design can be compared qualitatively or quantitatively with alternative refinements. For example, it is common to introduce a deadlock inadvertently when weakening the synchrony of several processes, so this must be checked for carefully. And introducing an intermediary process may or may not reduce bottlenecks, depending on assumptions about transaction rates. It may therefore be necessary to simulate the design to check its performance.

Using VERDI it is possible to execute or simulate a design visually. A process is animated by showing a colored token moving along its control

Figure 9. An alternative refinement. Now POS and m_bank communicate directly.

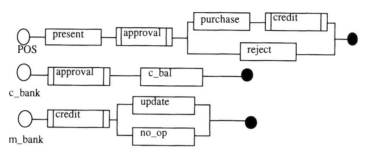

skeleton. Changes in color denote the state of each process (e.g., whether it is blocked, waiting to interact with another). When an interaction occurs, this too is shown visually by a "lightning bolt" between the interacting processes. A design may also be instrumented with delays, probabilities, and data collection code. Data collected during a performance simulation are displayed in real time and can be copied to a file.

An example of VERDI refinement is shown in Figure 8 and an alternative is shown in Figure 9. Designs eft2 and eft2a differ in how the original three-party transaction interaction has been thinned. Both designs protect the customer's privacy better than does eft1, in which the customer's balance was made available to POS and m_bank as well as to c_bank. In both refinements that information remains in c_bank. The two alternatives also have better performance and reliability than eft1, because there is less synchronization. Where they differ is in how m_bank learns it is to credit the merchant's account (i.e., the credit interaction). In eft2, it is informed of the credit by c_bank, whereas in eft2a, it is informed by POS. These alternatives differ by several criteria. Performance is likely to be superior in eft2a, because a POS terminal is installed in a store, and is therefore associated with a specific m_bank. This built-in connection can be exploited by a dedicated telephone line or a network connection that does not involve a switching overhead, whereas the connections between banks (any arbitrary m_bank, c_bank pair) must be switched. VERDI can be used to evaluate these performance predictions by instrumenting the design, running it, and collecting statistical data.

4.2. A Third Extension to IBIS: Design Analysis and Evaluation

The technical details of VERDI refinements are very different from the abstraction-based refinements of the Liskov and Guttag (1986) method, but if one stands back, there are remarkable similarities in the underlying

Figure 10. **Ontology for representing VERDI analysis and refinement and design rationale. IBIS is on the left; VERDI-specific information on the right.**

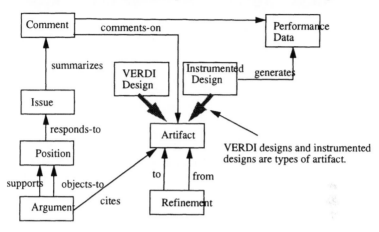

principles. In both cases, the designer creates artifacts that are method-specific in their form. The artifacts specify some features of the system: In the case of Liskov and Guttag, the artifacts specify procedures and data abstractions, whereas in VERDI an artifact specifies the interactions among concurrent, distributed processes. In both cases, the design is refined so that artifacts are supplemented or replaced by other artifacts: In the case of Liskov and Guttag this is done by identifying "helper" abstractions that encapsulate tasks; in the case of VERDI it is done by refining process interactions in such a way that the resulting design is closer to a feasible and efficient distributed system. Finally, in both cases there is choice and evaluation: In the case of Liskov and Guttag, abstractions may be selected and elaborated on grounds of ease of subsequent modification; in the case of VERDI, alternative refinements are chosen depending on their qualitative behavior and quantitative performance properties.

For an experimental VERDI Information Management environment (VIM), I developed the schema in Figure 10. In this data model there are two submodels: the IBIS component and a method-specific component. There are several differences from Figure 1, however:

1. The method-specific submodel contains VERDI-specific objects instead of abstraction-based objects (of course).
2. Refinements are reified as first-class objects, rather than relationships between artifacts. This enables information to be associated with a refinement, such as when it was performed, who did it, and so on.

Figure 11. **VERDI design history and superimposed design.**

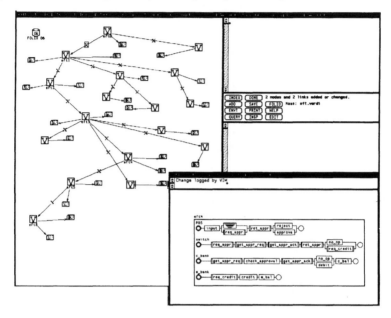

3. The deliberation submodel contains unstructured comment nodes, in addition to IBIS nodes.
4. The relationship between the two submodels is simpler, in that commentary is associated directly with artifacts or performance data, rather than mediating between an artifact and its refinement. This allows commentary to be represented that does not refer directly to a refinement step.

Figure 11 shows VIM in use. This shows a fragment of a design history and commentary using two automated tools: VERDI and GERM. GERM (Bruns, 1988) is a visual browser for entity-relationship databases. It is most often used as a node-to-node hypertext system—for example, it is the implementation substrate for gIBIS (Conklin & Begeman, 1989). It therefore serves the same function in this example that Planetext did in the Liskov and Guttag text formatter example of Potts and Bruns (1988). The "V" nodes represent VERDI designs. The node labeled "eft4" has been selected for closer scrutiny, and its contents, a collection of four interacting processes, are shown in the overlapping VERDI window. The "x" nodes represent refinements or instrumentation steps. And the small rectangular nodes represent either performance data derived from executing a design, informal commentary, or aggregated IBIS rationale.

4.3. Discussion of Third Extension to IBIS

The third ontology meets its goals in that it intertwines design synthesis and decision making with analysis in the form of simulation. It also relaxes the ontology to support informal commentary in addition to IBIS networks. In doing so, it reflects the intuition that method-specific IBIS networks, unstructured annotations and analysis, or simulation results are types of rationale that are useful for different purposes.

It is clear from using VIM, however, that a much tighter integration is necessary between the design technology (in this case, VERDI) and the decision support mechanism (in this case, GERM). The artifacts themselves (i.e., static VERDI diagrams and their animations) should "talk back" (Schön, 1983) more to the designer, rather than requiring the designer to switch attention from one tool to another and record the rationale manually. For example, when a deadlock is detected, a special kind of issue could be automatically raised for consideration by the designer either immediately or later. Gruber and Russell (1995 [chapter 11 in this book]) call such rationale recording *generative creation* of rationale. They explore the possibilities of utilizing device knowledge for generative creation of rationale in electronic design.

5. DISCUSSION

5.1. Implications

An issue-based analysis of different methods provides a comparatively objective basis for comparing software design methods quantitatively and qualitatively. One can ask several questions: Which methods have the most heuristics? Which have the closest correspondence between low-level, tactical commitments and high-level, strategic phases of analysis and design? Where, in a given method, does the practitioner have the most leeway, and where are decisions constrained by preordained alternatives? How rich is the correspondence between a method's tactics and strategy and its artifacts and evolutionary steps? These questions address the essence of software design methods—the degree to which they really help designers think.

Going further than analyzing existing methods, a principled design-theoretic approach should also be useful in the development of new methods, for example, by pointing out any gaps in the method. Unlike "process programming" (Osterweil, 1987), in which design processes are described in a programminglike formalism, an issue-based formulation avoids premature serialization into a sequence of stages or phases. Undoubtedly, some issues are considered before others, and some artifacts are produced before others (e.g., specifications before code), but these precedence relations arise from a network of design tasks and decisions and not from an a priori program.

One reason why new methods are needed is that software development is becoming increasingly domain-specific; and general-purpose design methods are increasingly being viewed as too vague and general. In some domains, such as some embedded real-time control applications and some financial applications, systems have standard architectures drawn from a limited repertoire of alternatives. As these domain-specific architectures and design options become increasingly codified, so the needs for methods that take account of domain specific details will increase.

Finally, we are interested in supporting the design process as a *process.* IBIS-based tools have been developed and applied to practical software design projects (Conklin & Burgess-Yakemovic, 1991 [chapter 14 in this book]; Burgess-Yakemovic & Conklin, 1990) but it is important that the deliberation components of design and the recording of rationale for decisions is not separated conceptually from working on the artifact itself. Fischer, Lemke, Mastaglio, and Morch (1991) and Fischer, Lemke, McCall, and Morch (1991 [chapter 9 in this book]) use the idea from Schön (1983) of the artifact "talking back" to the designer. If this tight loop from artifact to advice and back to artifact is to be supported by tools, it is essential that the deliberation and artifact components are more tightly integrated. The method-specific extensions to IBIS described in this chapter are a step in that direction.

5.2. Simplicity Versus Richness

Our experience has shown that IBIS is too simple and homogeneous a model to support decision making in the presence of change. IBIS networks tend to ossify and become impossible to revise or extend as soon as the design process moves beyond a certain stage. If the IBIS network does not refer to the issues inherent in the current design, as opposed to the design that was current when the issue base was started, then designers tend to work on the design and not maintain the now superseded issue base. Because the issue base consists of an amorphous web of only three types of object, distinguishable only by name or inspection of their content, it is impractical to change the structure of the network if the problem needs to be looked at in a new way.

One answer is to elaborate the model and use the extra categories, subtypes, and connections as semantic cues. This was the first way in which we extended IBIS (see Figure 1). Lee (1991; Lee & Lai, 1991 [chapter 2 in this book]) presents an even richer extension to IBIS. This extension strategy avoids some expressive shortcomings in IBIS and our first ontology. However, the richer representation of design rationale may become more difficult to understand as the design evolves. It remains to be seen how complex an ontology can be understood by users. To some extent, the complexity can probably be disguised by good interface decisions, but we know from using plain IBIS (Conklin & Burgess-Yakemovic, 1991 [chapter 14 in this book]; Burgess-Yakemovic & Conklin, 1990) that users cannot cope with too rich a set of node and link types.

Another approach is to keep IBIS as it is, but refine the categories in a context-specific way (e.g., specific to a design method) and add extra subschemas that refer to other aspects of the design process (e.g., artifact structure or design steps) (see Figure 5). An advantage to this second approach is that it is modular. It is possible for different users to view the design space according to different views. At one stage of the design, for example, the only artifact may be a list of desired features that the system should have. At that stage it is hardly necessary to go beyond the IBIS subschema. Later on, when there are many different artifacts with rich interconnections (specifications, detailed design documents, test plans, etc.) it will be necessary to use an explicit model of the artifacts and their interrelationships.

Another advantage of modularity is that it appears from our experience that users must understand the ontology underlying the model very quickly and be able to apply it to the higher levels of abstraction of their problem very quickly. In raw IBIS, for example, it is not always obvious to a naive user how to formulate his or her problem in issues, positions, and arguments. By keeping these extra constructs in separate subschemas, however, introducing them only when necessary, and only to users who need to consider them (e.g., a user who must understand the relationship between the resolution of issues and the project's deadlines or to quality assurance issues) the richness of the composite model can be kept manageable and easy to assimilate.

The third approach (see Figure 10) loosens the framework still further by allowing the designer to take shortcuts and not have to categorize all discussion into issues, positions, and arguments. We are currently exploring this idea in the context of software requirements specifications (Potts & Takahashi, in press).

5.3. Bird's-Eye View Versus Turtle's-Eye View

The software designer's job is to produce design artifacts and ultimately executable code, *not* primarily to produce a graphical network representing the history of design deliberations on a project—although see MacLean, Young, Bellotti, and Moran (1991 [chapter 3 in this book]) for an argument for coproducing artifacts and design rationale in human interface design. Although the integration of design artifacts and rationale presented in this chapter captures the rationale for design decisions, the graphs of the design histories (see Figures 2, 7, and 11) are views that the designer should rarely if ever need to inspect or work with. Such views can be regarded as *bird's-eye views*, because they look downward on an entire history of design argumentation and decision making from an abstracted and detached viewpoint.

An alternative way to view design rationale is *outward* from the context of an artifact that is currently under review. It is natural for a designer to raise issues about an artifact that is the current focus of design attention. This outward-focused type of view, we can call a *turtle's-eye view*, by analogy with turtle geometry. In such a view, the designer is always working with one or more artifacts (see Figures 3, 4, 8, and 9) and can create issues, positions, or arguments by annotating the artifact or annotating other annotations. The network of interlinked deliberation nodes is implicit in this view and must be inferred by navigating from node to node. In our current work, we are extending the second and third IBIS-based models described earlier but with tool support that is artifact centered rather than browser oriented (Potts & Takahashi, in press).

Bird's-eye and turtle's-eye views each have their advantages. We believe, however, that an artifact-centered or turtle's-eye view is much more useful in the practical work of designing.

NOTES

Author's Present Address.   Georgia Institute of Technology, College of Computing, Atlanta, GA 30332-0280. Email: potts@cc.gatech.edu

REFERENCES

Berzins, V., & Luqi (1991). *Software engineering with abstractions.* Reading, MA: Addison-Wesley.

Birrell, N. D., & Ould, M. A. (1985). *A practical handbook for software development.* Cambridge, England: Cambridge University Press.

Bruns, G. (1988). *GERM: A metasystem for browsing and editing* (Tech. Rep. No. STP-122-88). Austin, TX: Microelectronics and Computer Technology Corporation.

Burgess-Yakemovic, KC, & Conklin, E. J. (1990). Report on a development project use of an issue-based information system. *Proceedings of the Conference on Computer-Supported Cooperative Work,* 105–118. New York: ACM.

Conklin, E. J. (1987). Hypertext: An introduction and survey. *IEEE Computer, 20,* 17–41.

Conklin, E. J., & Begeman, M. (1989). gIBIS: A tool for all reasons. *Journal of the American Society of Information Science, 40,* 200–213.

Conklin, E. J., & Burgess-Yakemovic, KC. (1991). A process-oriented approach to design rationale. *Human–Computer Interaction, 6,* 357–391. Also in T. P. Moran & J. M. Carroll (Eds.), *Design rationale: Concepts, techniques, and use.* Hillsdale, NJ: Lawrence Erlbaum Associates, 1996. [Chapter 14 in this book.]

Fischer, G., Lemke, A. C., Mastaglio, T., & Morch, A. I. (1991). The role of critiquing in cooperative problem solving. *ACM Transactions on Information Systems, 9,* 123–151.

Fischer, G., Lemke, A. C., McCall, R., & Morch, A. I. (1991). Making argumentation serve design. *Human–Computer Interaction, 6,* 393–419. Also in T. P. Moran & J. M. Carroll (Eds.), *Design rationale: Concepts, techniques, and use.* Hillsdale, NJ: Lawrence Erlbaum Associates, 1996. [Chapter 9 in this book.]

Gruber, T. R., & Russell, D. M. (1995). Generative design rationale: Beyond the record and replay paradigm. In T. P. Moran & J. M. Carroll (Eds.), *Design rationale: Concepts, techniques, and use.* Hillsdale, NJ: Lawrence Erlbaum Associates. [Chapter 11 in this book.]

Jackson, M. A. (1983). *System development.* Englewood Cliffs, NJ: Prentice-Hall.

Lee, J. (1991). Extending the Potts and Bruns model for recording design rationale. *Proceedings of the 13th International Conference on Software Engineering,* 114–125. Los Alamitos, CA: IEEE Computer Society Press.

Lee, J., & Lai, K-Y. (1991). What's in design rationale? *Human–Computer Interaction, 6,* 251–280. Also in T. P. Moran & J. M. Carroll (Eds.), *Design rationale: Concepts, techniques, and use.* Hillsdale, NJ: Lawrence Erlbaum Associates, 1996. [Chapter 2 in this book.]

Liskov, B., & Guttag, J. (1986). *Abstraction and specification in software development.* Cambridge, MA: MIT Press.

MacLean, A., Young, R., Bellotti, V., & Moran, T. (1991). Questions, options and criteria: Elements of design space analysis. *Human–Computer Interaction, 6,* 201–250. Also in T. P. Moran & J. M. Carroll (Eds.), *Design rationale: Concepts, techniques, and use.* Hillsdale, NJ: Lawrence Erlbaum Associates, 1996. [Chapter 3 in this book.]

Osterweil, L. (1987). Software processes are software too. *Proceedings of the 9th International Conference on Software Engineering,* 2–13. Los Alamitos, CA: IEEE Computer Society Press.

Potts, C. (1989). A generic model for representing design methods. *Proceedings of the 11th International Conference on Software Engineering,* 217–226. Los Alamitos, CA: IEEE Computer Society Press.

Potts, C., & Bruns, G. (1988). Recording the reasons for design decisions. *Proceedings of the 10th International Conference on Software Engineering,* 418–427. Los Alamitos, CA: IEEE Computer Society Press.

Potts, C., & Takahashi, K. (in press). An active hypertext model for system requirements. *Proceedings of the 7th International Workshop on Software Specification and Design.* Los Alamitos, CA: IEEE Computer Society Press.

Rittel, H. (1984). Second-generation design methods. In N. Cross (Ed.), *Developments in design methodology* (pp. 317–327). New York: Wiley.

Schön, D. A. (1983). *The reflective practitioner: How professionals think in action.* New York: Basic Books.

Shen, V. Y., Richter, C., Graf, M. L., & Brumfield, J. A. (1990). VERDI: A visual environment for designing distributed systems. *Journal of Parallel and Distributed Computing, 9,* 128–137.

11

Generative Design Rationale: Beyond the Record and Replay Paradigm

Thomas R. Gruber
Stanford University

Daniel M. Russell
Xerox Palo Alto Research Center

ABSTRACT

Research in design rationale support must confront the fundamental questions of what kinds of design rationale information should be captured, and how rationales can be used to support engineering practice. This chapter examines the kinds of information used in design rationale explanations, relating them to the kinds of computational services that can be provided. Implications for the design of software tools for design rationale support are given. The analysis predicts that the "record and replay" paradigm of structured note-taking tools (electronic notebooks, deliberation notes, decision histories) may be inadequate to the task. Instead, we argue for a generative approach in which design rationale explanations are constructed, in response to information requests, from background knowledge and information captured during design. Support services based on the generative paradigm, such as

Thomas Gruber is a computer scientist interested in creating a collaborative knowledge medium; formerly a research associate at the Stanford University Knowledge Systems Laboratory, he is now chief technology officer for Colloquy Systems. **Daniel Russell**, a computer scientist with diverse interests, is the manager of the User Experience Research program for Apple Computer's Advanced Technology Group; previously he was at the Xerox Palo Alto Research Center, where this research was conducted.

CONTENTS

---

design dependency management and rationale by demonstration, will require more formal integration between the rationale knowledge capture tools and existing engineering software.

---

1. INTRODUCTION

When a product is designed, the primary output of the design process is a specification of the artifact, such as annotated CAD drawing or a circuit schematic. Additional information, about why the artifact is designed the way it is, is not captured in the artifact specification. Research in design rationale support is concerned with how this additional information might be effectively captured and used.

A design rationale is an explanation of why an artifact, or some part of it, is designed as it is. Such an explanation can include several kinds of information. A rationale may explicate tacit assumptions, such as the expected operating conditions for a device and its intended behavior under those conditions. It may clarify dependencies and constraints among design parameters, such as the set of modules that might need to be changed if a given module is replaced. A design rationale may also justify or validate a design, explaining why particular structures are chosen over alternatives, or how required functionality is achieved by device behavior.

Knowledge of the rationale for a design is needed for engineering tasks throughout the product life cycle, including the refinement of early conceptual designs, the realization of design specifications through manufacturing, and reengineering under changing requirements. Design rationale information can help engineers reuse, modify, and maintain existing designs. Unfortunately, the design rationale for complex artifacts mainly resides in the heads of the original designers, who are often not available or cannot remember the rationale. Existing methods of documentation do not adequately capture the information needed.

A technology for software support of design rationale is just beginning to be developed. A range of approaches has been proposed, mostly at the laboratory stage. Many of the approaches are represented in this book. Gruber, Boose, Baudin, and Weber (1991) offered a preliminary analysis of the design space for support tools. To evaluate the applicability and utility of these techniques, and chart the future direction of work in this area, we must address two fundamental questions: What kind of information should be captured, and how can that which is captured be used to support engineering practice. This chapter examines these two questions in depth, and offers prescriptive implications for the design of rationale support tools.

2. WHAT KIND OF DESIGN INFORMATION SHOULD BE CAPTURED TO SUPPORT RATIONALE?

To address the question of what information to capture for design rationale, we start by asking what kind of information is available and used in existing design practice. It would be of little use to build technology to represent and reason with knowledge that cannot be acquired, or is of little benefit to the information needs of engineers. To address this issue, we turn to data on the information that is available and used by designers when explaining the rationale for designs.

In a survey of design protocol studies (Gruber & Russell, 1992), we looked at how designs are explained in documents and live discussions. We examined the data from a set of empirical studies of designers requesting, communicating, and using design information. The data include protocols of individual designers thinking aloud during innovative design, teams of designers discussing a previous design during redesign, designers asking for information during an initial design, and designers negotiating with stakeholders about a design in progress. The studies covered designs in several domains: architecture (P. Pirolli & V. Goel, personal communication, December 1989), electromechanical devices (Baudin, Gevins, Baya, & Mabogunje, 1992; Baudin, Gevins, Baya, Mabogunje, & Bonnet, 1992; Baya et al., 1992; Kuffner, 1990; Kuffner & Ullman, 1990), heating and air conditioning systems for buildings (Garcia & Howard, 1992; see also A. C. B. Garcia, personal communication, November 1991), user interfaces in software and hardware (Bellotti & MacLean, 1989), and the design of instructional courses (P. Pirolli & V. Goel, personal communication, December 1989).

As an example, consider the following protocol fragment from one of the studies surveyed (Baudin, Gevins, Baya, Mabogunje, & Bonnet, 1992). A mechanical engineer has been given the task of redesigning a computer-controlled automobile shock absorber. The engineer doing the redesign

(R) is asking one of the designers of the original device (D) about the design of the damper mechanism:

R: The existing equipment . . . What is the problem with it?
     Why can't it go up to 750 pounds [resistive force]?
D: There are related problems . . . heat transfer . . . the
     solenoid is . . . exposed to a lot of heat. And solenoid
     is a heat sensitive device, and cannot function above
     a certain temperature. . . From the vendors we got a
     limit of 120 degrees centigrade for the solenoid . . .
     This is a critical point in how high a force you can get.
     . . . If you are dissipating at high force and high
     stroke lengths and high speed, then a lot of heat is
     generated.
R: Did you anticipate that problem?
D: Yes . . . The ideal thing would be for us to have the
     solenoid outside and to get heat convected out by air.
     But once we put it outside . . . we were using up more
     volume than we had, so we chose to put it inside and to
     insulate the solenoid.
R: Okay . . . The heat problem is associated with the so-
     lenoid. If you used a different device, then may not
     . . .
D: We looked at various different kinds of actuators. So-
     lenoid is just an actuator which gives you a certain
     force when passed it a certain current and voltage.
     . . . There's a trade-off that we have so little power
     and we have to generate high force. The solenoid was
     chosen because with less power it can generate high
     force at various small stroke length. . . .
R: If I get a solenoid with very good insulation. That
     would be too expensive[?]
D: No, the best you can get can go up to 175 degrees centi-
     grade.

The designer was given a changed requirement for the device to produce a higher resistive force. In the previous design, driving the device to produce the higher force would have resulted in extra heat. He is trying to understand why the previous damper cannot be run at a higher temperature, and is told that the solenoid would not tolerate the heat that would be generated. In exploring alternatives, he wonders why the previous design did not use a different kind of actuator or a solenoid that is adequately insulated. He hypothesizes incorrectly that such a solenoid might have been too expensive. In fact, even the best insulated solenoid available could not tolerate the extra heat.

We analyzed the questions, conjectures, and statements of designers in settings like the aforementioned, noting the types of rationales requested, generated, and used by designers. Figure 1 summarizes the categories of information requested, in the form of generic questions asked about a

Figure 1. Categories of information requested about designs. These are a set of generic questions one might ask about a design, derived from an analysis of designers talking about designs in the protocol studies surveyed (Gruber & Russell, 1992). Each generic question is derived from segments of design protocols in which a subject requested the information directly, or made a conjecture or statement about the answer to the question in a think-aloud setting. Given the nature of the protocol data, the relative frequency of each question type is not reported (see Kuffner & Ullman, 1990, for such data). Instead, the generic questions characterize the broad range of knowledge used to create a design rationale statement.

---

1. Requirements
   What are the given requirements?
   Is this constraint a requirement?
   Give more detail about this parameter of the operating
       environment.
   Can I assume this fact about the operating environment?
   What are the requirement constraints on this parameter?
   Is this parameter constrained by external requirements?
   What is the expected behavior of this artifact in the scenario
       of use?
   Should I assume that this functionality is required?
   Can I modify this requirement?
2. Structure/Form
   What are the components?
   What class of device or mechanism is this part?
   What is the geometry of this part? (qualitative)
   What material is this part made of?
   How do these components interface?
   What are the locations of parts, connections, etc. (for
       constraint checking)?
   What are the known limitations (strengths) of this
       part/material class?
   What affects the choice of artifact components?
3. Behavior/Operation: What Does It Do?
   What is the behavior of this parameter in the operating condi-
       tions?
   What is the behavioral interaction between these subsystems?
   What is the range of motion of this part?
   What is the cause of this behavior?
   What are the expected failure modes in the scenario of use?
4. Functions
   What is the function of this part in the design?
   What is the function of a feature of a part in the design?
5. Hypotheticals
   What happens if this parameter changes to this new value?
   What is the effect of this hypothetical behavior on this
       parameter?
   Adapt equation to this changed parameter and recompute?
   What will have to change in the design if this parameter
       changes to this new value?

(Continued)

Figure 1. Continued

6. **Dependencies**
   What are the known dependencies among the parts?
   What are the constraints on this parameter?
   Is this parameter critical (involved in a dominant constraint)?
   How is this subassembly related to this parameter?
   What is the source of this constraint?
7. **Constraint Checking**
   Is this constraint satisfied?
   Does this structure have this behavior that violates this
     constraint?
   What are the known problems with this design?
   Would a part with this functionality satisfy this constraint?
8. **Decisions**
   What are the alternative choices for this design parameter?
   What decisions were made related to this parameter?
   What was an earlier version of the design?
   What decisions were made related to satisfying this constraint?
   Which parameter, requirement, constraint, or component should be
     decided first?
   What design choices are freed by a change in this input
     parameter?
   What alternative parts that satisfy this constraint could
     substitute for this part?
   Where did the idea for this design choice come from?
9. **Justifications and Evaluations of Alternatives**
   Why this design parameter value?
   Why is design parameter at value V1 instead of normal value V2?
   Why was this alternative chosen over that alternative?
   What is person P's evaluation of these alternatives?
   Why not try this alternative?
10. **Justifications and Explanations of Functions**
    Why is this function provided?
    Why is this function not provided?
    Why can't the current design achieve this new value of this
      functional requirement parameter?
11. **Validation Explanations**
    How is this requirement satisfied?
    How is this function achieved?
    How is this functional requirement achieved?
    How will this part be maintained?
12. **Computations on Existing Model**
    Compute a parameter value given other parameters.
    What are the trajectories of parameters?
13. **Definitions**
    What does a term in the documentation mean?
14. **Other Design Moves**
    Search for information expected to be in documentation, e.g.,
      equation or diagram.
    Change this requirement constraint and update design.
    Have all the arguments for/against this alternative been
      checked?

design. Each generic question is abstracted from one or more explicit information requests or think-aloud conjectures made by designers as recorded in the protocols. The data indicate the range of questions that are asked about why a design is as it is.

The survey resulted in a set of general observations about the information underlying design rationale. Each observation motivates a conclusion about the information that should be captured to create design rationales like those we see designers using.

Observation 1. Rationales Are Based on Many Kinds of Underlying Information. The questions designers asked about the rationales for existing designs included a broad range of sources and subjects, as listed in the Appendix. The information sources included people, documents, databases, CAD tools, spreadsheets, and notes. The subjects included requirements, constraints, decisions, artifact structure, expected behavior, and intended function.

We found that the scope of design rationale information is very broad—broader than the current literature in design rationale might suggest:

```
D:  No single model of the design process (e.g., design as
    argumentation, design as decision making, design as
    constraint satisfaction) accounted for a dominant
    share of the questions. Furthermore, rarely did the
    language of a rationale explanation in a protocol sug-
    gest a unique process model; the protocol reader must
    reinterpret the natural explanations using the terms
    provided by the design process model (e.g., "issue,"
    "option," "constraint," etc.). . . .
R:  What of these pieces already existed? Was it just,
    what'd I do with them?
D:  Right. Yeah. See, these disks are alternate disks. You
    see, one is friction disk, and the other is a metal disk
    . . . and they're made together to get friction. And the
    concept was used from a motorcycle clutch, although we
    got these manufactured to our dimensions. . . .
```

This fragment of discussion between designers gives an illustration of a nonissue, nonrationale conversation. D begins the answer by giving some background context before answering. Rationales are more like ongoing conversations: They exist in a problem-solving context that may not be directly modeled by design information alone.

Conclusion: Capture the information that is used to answer designers' questions, not the information that fits a preconceived model of the design process. The observation that designers draw on many kinds of information in explaining designs is not surprising. However, it suggests that design rationale support tools should not limit the captured information to that which fits a particular view of the data. Many of the tools described in the current design rationale literature do just that: They are based on models of the design process that prescribe what information is relevant and provide a representation for capturing it. For example, a tool based on a view that design is constraint satisfaction would only capture formal design constraints among mathematical parameters. However, this tool will not support explanations drawing from natural language documents, CAD drawings, and other relevant sources. Similarly, only capturing information that can be labeled as an argument for or against a design decision may preclude other information that is relevant to the rationale in unforeseen ways, such as descriptions of expected artifact behavior.

Observation 2. Rationales Are Constructed and Inferred. Rationale explanations are often *constructed* and *inferred* from information that is stored, rather than being stored as complete answers. For example, even though the original damper design was carefully documented using a design notebook, the rationale for rejecting the insulated solenoid alternative was constructed in response to a hypothetical question ("If I get a solenoid . . . No, the best you can get . . ."). Often the explanations are constructed in response to *new* questions that arise in the context of redesign. ("[Can I] get a solenoid with very good insulation?") Answering these questions requires inference from stored information, rather than a database lookup for the answer.

Although the original designers in the protocol studies did not often record complete explanations, they did record many of the constraint equations and component specifications that were used to construct explanations. For roughly half of the information requests in the protocols, answers were computed or inferred from information that was stored or could have been accessed; the answers to these rationale questions could not be retrieved directly from a design record.

Conclusion: Acquire data rather than answers. If many rationale explanations are inferred from data, it is more important to capture the data that might be used to infer answers to later questions than to try to anticipate the questions and capture preformulated answers. In particular, it is important to identify the parameters of the design that are used in standard inference procedures, such as predicting behaviors from structures, verifying functions, and searching for parts meeting some specifications.

Observation 3. Rationales Are Not Just Statements of Fact, But Explanations About Dependencies Among Facts. Design rationales are often *explanations*, rather than simple statements of fact. An argument consisting of a set of reasons for and against a position is an explanation, whereas the individual facts offered as evidence are not. In the damper example, the explanation of a choice among alternatives is found in the relationships among several pieces of information: a requirement ("we have to generate high force"), the functional role of a component ("solenoid is just an actuator which gives you a certain force"), and a limiting constraint ("This is a critical point in how high a force you can get. . . . If you are dissipating at high force and high stroke lengths and high speed, then a lot of heat is generated."). The rationale explanation is the statement that these particular facts, among all the data available on the design, are combined to explain the decision to choose the solenoid.

Frequently rationale explanations describe *dependencies* among decisions or design parameters. Dependency relations are important for *managing change* in designs. In the protocols surveyed, designers often ask questions or make conjectures about the effects of changes to the design. Here are a few examples from several of the protocols covered by the survey:

"Now if we changed GH, would we have to change anything else in the frame?"

"What will I have to do if I try to extend the force range from 500 to 750?"

"What can I do with more current which could not be done with less current?"

"The other thing that would have to change would be the spring, like I said, you adjust the spring constant if you are damping this."

"Supply chilled water temperature should be changed to 45 degrees F. All coils need to be recalculated at 45 degrees F."

Conclusion: Capture dependency relationships. Answering questions such as those just listed requires knowledge of the dependency relationships among elements of the design (i.e., physical components, model parameters, functional requirements, other requirements, evaluation criteria, and decisions). Thus to support rationale, it would be useful to capture these relationships. It would not be necessary to acquire a formal theory capable of *inferring* the dependencies. Dependency information can be captured as semiformal links (e.g., between the water temperature parameter and the coil parameters). Although these semiformal links represent relationships among design elements that are not formally modeled, the links can be used by tools that perform simple dependency management services. For example, software engineering tools called source management systems are used by program developers and document writers to maintain complex dependencies among modules, yet the source management tool does not know what the modules represent.

Observation 4. Rationale Explanations Refer to Real Engineering Data and Models. In the protocol review study, we examined the source of each piece of information mentioned in designer questions or used in answers. We found that much of the information used in rationale explanations is either available from information sources already available to the practicing engineer, or could be made available if the documentation were brought online. For example, information about the availability of solenoids may be found in manufacturer databases, which are accessed routinely by mechanical engineers. The functional description of the solenoid as a force actuator is an example of standard engineering knowledge found in textbooks and handbooks. In addition, many of the more difficult questions asked about designs could be answered by reference to engineering *models*, such as structure models in CAD databases and mathematical behavior models used for simulation and analysis. Such models are used routinely in traditional engineering disciplines. For instance, equations describing relationships among power, stroke length, force, and temperature were maintained online in a spreadsheet. The damper designer could then point to those equations when explaining the relationship between a component and design constraints. Although formal models are not available in all domains, where models exist they can play integral roles in constructing rationale explanations.

Conclusion: Capture data and models used in engineering practice. Engineering design is supported by software tools and on-line information sources, and will become increasingly so in the future. Because the data and models manipulated by these tools and databases are constituents of rationale explanations, this information could be captured as a by-product of using these tools. For instance, the annotation facilities in CAD drawing tools could be extended to allow for hypertext-style linking into other design information. This sort of data capture is not possible unless rationale information capture tools are integrated with mainstream engineering tools. This conclusion is elaborated in Section 4.3.

Observation 5: Rationales Can Be Reconstructed From the Relevant Data. Although many of the facts that are mentioned in rationale explanations come from formal models and engineering data, justifications for design decisions in the protocols studied were usually informal. Decision rationales were often lists of relevant data of the form, "The following factors were considered: . . ." We call such a statement a *weak explanation.* A strong explanation would show *how* the factors led to the decision. In an automated design system that operates on a complete domain model, a strong explanation can be captured for each design step. For example, in explanation-based learning systems, an "explanation" of a design decision is a trace of the formal reasoning done, such as deductive proof of goal satisfaction (Mitchell, Keller, & Kedar-Cabelli, 1986). More generally, one can think of the trace of a design problem solver as a "derivation" that can be "replayed" to justify existing designs and create solutions for similar design problems (Mostow, 1989).

It is not surprising that most justifications were weak explanations, because the designers in the protocols were speaking or writing in natural language to other humans, rather than writing models or proofs for mechanical verification. In the domains we studied, the design process was not automated, nor were the models complete enough to generate strong explanations for most decisions. Fortunately, human engineers can make use of weak explanations, without the ability to simulate the mental processes of the original designer. For example, it is useful to a designer to know that the original designer of the damper "anticipated that [heat transfer] problem," without knowing exactly what the designer thought about the heat problem.

Conclusion: Capture weak explanations (just the relevant data), when a complete justification for a decision is not available. From the first four observations we conclude that rationale support tools should focus on capturing the variety of data and models that are relevant to designers, from existing sources where possible, and manipulated by existing tools and databases. We found that the dependency relations among these data, models, and other information are especially relevant. The emphasis is on capturing the information, but what about rationale—the explanation? The final observation suggests that it is more important to acquire from the designer the relevant set of facts (to be able to reconstruct a rationale) than it is to assemble them into a coherent argument at the point of capture. For example, instead of asking the designer to document the deliberation about how to produce force without generating too much heat, a capture tool might only require the user to *associate* the requirement (perhaps by selecting a region of text in a document), the alternatives under consideration (solenoids, other force actuators, perhaps by references to a manufacturer's database), and their features that are relevant to the deliberation (force, current, heat). The explanation of how these elements justify a design decision could be left to the reader.

This conclusion is not a call to abandon active support for design rationale. When the information is available, a design rationale tool should be able to help generate the explanation. This idea is explored in Section 4.2.

3. HOW CAN DESIGN RATIONALE SUPPORT ENGINEERING PRACTICE?

The purpose of design rationale support is to improve design quality and the productivity of product development. Schemes for capturing, representing, and operating on design rationale should be evaluated with

Figure 2. Categories of use for information about designs.

- *Clarifying requirements or assumptions* about the operating environment. ("Does [the pivot arm] flip all the way out, or [are there] two positions?"—manufacturing tool design)
- *Formulating a decision among alternatives.* ("[What materials could be used for the case?] ABS . . . Polycarbonate . . ."—battery case design)
- *Understanding the artifact specification itself.* ("Is this [points to drawing] steel or aluminum?"—manufacturing tool design)
- *Explaining the effects of changes* to the design or a requirement. ("What will happen if I try to extend the force range from 500 to 700 pounds?"—damper design)
- *Explaining the expected operation of the device, often in hypothetical situations.* ("I don't know if you expect some receipt of something like that for your transaction . . . for parcels and registered letters."—postal ATM design)
- *Verifying and validating the design.* ("The area over the . . . loading dock is covered. How do you plan to exhaust truck fumes from that area?"—building design)

respect to their impact on engineering practice. In other words, to decide what design rationale information to capture, we should consider how that information might be used. In this section, we examine the tasks that might be supported by design rationale. We start by summarizing how the information was used in the verbal design protocols, and then consider the tasks that might be supported by software tools.

3.1. Uses of Design Rationale Information in the Design Protocols

In our survey of design protocol studies, we classified the designers' questions and answers to identify a set of general categories of information use. Figure 1 summarizes these categories, each representing a way that information about an existing design is used in the design process. Figure 1 summarizes the kinds of questions asked about designs. The categories in Figure 2 describe why such questions are asked and how the answers are used in design.

3.2. Computational Services That Use Design Rationale Information

Although human designers can use rationale information in a variety of ways, the focus of this analysis is on computational uses: how design rationale information can be used by software tools to support engineering tasks. We survey four kinds of such services: information retrieval, decision support, dependency management, and rationale by demonstration.

Documentation and Information Retrieval. The service provided by many proposed design rationale tools is documentation, or more generally, information retrieval. The idea is simple: If we can capture design information online, we can retrieve the information when it is needed. The efficacy of the information retrieval service depends on the quality of the indexing. Design notebooks provide traditional indexing based on textual search or user-supplied indices (Hwang & Ullman, 1990; Lakin, Wambaugh, Leifer, Cannon, & Sivard, 1989; Uejio, 1990; Uejio, Carmody, & Ross, 1991). Semiformal tools for argumentation, deliberation, design space, and design process history provide an ontology of concept and relation types, which structure the information for graphical browsing (Brown & Bansal, 1991; Chen, Dieterich, & Ullman, 1991; Conklin & Yakemovic, 1991 [chapter 14 in this book]; Fischer, Lemke, McCall, & Morch, 1991 [chapter 9 in this book]; Lee & Lai, 1991; MacLean, Young Bellotti, & Moran, 1991 [chapter 3 in this book]; Ramesh & Dhar, 1992).

A major obstacle to the practical use of these tools is the burden of structuring the information to rationalize design choices. For example, a design support tool called IDE (Russell et al., 1990) elicited design rationale at design time together with the specification of the designed artifact (instructional courses). Experience of the second author with IDE users over several years revealed that even highly motivated and disciplined designers suffer from "rationale fatigue." Users who started a design by including rationale notes to justify design choices soon began to bypass the rationale capture interface to focus on their primary design task.

Our examination of the design protocols suggested that whereas some of the design rationale dialog flowed naturally along the lines of argumentation or deliberation, much of the information available to the original designers and needed by others was not structured in this manner. This is an example of a mismatch between the knowledge as available and the knowledge as intended to be used in a rationale.

However, semiformal design documentation can be used for more than information recording. The effort of categorizing and relating notes on the issues, arguments, and so forth, can pay dividends if the capture tool can provide project management services, such as checklists. If all design issues are labeled as such, and all alternatives are enumerated and categorized, then the support tool can do a simple kind of completeness analysis, looking for unresolved issues and questions. Indeed, the main purpose of tools such as QOC (MacLean et al., 1991 [chapter 3 in this book]) is to guide the designer through a systematic exploration of the design space, producing a better design document and presumably a better design.

Decision Support. Decisions are so ubiquitous in design activity that design is sometimes *defined* in terms of decision making. We found two basic uses for information explicitly related to decisions. First, decision points serve as loci for considering alternatives and linking dependent elements in the design. For example, a designer will ask what decisions in the previous design were related to a specific parameter. The answer is a set of clusters of information (alternatives, criteria). These clusters identify places in the search space where the new design might differ, and link parameters that should be reexamined if the design took a different path (Lee, 1990; Mark & Schlossberg, 1991). This use of decision representations is similar to semistructured note taking, and can provide similar information retrieval services.

A second use of decision information is design evaluation. In this context, decision making means choosing alternatives based on evaluation against criteria. Formal tools can help the designer be consistent with normative ideals (i.e., overcoming human bias in judgment). For example, some tools offer spreadsheet-like services for experimenting with trade-offs among alternatives and criteria in design (Boose, Bradshaw, Koszarek, & Shema, 1993; Boose, Shema, & Bradshaw, 1991; Shema, Bradshaw, Covington, & Boose, 1990). Tools can also help with evaluation tasks such as calculating resource budgets and checking constraints. The most structured formalism for decision making is based on the mathematics of maximizing expected utility, given a network of conditional probabilities (see Howard, 1968, for an introduction). Such a network can determine *exactly how good* a decision would be given information about the probabilities and utilities of consequences of the decision. Of course, greater precision in the decision model comes at the expense of more work in developing the model; in many cases, approximate representations such as partial preference orderings are adequate for documenting important design decisions.

None of the information questions or answers in the protocols we surveyed would have required or used this degree of precision in decision support. However, even seemingly small design decisions can have amplified monetary consequences in products designed by large teams. In such situations, the costs of representing design decisions with the rigor and completeness required of normative formalisms might be outweighed by the benefits.

Dependency Management. A primary reason why designers are interested in the rationale for an existing design is to change it. They want to reuse the good parts of the existing design and modify the bad parts such that the changes do not undo the careful trade-offs and choices that make up the design. We have already pointed out the need to capture dependency relationships. The rich webs of interconnected dependencies in a realistic design are a source of complexity. Managing this complexity is a problem for designers and an opportunity for software support.

The ubiquitous question "what is affected by a change in this design element?" can only be answered by a program to the extent that the program can *represent* the dependencies and invoke the *reasoning* services that correspond to dependency paths.

Some dependencies are operational, that is, formulated in a representation in which they can be automatically calculated, propagated, or checked. In the simplest case, semiformal links can be followed when the contents of nodes change (Ramesh & Dhar, 1992). More information can be represented by dependencies that are expressed as *constraints*. For instance, in the damper design, the force produced by the solenoid can be computed by an operational constraint—a closed-form function of current and stroke length. Another kind of constraint is a limit on possible behavior, such as the maximum stroke length produced by the solenoid. Given constraints in this form, programs can help verify the intended function and behavior of components by following the paths of constraints from components to requirement specifications (Baudin, Sivard, & Zweben, 1990; Bowen & Bahler, 1993). Another class of dependencies is called *commitments*. Commitments are constraints that must be satisfied for a particular software module to be incorporated into a larger design (Mark, Tyler, McGuire, & Schlossberg, 1992). Capturing commitments in operational form enables tools that help developers determine what needs to change in a design when a requirement or assumption changes. Another kind of dependency relation is the functional role of a component in a larger system. Explicit descriptions of functional roles can support the retrieval and modification of cases from a design case library (Goel & Chandrasekaran, 1989) and the description of design intent (Chandrasekaran, Goel, & Iwasaki, 1993).

When dependencies are not operational, tools can help elicit them. The deliberation and decision support tools are designed to help the user enumerate and record the dependencies between alternatives and justifying information. The ontologies of argumentation (Kunz & Rittel, 1970; McCall, 1986; Newman & Marshall, 1990) design space possibilities (MacLean et al. 1991 [chapter 3 in this book]), and decision making (Lee & Lai, 1991) are explicit representations of dependencies that are often implicit. When a designer wants to reconsider a decision or design option, he or she can browse the links in the dependency network. An important future direction for such tools is to augment them with more operational dependency management techniques, such as rule-based constraint checkers (Fischer, Lemke, Mastaglio, & Morch, 1990; Fischer et al., 1991 [chapter 9 in this book]), truth maintenance (Lubars, 1991; Petrie, 1990), and mathematical modeling tools (Wolfram, 1991).

Rationale by Demonstration. One of the primary uses of design rationale information is to communicate the intended purpose or expected behavior of the design. Although it is valuable to capture information about expected behavior and contexts of use, it is difficult to specify the *dynamic* and *situational* aspects of a design in static artifact specifications. CAD drawings, program source code, and text-graphic documents are not well suited to the task of communicating time-varying (dynamic) behaviors. Furthermore, specifying the situation in which the artifact interacts with the environment (including the human contexts of use) requires envisioning in a hypothetical world with detail that is difficult to anticipate. A passive medium offers no assistance with this cognitive task.

We are exploring the idea of using interactive simulation as a communication medium for conveying the intended function of an artifact in an expected operating environment (Gruber, 1990, 1991). The technique is called *rationale by demonstration* because one demonstrates the phenomenon of interest, using packaged simulation *scenarios.* An engineer sets up a simulation scenario by specifying structures and behaviors of interest, and the expected operating environment in terms of initial conditions and exogenous variables. The system performs a simulation and generates an explanation of the scenario. The process is illustrated in Figure 3. An intelligent model formulation system actively elicits the engineer's specification, checking that parameters are not under- or overconstrained and ensuring that component connection topologies meet structural constraints. The model formulation system then generates an operational simulation model, drawing from a library of standard model fragments. The simulation model is given to a simulation system that predicts behavior and generates explanations for human consumption (tracing causal pathways, filtering extraneous detail, etc.). The explanations, possibly including animations of the artifact, serve as demonstrations of the designer's intent. The models and initial conditions can also be saved, so that the demonstration can be regenerated later with slightly different parameter settings to explore alternative scenarios.

The basic technology required for rationale by explanation is model formulation assistance (Falkenhainer & Forbus, 1991; Iwasaki & Low, 1993; Nayak, 1992; Palmer & Cramer, 1991) and explainable simulation (Falkenhainer & Forbus, 1992; Gruber & Gautier, 1992). It only works in domains where formal simulation models exist, and is only practical if the model formulation process can be made efficient. (Research on both fronts is progressing rapidly.)

In any case, the potential payoffs to the user are significant. Not only is design rationale information captured as a by-product of normal engineering, but the explanation is guaranteed to reflect the assumptions and input data specified *unambiguously* by the designer. Furthermore, the act of specifying a scenario—particularly in the interactive, qualitative modeling systems—*prompts* the designer to explicate assumptions and enumerate relevant conditions.

Figure 3. Capturing demonstrations of design intent. The human designer specifies a simulation scenario, which is used to formulate a simulation model, which is used to predict behavior. The output of the simulation is presented in text and graphic explanations for human consumption. The designer can use the entire mechanism as a medium for capturing demonstrations of expected behavior and assumed operating conditions. The architecture shown is implemented in the DME system (Iwasaki & Low, 1993).

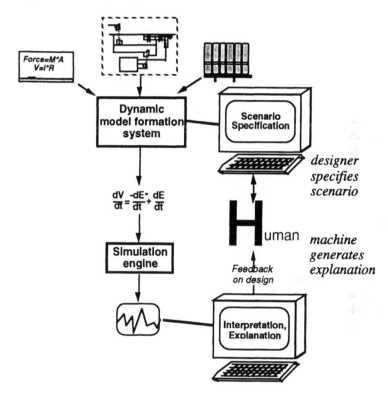

Rationale by demonstration shares the intuition of the scenario-driven design, that one can help communicate the intent of a design by describing the context of use and expected behavior (Rosson & Carroll, 1992). However, because rationale by demonstration is based on predictive behavior models and simulation software, it is possible to actively prompt for the information needed to specify a scenario and to verify that the specification is sufficient to model the scenario. The active prompting of input needed for a model coupled with feedback about predicted behavior is similar to what is done by decision support tools (Boose et al., 1991; Shema et al., 1990), which help the user specify the relevant factors that go into a decision and provide feedback on the consequences of alternative decision outcomes.

4. IMPLICATIONS FOR THE DESIGN OF SUPPORT SOFTWARE

Our analysis of the information available for design rationale and the possible computational services that can operate on that information can inform the design of rationale support software. Our major conclusion is that the record and replay paradigm is inadequate. In this final section, we argue for this conclusion, offer an alternative paradigm, and discuss practical implications for the development of tools.

4.1. "Record and Replay" Is Incomplete

Many proposals for design rationale support are based on the strategy of recording a complete rationale explanation (argument, design space exploration, decision justification) so that it can be easily replayed (retrieved, presented, browsed) by consumers of the information. We call this the *record and replay* paradigm. The research within this paradigm emphasizes the interesting problem of eliciting knowledge that is often tacit, difficult to formalize, and generally incomplete. However, our analysis suggests that for many types of design rationale information and several potential applications, the capture of complete explanations is not an adequate solution.

A record and replay approach is incomplete with respect to the information needs of designers for several reasons. First, rationale explanations cover a broad range of information requests (see Section 2 and Figure 1). If design rationales are to be captured and played back, then the design rationale author has to anticipate the space of questions that might be asked by the reader, and formulate possible rationale explanations in advance. A semiformal capture tool can prompt the designer to construct explanations along some dimension, such as deliberation, decision making, or design space exploration. However, the particular nodes and links used in explanations captured at design time will answer only a fraction of the questions asked by designers about existing designs.

Of course, the main objective of semiformal rationale approaches such as QOC (MacLean et al., 1991 [chapter 3 in this book]) is to help the author systematically generate and explore the questions that might be of interest to the downstream reader. We are not asserting that the resulting rationales are not useful (see Yakemovic & Conklin, 1990, for evidence on this question). We are questioning whether it is worth asking the designer to construct *complete* explanations at design time, given the cognitive burden of this task.

Second, many uses of rationale information, such as tracking dependencies and demonstrating design intent, require *inference* from available information to answer questions. In the damper protocol, the original designer explained the reasoning that links the heat transfer problem to the availability of solenoids with desired properties. The original designer explored a space of several competing factors, including functional alternatives to solenoids. The path from overall functionality to solenoid properties is just one of many possible inferences. The space of inferences (and corresponding explanations) that might be made from available design data is much larger than the space of the design data themselves.

Third, the inferences underlying rationale explanations are based on assumptions and design state that can change after a rationale is constructed. This would invalidate "snapshot" explanations. To allow for change, the person recording the rationales would have to anticipate all the assumptions that might change, and explicitly instantiate them for each case (e.g., identify the availability of high-temperature solenoids as a supporting argument for some position). If the rationale only included the history of decisions as they actually occurred, the resulting record could end up reflecting an opportunistic hopping among constraints and alternatives (as has been observed in software design; see Guindon, 1990), rather than the structure of dependencies between decisions and the data on which they are based.

4.2. Generating Rationale Explanations From What Is Captured

We conclude that design rationale tools must go beyond the record and replay paradigm if they are to support the kinds of computational services we have described. In particular, we argue for a paradigm in which design rationale explanations are *generated*, in response to information requests, from relevant stored information. The knowledge capture task, then, is to elicit the information that is potentially relevant to explaining a rationale, rather than the explanations themselves. The generation task is to construct an explanation that answers a given query, possibly by way of automated inference or other computation from the captured information and other background knowledge.

A Conceptual Framework for Generative Rationale. The generative view of design rationale can be explained in the terms of a conceptual framework illustrated in Figure 4 (Gruber & Russell, 1990). In this framework, the basic element of information is a design description, which we assume can be captured and represented. Common classes of design descriptions are *structure* (the physical and/or logical composition of an artifact, typically in terms of the composition of parts, and connection topologies), *behavior* (something an artifact might *do*, in terms of observable states or changes), *function* (effect or goal to achieve by artifact behavior), *requirements* (prescriptions concerning the structure, behavior, and/or function that the designed artifact must satisfy, typically specified as constraints), and *objectives* (specifications of desired properties of the artifact other than pure

Figure 4. Relationships among design descriptions. Nodes labeled *S, B, F, R,* and *O* are classes of design descriptions. Arcs correspond to *inferences* between descriptions. For example, $S \rightarrow B$ inferences are performed by simulation and analysis tools. *Explanations* follow paths of inference. An explanation of why a component was chosen, for instance, might show how the component structure (*S*) implements a behavior (*B*), achieving a function (*F*), which satisfies a requirement (*R*).

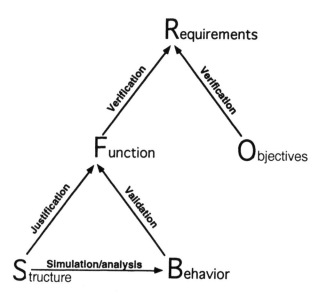

functions, such as cost and reliability, often given as decision criteria). Some of the information used by the designers in the protocol survey are simple facts of one of these types. Answers to many rationale questions involved explanations (see Section 2). Explanations follow paths of inference among design descriptions, shown as directed arcs in the figure. An explanation of why a component was chosen, for instance, might show how the component structure (*S*) implements a behavior (*B*), achieving a function (*F*), which satisfies a requirement (*R*). The generative view of rationale says that such explanations are generated in response to questions about the relationships among design descriptions. The inferences may be performed by engineering tools and/or human engineers.

A record and replay approach would capture completed explanation graphs (perhaps structured as decisions among alternatives). In contrast, the generative approach captures the nodes involved in a decision (the "relevant facts") and the inference processes that link them (if they are available on-line). The explanations are generated when a consumer of the rationale information makes a query, such as "what functions depend on this structure?" or "how is this requirement met?"

Application of Generative Design Rationale. The generative approach can be applied to several of the computational uses of design rationale described in Section 3. The cases of dependency management and rationale by demonstration illustrate the idea.

For dependency management, in the record and replay paradigm one would explicitly capture the dependencies between each pair of related design elements. Perhaps the dependency would be couched in the terminology of evidential support for a position. In the generative approach, one would capture only the local dependencies. At explanation time a process would traverse the dependency network, uncovering unanticipated relationships among design elements. If the value of one of the nodes in a dependency network changes, a new explanation can be generated.

The technique of rationale by demonstration is an inherently generative approach. The analog in the record and replay paradigm is the handcrafted mock-up or storyboard animation. This is a known technique for flushing out assumptions and documenting intent, especially in domains such as user interface design. However, the medium is essentially static and passive. In the generative approach, one captures an operational specification of a scenario, which can be run. The specification is given as the inputs to the model formulation system: the basic device model, input conditions, behavior of interest, and operating conditions. The explanation of how the device operates in the specified scenario is *generated* by the simulation program. This can be used to produce an active, interactive document. Textual and graphical explanations are generated in response to user queries. Users can ask questions such as "what caused this behavior to occur?" and "what else might have happened at this point?"

When the knowledge consumer views the demonstration, he or she has access to all the detail specified by the original designer. Note that the "author" had to specify an *operational* behavior model and an *unambiguous* set of initial conditions to run the simulation. This information is more useful than a static, textual summary of expected behavior, for two reasons. First, because the explanation is generated from the specification, it is guaranteed to be consistent with the engineering models. Second, when the models or assumptions change, the explanations are automatically kept up to date and accurate. Of course, the cost to the user of specifying operational models is higher than jotting down notes. However, if model formulation can be made easier and useful in earlier stages of the design process, the capture of operational demonstrations of design intent can be a by-product of the normal use of simulation and analysis tools.

Another example of generative rationale support is Garcia's ADD system (Garcia, 1992; Garcia & Howard, 1992). ADD supports the documentation of parametric designs. During design, designers make choices for parameters in light of requirements and constraints. ADD monitors the decision-making process, as an apprentice. The program has a strong a priori model of the domain (design parameters for heating and air-conditioning systems), design process (the ordering dependencies among design decisions), and decision making (how to evaluate alternatives considering multiple criteria and constraints). The designer enters requirements, constraints, and decision criteria for a particular case. Then, as the designer makes design decisions, the system uses the domain, process, and decision models to predict expected choices for parameters. If the designer's choices do not match the system's expectations, ADD asks for more information from the user to help justify the choices. If they do match, ADD can justify the choices. As a result, all decisions in a design can be explained by ADD.

ADD is an excellent example of the generative approach. It acquires rationale information from the designer that is used to generate explanations to the user of the document. It produces focused explanations in response to user queries, rather than asking the designer to anticipate the reader's needs. The explanations are constructed from models and acquired information (parameters, requirements, constraints, criteria). As with simulation-based explanations, the ADD's explanations are guaranteed to be consistent and up to date with the design choices.

In sum, the generative approach enables more powerful computational services yet requires the capture of more operational information. This is practical if the capture of dependency relations and scenario specifications results from the routine use of engineering support tools. This brings us to a final issue: the impact of this perspective on the design of software.

4.3. Architectural Implication: Integration With Existing Engineering Tools

We believe that much of the information to be captured for design rationale is either available now from existing engineering tools and information sources, or will be. Knowledge-based CAD, advanced programming environments, on-line engineering handbooks, simulation and modeling programs, design-checking tools, and electronic mail are all to be found in the designer's toolbox. If more and more of design will be done in the context of such tools, then there will be information freely available in machine-readable form. However, "freely available" does not imply "ready to use."

To bring these tools into the service of design rationale support requires *integration*. Consider again the damper example. The information on properties of available solenoids might come from a database. The equations relating voltage, current, and force might be in a model library. The decision to use a short stroke length might be found in a design history, and hyper-linked to a diagram in the requirements document. To construct complete explanations in response to user questions, a design rationale tool needs to reason about the *dependencies* among these heterogeneous information sources. This requires a common conceptualization of the shared data and a mechanism for representing relations that cross tool boundaries. In addition, because some of the relationships among design elements are *computed* by engineering tools, reconstructing the explanation when assumptions change requires reinvoking the tools in a coordinated fashion.

Fischer and his colleagues have shown the value of integrating critics (design checker tools) with artifact specification (e.g., domain-specific CAD drawing) and hypertext documentation (Fischer et al., 1991 [chapter 9 in this book]). This work showed that the explanations given by critics could serve the same kinds of information roles as the hypertext documentation (i.e., giving reasons for rejecting alternatives). Furthermore, the explanations seem better grounded and more informative when presented in the same context as the textual annotations and CAD drawings.

Critics that generate their explanations from engineering models, such as the industrial design checkers used to validate the manufacturability and testability of electronic circuits, could be incorporated into the documentation of designs in a similar manner. One can view the output of the generative tools as part of the documentation. Because the tools can be rerun with different input conditions (alternative designs), the documentation can answer hypothetical questions not covered by the original document writer and survive change in the life of the design.

However, achieving this vision of the generative documentation of rationale requires greater integration of tools than exists today. The tightest integration in existing engineering tools occurs in *CAD frameworks*—families of programs designed together to share data in standard formats and common domain models. However, we believe that large-scale rationale capture and generation will require the integration of design rationale capture tools with sets of *independently* developed, *heterogeneous* engineering tools of the designers' choosing (across CAD frameworks if desired). One should be able to specify dependencies between a section of text in a requirements document, a drawing in a CAD database, textual annotations to CAD objects, models of the components represented by those objects, constraint equations and simulation models specifying the physical behavior of those objects, and plans generated by a process planner for manufacturing the components. An architecture called SHADE is under development that will provide this capability (Gruber, Tenenbaum, & Weber, 1992).

NOTES

Acknowledgments. We would like to thank the participants in the Stanford design rationale seminar in the fall of 1991 and the members of the Palo Alto Collaboration

Testbed consortium (Cutkosky et al., 1993) for many discussions that contributed to this work. We are grateful to the researchers who generously allowed us to analyze their hard-earned protocol data, much of which was previously unreleased. They include Catherine Baudin, Vinod Baya, Jean-Charles Bonnet, Jody Gevins, Larry Leifer, and Ade Mabogunje at NASA Ames and the Stanford Center for Design Research; Ana Cristina Bicharra Garcia from the Stanford Civil Engineering Department, Thomas Kuffner and David Ullman at Oregon State University (NSF grant DMC.87.12091); Pete Pirolli and Vinod Goel at the University of California at Berkeley School of Education; and Victoria Bellotti and Alan MacLean at EuroPARC. Conversations with Tom Dietterich, Bill Mark, Richard Pelavin, Mark Stefik, Marty Tenenbaum, and Jay Weber were particularly enlightening. Reviews by Tom Moran and Jack Carroll were extremely helpful.

Support. Work by the first author is supported by NASA Grant NCC2-537, NASA Grant NAG 2-581 (under ARPA Order 6822), and DARPA prime contract DAAA15-91-C-0104 through Lockheed subcontract SQ70A3030R, monitored by the U.S. Army BRL.

Authors' Current Addresses. Thomas R. Gruber, 147 Lakeview Way, Redwood City, CA 94062. Email: Gruber@hpp.stanford.edu; Daniel M. Russell, Apple Computer, Advanced Technology Group, 1 Infinite Loop, MS 301-3D, Cupertino, CA 95014. Email: DMRussell@Apple.com.

REFERENCES

Baudin, C., Gevins, J., Baya, V., & Mabogunje, A. (1992). Dedal: Using domain concepts to index engineering design information. *Proceedings of the 14th Annual Conference of the Cognitive Science Society*, 702–707. Bloomington, IN: Cognitive Science Society.

Baudin, C., Gevins, J., Baya, V., Mabogunje, A., & Bonnet, J.-C. (1992). *The variable damper experiment* (Tech. Rep. No. FIA-TR-92-08). Moffett Field, CA: NASA Ames Research Center.

Baudin, C., Sivard, C., & Zweben, M. (1990). Recovering rationale for design changes: A knowledge-based approach. *Proceedings of the IEEE International Conference on Systems, Man, and Cybernetics*, 745–749. Los Angeles: IEEE.

Baya, V., Gevins, J., Baudin, C., Mabogunje, A., Toye, G., & Leifer, L. (1992). An experimental study of design information reuse. *Proceedings of the 4th International Conference on Design Theory and Methodology*, Scottsdale, AZ, 141–147. New York: American Society of Mechanical Engineers.

Bellotti, V., & MacLean, A. (1989). *Transcription of: Two designers discussing a proposed design for a Fast Automated Teller Machine* (Internal Tech. Rep. No. AMODEUS RP6/DOC1). Cambridge, England: Rank Xerox EuroPARC.

Boose, J. H., Bradshaw, J., Koszarek, J. L., & Shema, D. B. (1993). Knowledge acquisition techniques for group decision support. *Knowledge Acquisition Journal*, 5(4), 405–448.

Boose, J. H., Shema, D. B., & Bradshaw, J. M. (1991). Knowledge-based design rationale capture: Automating engineering trade studies. In M. Green (Ed.), *Knowledge aided design*. London: Academic Press.

Bowen, J., & Bahler, D. (1993). Constraint-based software for concurrent engineering. *IEEE Computer, 26*(1), 66–68.

Brown, D. C., & Bansal, R. (1991). Using design history systems for technology transfer. In D. Sriram & R. Logcher (Eds.), *Computer-aided cooperative product development* [lecture notes] (Series No. 492). Berlin: Springer Verlag.

Burgess-Yakemovic, KC, & Conklin, J. (1990). Report on a development project use of an issue-based information system. *Proceedings of the Conference on Computer-Supported Cooperative Work.* New York: ACM.

Chandrasekaran, B., Goel, A. K., & Iwasaki, Y. (1993). Functional representation as design rationale. *IEEE Computer, 26*(1), 48–56.

Chen, A., Dietterich, T. G., & Ullman, D. G. (1991). A computer-based design history tool. *NSF Design and Manufacturing Conference,* 985–994. Austin, TX.

Conklin, E. J., & Yakemovic, KC B. (1991). A process-oriented approach to design rationale. *Human–Computer Interaction, 6,* 357–391. Also in T. P. Moran & J. M. Carroll (Eds.), *Design rationale: Concepts, techniques, and use.* Hillsdale, NJ: Lawrence Erlbaum Associates, 1996. [Chapter 14 in this book.]

Cutkosky, M., Engelmore, R. S., Fikes, R. E., Gruber, T. R., Genesereth, M. R., Mark, W. S., Tenenbaum, J. M., & Weber, J. C. (1993). PACT: An experiment in integrating concurrent engineering systems. *IEEE Computer, 26*(1), 28–37.

Falkenhainer, B., & Forbus, K. D. (1991). Compositional modeling: Finding the right model for the job. *Artificial Intelligence, 51,* 95–143.

Falkenhainer, B., & Forbus, K. (1992). Self-explanatory simulations: Scaling up to large models. *Proceedings of the Tenth National Conference on Artificial Intelligence,* 685–690. San Jose, CA: AAAI Press/MIT Press.

Fischer, G., Lemke, A. C., Mastaglio, T., & Morch, A. I. (1990). Using critics to empower users. *Proceedings of the ACM Conference on Human Factors in Computing Systems (CHI '90),* 337–347. New York: ACM.

Fischer, G., Lemke, A. C., McCall, R., & Morch, A. I. (1991). Making argumentation serve design. *Human–Computer Interaction, 6,* 393–419. Also in T. P. Moran & J. M. Carroll (Eds.), *Design rationale: Concepts, techniques, and use.* Hillsdale, NJ: Lawrence Erlbaum Associates, 1996. [Chapter 9 in this book.]

Garcia, A. C. B. (1992). *Active design documentation.* Unpublished doctoral dissertation, Stanford University, Stanford, CA. (Available as CIFE Tech. Rep. No. 82, Department of Civil Engineering)

Garcia, A. C. B., & Howard, H. C. (1992). Acquiring design knowledge through design decision justification. *Artificial Intelligence for Engineering, Design, and Analysis in Manufacturing, 6*(1), 59–71.

Goel, A., & Chandrasekaran, B. (1989). Functional representation of designs and redesign problem solving. *IJCAI-89,* 1388–1394. San Mateo, CA: Morgan Kaufmann.

Gruber, T. R. (1990). *Model-based explanation of design rationale* (Tech. Rep. No. KSL-90-33). Stanford, CA: Stanford University, Knowledge Systems Laboratory.

Gruber, T. R. (1991). Interactive acquisition of justifications: Learning "why" by being told "what." *IEEE Expert, 6*(4), 65–75.

Gruber, T. R., Boose, J., Baudin, C., & Weber, J. (1991). Design rationale capture as knowledge acquisition: Tradeoffs in the design of interactive tools. In L. Birnbaum & G. Collins (Eds.), *Machine Learning: Proceedings of the Eighth International Workshop,* 3–12. Chicago: Morgan Kaufmann.

Gruber, T. R., & Gautier, P. O. (1992). Machine-generated explanations of engineering models: A compositional modeling approach. *Proceedings of the 13th International Joint Conference on Artificial Intelligence*, Chambery, France, 1502–1508, San Mateo, CA: Morgan Kaufmann.

Gruber, T. R., & Russell, D. M. (1990). *Design knowledge and design rationale: A framework for representation, capture, and use* (Tech. Rep. No. KSL 90-45). Stanford, CA: Stanford University, Knowledge Systems Laboratory.

Gruber, T. R., & Russell, D. M. (1992). *Derivation and use of design rationale information as expressed by designers* (Tech. Rep. No. KSL-92-64). Stanford, CA: Stanford University, Knowledge Systems Laboratory.

Gruber, T. R., Tenenbaum, J. M., & Weber, J. C. (1992). Toward a knowledge medium for collaborative product development. In J. S. Gero (Ed.), *Artificial intelligence in design '92* (pp. 413–432). Boston: Kluwer Academic Publishers.

Guindon, R. (1990). Knowledge exploited by experts during software system design. *International Journal of Man–Machine Studies, 33*, 279–304.

Hwang, T. S., & Ullman, D. G. (1990). The design capture system: Capturing back-of-the-envelop sketches. *International Conference on Engineering Design (ICED-90)*, Dubrovnik, Yugoslavia. New York: American Society of Mechanical Engineers.

Iwasaki, Y., & Low, C. M. (1993). Model generation and simulation of device behavior with continuous and discrete changes. *Intelligent Systems Engineering, 1*(2).

Kuffner, T. A. (1990). *Mechanical design history content: The information requests of design engineers*. Unpublished master's thesis, Oregon State University, Corvallis.

Kuffner, T. A., & Ullman, D. G. (1990). The information requests of mechanical design engineers. *Design Studies, 12*(1), 42–50.

Kunz, W., & Rittel, H. (1970). *Issues as elements of information systems*. Berkeley: University of California, Center for Berkeley, Planning and Development Research.

Lakin, F., Wambaugh, H., Leifer, L., Cannon, D., & Sivard, C. (1989). The electronic design notebook: Performing medium and processing medium. *Visual Computer: International Journal of Computer Graphics, 5*(4), 214–226.

Lee, J. (1990). SIBYL: A tool for managing group decision rationale. *Proceedings of the Conference on Computer Supported Cooperative Work (CSCW-90)*, 77–92. New York: ACM.

Lee, J., & Lai, K-Y. (1991). *A comparative analysis of design rationale representations* (Tech. Rep. No. 121). Cambridge, MA: MIT Center for Coordination Science.

Lubars, M. D. (1991). Representing design dependencies in an issue-based style. *IEEE Software, 8*(4), 81–89.

MacLean, A., Young, R., Bellotti, V., & Moran, T. (1991). Questions, options and criteria: Elements of design space analysis. *Human–Computer Interaction, 6*, 201–250. Also in T. P. Moran & J. M. Carroll (Eds.), *Design rationale: Concepts, techniques, and use*. Hillsdale, NJ: Lawrence Erlbaum Associates, 1996. [Chapter 3 in this book.]

Mark, W., & Schlossberg, J. (1991). Interactive acquisition of design decisions. In J. Boose & B. Gaines (Eds.), *Proceedings of the 6th Knowledge Acquisition for Knowledge-Based Systems Workshop*, Banff, Canada. Calgary, Alberta, Canada: SRDG Publications.

Mark, W., Tyler, S., McGuire, J., & Schlossberg, J. (1992). Commitment-based software development. *IEEE Transactions on Software Engineering, 18*(10), 870–885.

McCall, R. (1986). Issue-serve systems: A descriptive theory for design. *Design Methods and Theories, 20*(8), 443–458.

Mitchell, T. M., Keller, R. M., & Kedar-Cabelli, S. T. (1986). Explanation-based generalization: A unifying view. *Machine Learning, 2*(1), 46–80.

Mostow, J. (1989). Design by derivational analogy: Issues in the automated replay of design plans. *Artificial Intelligence, 40,* 119–184.

Nayak, P. P. (1992). *Automated modeling of physical systems.* (Tech. Rep. No. STAN-CS-92-1443). Stanford, CA: Stanford University, Computer Science Department.

Newman, S. E., & Marshall, C. C. (1990). *Pushing Toulman too far: Learning from an argumentation representation scheme* (Tech. Rep. No.). Xerox PARC, Palo Alto, CA.

Palmer, R. S., & Cramer, J. F. (1991). *SIMLAB: Automatically creating physical systems simulators* (Tech. Rep. No. TR 91-1246). Ithaca, NY: Cornell University, Department of Computer Science.

Petrie, C. (1990). *REDUX: An overview* (Tech. Rep. No. TR ACT-RA-314-90). Austin, TX: Microelectronics and Computer Technology Corporation.

Ramesh, B., & Dhar, V. (1992). Supporting systems development using knowledge captured during requirements engineering. *IEEE Transactions on Software Engineering, 18*(6), 498–510.

Rosson, M. B., & Carroll, J. M. (1992). *Extending the task-artifact framework: Scenario-based design of Smalltalk applications* (Research Rep. No. RC 17852 [No. 78516]. Yorktown Heights, NY: IBM Research Division.

Russell, D. M. Burton, R. R., Jordan, D. S., Jensen, A. M., Rogers, R. A., & Cohen, J. (1990). Creating instruction with IDE: Tools for instructional designers. *Intelligent Tutoring, 1*(1), 3–16.

Shema, D., Bradshaw, J., Covington, S., & Boose, J. (1990). Design knowledge capture and alternative generation using possibility tables in CANARD. *Knowledge Acquisition, 2*(4), 345–364.

Uejio, W. H. (1990). Electronic design notebook for the DARPA initiative in concurrent engineering. *Proceedings of the Second Annual Concurrent Engineering Conference,* 349–362. Los Alamitos, CA: IEEE Computer Society Press.

Uejio, W. H., Carmody, S., & Ross, B. (1991). An electronic project notebook from the electronic design notebook (EDN). *Proceedings of the Third National Symposium on Concurrent Engineering,* 527–535. New York: American Society of Mechanical Engineers.

Wolfram, S. (1991). *Mathematica: A system for doing mathematics by computer.* Menlo Park, CA: Addison-Wesley.

USING DESIGN RATIONALE
FOR TEACHING

12

Rationale in Practice: Templates for Capturing and Applying Design Experience

George Casaday
Digital Equiptment Corporation

ABSTRACT

This chapter presents a practical method useful in design of human–computer interaction. The method was developed pragmatically in a consulting practice to satisfy a need for reliable, reusable, and cost-effective problems and design-meeting situations.

Design is viewed as a process of constructing and evaluaing a succession of more and more complete design models until a final design is produced. In this view, an important part of support for design is a powerful method for constucting design models. The chapter offers generic forms, called templates, that can be instantiated to produce specific design models, greatly facilitating construction. Three types of template are described. The first group supports modeling of the form of a software user interface. The second group helps in establishing design goals. The third group facilitates managing the design process. Each template is based on actual practice, and each is supported with a procedure for using it and an example of its use in practice.

George Casaday is a software engineer with Digital Equipment corporation where he designs, consults, and teaches in the area of human–computer interaction.

CONTENTS

1. INTRODUCTION

This chapter is written from the point of view of a user-interface designer and former software developer. But my work includes less hands-on design than facilitating user-interface design meetings and coaching others as they learn to design. To be effective I need good, practical, transferable techniques for design and facilitation. I try to adapt and apply promising techniques from the literature on human–computer interaction, keeping the successes. I invent techniques, evaluate their effectiveness, and develop the promising ones. So my work includes an inquiry that is not aimed at scientific rigor; nor is it simply aimed at developing a personal craft skill. Rather it is aimed at creating useful techniques, grounded in successful practice and sound theory, and transferable to other designers. The chapter is written very much from the point of view of an engineer trying to make sense of a work area that is still mainly practiced as a craft. I think of myself as a reflective practitioner, to borrow the phrase of Donald Schön (1983).

In my experience, clear and powerful reasoning about design, that is design rationale, is one of the main ingredients of cost-effective design for computer–human interaction. How to develop useful design rationale is very unclear, as the wide range of approaches in this volume demonstrates. However, experience shows that reasoning from first principles during the course of design is too burdensome to be practical. This leads me to develop forms—I call them *templates*—that can guide development of useful parts of rationale by instantiation of the template. Templates can be extracted from designs, from artifacts, and from design process. They can then be evolved, fine-tuned, and adapted. I find that after a template is in use for a few months and has been critiqued and modified in several contexts, it can become a very powerful job aid for both designing and explaining.

Because templates are more naturally aids for constructing than for evaluating, I find them most effective as support for the kind of design that relies upon evolving a sequence of models toward a final result. The next section briefly outlines such an approach to design, one that works well with templates. The following section then explains several templates from my practice.

2. DESIGN

Teaching design, facilitating design, and teamwork in design require a clearly articulated statement of what design is, that is what kinds of activities and objects will be involved. That statement, even if it is only a useful metaphor, serves as the deep foundation for communication and cooperative work. In addition, the form of design rationale, the statements that designers accept as convincing arguments for design choices, depend on the statement of the nature of design. This section is the statement I use in practice.

Design for human–computer interaction is an inquiry rather than a computation. It is not the kind of inquiry that has a fixed answer uniquely determined by requirements or constraint. There is as yet no way to compute a user-interface design. Also, design for human–computer interaction is not exactly a search, at least not in Simon's (1969) sense of looking for a path through a problem space. It is much more like an exploration of possibilities. To use a metaphor, the raw materials for design are active: They actively, sometimes aggressively, respond and the wise designer listens carefully. In effect, the designer asks the materials just what they are capable of in the given situation and negotiates with them by constructing and evaluating design models.

2.1. Design Includes Construction and Evaluation

My view of design is that designers iterate through a loop in which they construct design models and evaluate them (Figure 1). The practice of design in this manner is explored by Donald Schön in two influential books, *The Reflective Practitioner* (1983) and *Educating the Reflective Practitioner* (1987). These books examine the actions of skilled practitioners—architects, city planners, psychiatrists, musicians—in which individual creativity is required beyond standard procedures. Schön's analysis applies to the design of human–computer interaction because it is also a skilled practice that so far has defied reduction to standard procedures. Schön's study is reinforced with examples from practice, but others have also described a similar design loop (e.g., Cohill, 1991).

Schön (1983, 1987) found the same process of investigation by informed trial for many examples of skilled practice. First, engage the task. Sketch a building, meet with real estate developers, counsel a patient, perform

Figure 1. Construction and evaluation of models.

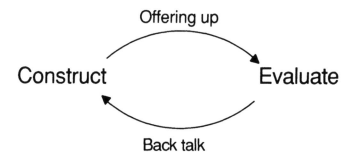

music. In the case of design this means to construct a physical representation, called a design model, of some object that is to be built. For
human–computer interface design, an example would be attempting a
screen layout sketch to get as much information displayed on one screen
as practical. Second, "offer up" the model to get the results of the engagement. Sometimes the results are expected, sometimes surprising. In
Schön's words, the model "talks back." In the screen layout example, the
sketch could show that much information can be displayed but that the
visual density makes items hard to find. Third, reflect upon the "back talk."
This is in preparation for the next round of engagement with this task, to
prepare for engagement with some new task, or to form a concept for
teaching. In the screen layout example, the difficulty of finding items could
trigger reorganization of the display into visually distinct subsections.

The construct-evaluate loop occurs on various time and size scales. The
individual designer constructs and evaluates the most preliminary design
models in minutes. Design teams may review more elaborate design models
together on a time scale of days. Users of a system may be involved in
evaluation on a longer time scale, perhaps with a test cycle of weeks. The
longest commercially interesting cycle involves the market and may take
years. An even longer cycle fills historical time spans, perhaps decades.

In this view, there must be a design object, which we call a design model.
There is no purely conceptual design; there must be something concrete to
be offered up to evaluation in order to get the needed back talk. For
human–computer interaction, the basic elements of design are not individual decisions but are design models. Each design model captures a syndrome
of related and interdependent decisions. It seems that the level of individual
decisions is usually too fine-grained for practical work because there are too
many individual decisions and because decisions pass through the construct-
evaluate cycle packaged in models rather than individually.

Figure 2. **Design driven mainly by evaluation or construction.**

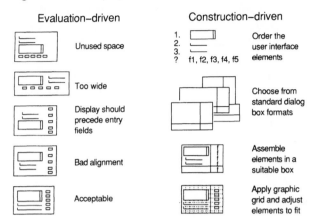

Rationale, in this framework, is built on the history of design models. In practice, the evolving sequence of design models is *very* useful to a design team; once a team begins to explicitly model and to save each generation of attempts, they generally find it cost-effective to continue creating, saving, and referring to this record. As a result, the raw material for rationale becomes available without extra work as a coproduct of design.

2.2. Design Can Be Evaluation Driven or Construction Driven

Good designs are probably always strongly influenced by design principles, the accumulated collective wisdom that is more general than the design choices of a particular artifact. In the view of design I am using here, principles can influence design either at the point of evaluation or the point of construction. In human–computer interaction, the main medium for influencing evaluation has been the guideline, whether applied heuristically or as a basis for testing. The main medium for influencing construction has probably been existing, successful user interfaces. Imitation is not powerful enough, so in this chapter I offer ways of presenting principles and abstractions from successful practice as templates, forms that are specifically crafted to help guide the construction side of the design loop (see also Carey, McKerlie, & Wilson, 1995 [chapter 13 in this book]).

To illustrate the distinction between evaluation-driven and construction-driven design, I examine one design exercise from both viewpoints. Consider designing a simple dialog box that must contain five action buttons, a vertically scrolling display area, and two single-line text-entry fields (Figure 2).

The first approach to this design emphasizes evaluation, which drives design changes. The first trial sketch appears as a designer's intuitive leap.

This version is evaluated as having too much unused space in the upper right quadrant of the box. This back talk triggers a rearrangement to squeeze out dead space. The second version is evaluated as too wide and so a different arrangement is triggered. The third version looks like it may lead to an acceptable use of space. But the designer knows that the user must read the scrolling list first upon entering the box and so moves it into the upper left corner in the next version. The fourth version looks promising. It needs adjustment of the haphazard alignment. The fifth version is evaluated as acceptable and design stops.

The second approach arrives at the same result, but in a construction-driven way that relies on known forms or procedures for creating versions and uses evaluation only for selection and tuning. The elements are defined and organized. A format is chosen from a known set of dialog box formats, which I would call templates. The elements are assembled in the box. There may be some backtracking here if earlier choices were ill advised. A graphic grid is applied to guide adjustment of the elements to fit and harmonize.

I generally prefer construction-driven design for several reasons. Experienced designers usually do construction-driven design, even if the templates or procedures they use are not explicitly stated. I suspect that construction-driven techniques may be faster. Construction techniques using suitably focused templates are transferred more easily than evaluation techniques, which may depend on a very broad background of knowledge. That is, it is easy to teach guidelines but very difficult to teach when and how to apply them. This also has the side effect of quicker convergence on shared understandings for a design team. It is easier to demonstrate progress with construction, a political consideration. Assistance with construction is simply more acceptable than evaluation for clients in a consulting relationship, a pragmatic consideration. Finally, construction-driven design automatically leaves behind a historical record that is useful for the designers in understanding rationale.

The best support I know of for guiding construction are the kinds of forms that I call templates. Because I choose to do this style of design and to teach it to my clients, I have begun creating templates for user-interface design. A few of these templates are presented in the next section.

3. TEMPLATES FOR DESIGNING HUMAN–COMPUTER INTERACTION

My work includes more teaching, advising, and explaining than direct designing for human–computer interaction. I find the most useful aids for that work to be what I call templates, essentially, forms that guide construction. In each of the eight examples that follow, the template guides either the form of a user-interface design, the articulation of design goals,

Figure 3. **Three important distinctions of design.**

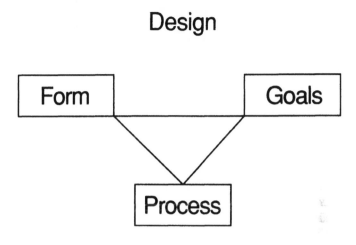

or the course of the design process (Figure 3). The equivalents of templates appear in other design areas under other names. For example, designers of training call them job aids (Rossett & Gautier-Downes, 1991). Organizational system modeler Peter Senge (1991) called them organizational archetypes. Architect Christopher Alexander referred to them as patterns (Alexander, Ishiwaka, & Silverstein, 1977). Carey et al. (1995 [chapter 13 in this book]) refer to similar objects as exemplars. The key is that templates lay out a form in advance, freeing the designer to concentrate on the content specific to the case under consideration.

I find templates more useful than guidelines, design principles, or specific advice because they are aimed at guiding designers on the construction side of the construction-evaluation loop. It is usually much easier to supply good answers than to ask the right questions, and templates are ways of asking the right questions in a way that is known from experience to elicit answers useful for construction. That intervention often enables design teams to do most of the design work on their own.

Good templates can be taught to development teams because they are at an appropriate level of abstraction; they must be specific enough to be illustrated with examples but general enough to be adapted to individual design problems. They can often be used immediately because they provide a fairly specific structure, sometimes as simple as a checklist, and they are fairly close to specific application requirements. They help design teams composed of varied skill sets by focusing attention and providing a common language. They also provide a tested framework for understandable and believable explanations for those outside the design team.

I believe that much of my own design ability consists of having a repertoire of applicable templates and some skill at calling up the ones

Figure 4. Mapping conceptual hierarchy to user interface hierarchy.

Template 1

Concepts		User Interface		Example
Concept	◁----▷	Object	◁----▷	Program entities (2)
Subconcept	◁----▷	Subobject	◁----▷	Data (3)
Sub–subconcept	◁----▷	Sub–subobjects	◁----▷	Terminal
⋮		⋮		Variable
				File
Subconcept	◁----▷	Subobject	◁----▷	Processes (3)
Sub–subconcept	◁----▷	Sub–subobjects	◁----▷	Move data
⋮		⋮		No–op
				Compute
Concept	◁----▷	Object	◁----▷	Program relationships (2)
Subconcept	◁----▷	Subobject	◁----▷	Flow (2)
Sub–subconcept	◁----▷	Sub–subobjects	◁----▷	Execution flow
⋮		⋮		Data flow
Subconcept	◁----▷	Subobject	◁----▷	Control structures (2)
Sub–subconcept	◁----▷	Sub–subobjects	◁----▷	Iteration
•		•		Alternative

that fit a circumstance. The repertoire is not a small set of general design principles. Instead, it is more like a chess player's memory of familiar board positions. The templates are supported, made concrete in my mind or in my notebooks, by specific instances that I have seen or created. When I study designs, rationales, or analyses, one of my main goals is to find new templates or new instances of templates that I already know.

This section is a sample from the templates that I use in design for human-computer interaction. All the supporting examples are drawn from real cases, although they are disguised to avoid problems with disclosure of proprietary information. None of the templates is surprising. The point is that they all have a compact, easily accessible form. They are primarily aids that make it less necessary for designers to invent everything from scratch or to remember every relevant feature of each design task.

3.1. Templates for Form of the User Interface Design

The two templates in this section supply specific ways for designers to approach the form and organization of the user interface.

Template 1: Mapping Conceptual to Physical Hierarchy

One characteristic of a good user-interface design is a close correspondence between the physical user interface; that is, the *look and feel* level, and the concepts that the user forms about the system. This template directs attention to that desirable correspondence for one kind of conceptual structure.

The template concerns only hierarchical structure, although it is easily modified for other structures such as grids or lists (Figure 4). To fill in the template, the designers construct an explicit hierarchical conceptual

Figure 5. Conceptual hierarchy expressed in user–interface hierarchy.

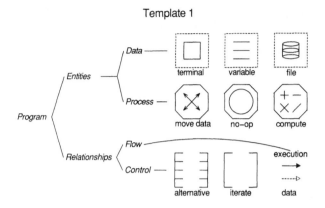

structure that they intend for the user to form as indicated in the left column of the figure. They then construct a corresponding hierarchy of user interface objects as shown abstractly in the middle column and by an example in the right column. This leads naturally to the physical features, such as graphics, that distinguish and structure the interface objects.

This general kind of mapping of intended user concepts to user-interface design structure is obviously useful; the designer's concern is, after all, more with the user's concepts than with the physical user interface. So I find it surprising how seldom explicit modeling of intended user concepts is done. The template exists to encourage and support the often neglected design move of modeling intended user concepts and using that result to drive physical interface design.

For example, Figure 5 shows a small illustrative piece of a graphic interface for a visual, general-purpose programming language. Programs are composed of *entities* and *relationships*. Entities are either *data* or *process*, whereas relationships are either *flow* or *control*. (More detailed structure is in the figure.) So the conceptual structure is set up to be hierarchical, and the conceptual structure is reflected in the physical structure as follows: The entity icons are grouped together spatially as are the relationship icons. The two are separated from each other by extra white space between the second and third rows. Data and process icons are distinguished by square dotted outlines for the data icons and octagonal solid outlines for process icons. Control icons are grouped by being horizontally adjacent and by having similar forms. The flow icons are separated from the control icons by shape, by label arrangement, and by placement of the arrows slightly off the graphic grid. The execution flow arrow is solid to indicate its logical connection with process icons; the data flow arrow is dotted to show its logical connection with the data icons.

Figure 6. Organizing the interface by system-subsystem, by task-subtask, or by a combination.

Template 2

Organization by System–Subsystem	Organization by Task–Subtask
Document tools	Document usage
Word processor Text editor Text formatter Spell checker	Memos Department Company External
Graphics editor Drawing tools Color mapper Print facility	Presentations Formal Informal
	Reports Monthly Quarterly Annual

As an example of how this template might aid development of rationale by focusing discussion, consider the plausible suggestion that the execution flow and data flow arrows should be separated and placed respectively with the data and process icons. With the explicit mapping done, this can lead to a fruitful design discussion about the conceptual structure for the program elements that will be most useful to the user. Without the explicit mapping, the discussion is likely to revolve around some less critical issues of graphic form or style.

Template 2: Organizing by Task-Subtask or System-Subsystem

It is helpful to define the overall organizing themes of a user interface very early in design. This template offers a choice of two such themes (Figure 6).

Two convenient ways to organize an application are: (a) according to a user's tasks as expressed in the language of the work, that is, by task and subtask, and (b) according to the applications functional components, that is, by system and subsystem. To use the template, the designers simply choose one and then use the choice to guide design. Also, it is easy to extend this template by defining other organizing themes. Examples are organization by predefined sequence versus organization by free access, or organization by objects versus organization by functions.

Each of these is appropriate at some time, and of course it is common to see both organizations present in a single good design but clearly separated. A problem results from mixing the two in a haphazard way so that the user cannot develop a clear organizing concept.

A task-subtask organization may be more appealing to users. However, to design task-subtask organization requires much more knowledge of the

Figure 7. **Criteria for type and level of user learning and for rate of skill development.**

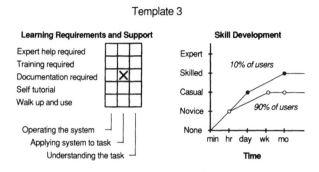

application domain. If the fit is good the application will appear powerful; if the fit is poor, it will appear inflexible, or will appear to deny the user access to underlying capability.

For example, an interesting real application of this template concerned online operating documentation for a large chemical plant. The development team, immersed in the detail of creating, organizing, and navigating a large image database, struggled with a user-interface organization for this large database system that users would accept. The solution came fairly easily with the realization that the data could be organized by department and job, using a volume, chapter, and page metaphor adapted from existing paper documentation (i.e., by task concepts rather than by system concepts). A consistent organization required some real work to avoid expressing database organization inappropriately. But with a clear organizing theme as a basis for design rationale, the developers were able to detect unwanted intrusions of database and system concepts into the user interface and to replace them with task concepts.

3.2. Templates for Establishing Design Goals

A clear statement of design goals or requirements is an essential component of rationale and an essential guide for design. The three templates in this section pose questions or identify distinctions concerning design goals.

Template 3: Specification of Learnability Goals

A carefully framed statement of learnability goals can be very helpful in guiding design choices. This template provides a framework for such a statement (Figure 7).

To specify *learning requirements and support,* the designers choose the level of learning support that the system will supply. If the system is *walk-up-and-use,* then everything the user needs to know is *immediately* available. If it is self-tutorial, then adequate help and online instruction must be supplied. If the system supplies less support, then external documentation, training, or even expert coaching may be required. The designers also choose the kind of learning supplied. Using a word processor as an example, operating the system would involve such operations as reading and writing files, printing, and operating a menu system. Applying the system to a task would involve composing documents, editing paragraphs, and checking spelling. Understanding the task would probably mean understanding how to be a writer.

To specify the planned course of *skill development,* the designers specify the practice time in six increments (none, 1 hour, 1 day, 1 week, 1 month, plateau) needed to reach each of five skill levels (none, novice, casual, skilled, expert). Part of the process is to reach a shared understanding among the designers of what each of those levels means in the context of their specific project.

The explicit statement of goals is very much like the specification of goals in classic usability engineering (Whiteside, Bennett, & Holtzblatt, 1988); this template differs in its focus on construction rather than evaluation and in its briefer format. The levels, categories, and increments were chosen through experience as useful guideposts at a very early stage of design. They are numerous enough to define a continuum or range, not just a binary choice. Yet they are far enough apart to suggest qualitatively different approaches to design. For example, specifying a self-tutorial system requires building additional tutorial subsystems, not just enhancements. As an additional value, if these learning goals are specified, they not only guide design, they also serve as the basis for user testing at a later time.

For example, a software tool was needed to capture knowledge of fault diagnosis of computer hardware for inclusion in an expert system. The tool was required to be accessible for one set of occasional users who were diagnosis experts but unfamiliar with expert systems and knowledge capture. Furthermore, these users were known to have little patience with new systems not directly related to their main diagnostic work. Also, they would never spend time in training. The following design decisions were made:

1. The tool was designed to require documentation but no training.
2. Documentation would explain how to use the system in support of diagnosis, but would not explain the actual diagnostic work.
3. Plateau skill was set at casual.
4. Time to reach plateau was set at 1 week.
5. Novice level was required by the end of 1 hour of exploration of the system. As defined by this team, novice meant an ability to do at least some useful work, no matter how inefficiently.

Figure 8. **Three main distinctions of human–computer interaction.**

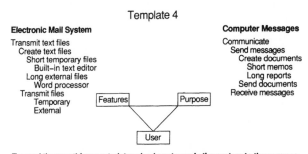

Template 4

Electronic Mail System		Computer Messages
Transmit text files		Communicate
Create text files		Send messages
Short temporary files		Create documents
Built-in text editor		Short memos
Long external files		Long reports
Word processor		Send documents
Transmit files		Receive messages
Temporary	Features	Purpose
External		

User

To send the monthly report, determine how to apply the system to the purpose
1. Create and revise the draft in the system word processor.
2. Bring the final file into the e–mail system.
3. Create a short cover message with the built–in editor.
4. Send the message using the transmit command.
5. Specify the external file name of the distribution list.

This clearly stated requirement guided (in fact, severely constrained) the construction side of design. It also provided conditions for passing the user evaluations that were conducted as soon as a prototype system could be field tested. Additional support was required for a set of skilled users, and that produced a very different set of entries in the template. The new understanding of the requirement to support the two user types guided the developers to divide the tool into two subsystems, one for skilled use and one for occasional use. In addition, they developed a learning path from one subsystem to the other. So, simple and rough answers to the questions of this learning and skill development template provided much of the rationale that guided the design.

Template 4: Three Parts of Human–Computer Interaction

This template (Figure 8) divides the concerns of human–computer interaction into three familiar components that have been recognized for years under many different names (Card, Moran, & Newell, 1983; Carroll & Rosson, 1991 [chapter 4 in this book]; Hewett et al., 1992; Laurel, 1991; Norman, 1986). I have included it as a template because this simple statement of the distinction is useful for keeping design teams representing many different skills oriented and goal directed in the exceedingly complex user-interface design space. The real-world example at the end of the section should support this contention.

To use this template, designers need to be be very careful to distinguish and explicitly state three main concerns: (a) the purpose for the system as expressed in the language of the application domain, (b) the system features that make it possible for a user to achieve the purpose using the system, and

(c) the user's concept of how to achieve the purpose by using the system as expressed from the user's subjective point of view. This is easy to say, but in my experience it requires care, discipline, and thoughtful judgment to accomplish in practice. The figure shows a very simple example.

Even though numerous published analyses show essentially the same division, many of the design discussions I witness become confused because of failures to distinguish these different topics. For example, implementation-oriented developers sometimes neglect to state the purpose of the system, but jump directly to features. Sometimes a user interface becomes very muddled because system concepts expressed as features are mixed indiscriminately with task concepts. (See Template 2.) Even though the general distinction is well known and well established, even though it is fairly obvious, many developers who need to use it have not yet heard of it. Many who learn the words are still unable to apply the distinction in practice. This is a simple template that can be used to raise the quality of many design discussions, especially when there are participants who do not specialize in human–computer interaction.

For example, the manager of a group, justifiably impressed by the usability and learnability of a popular family of personal computers, ordered that all future applications in the organization should have user interfaces essentially identical to the GUI model that had been so impressive. The real purpose, that is usable and learnable applications, was somehow lost. Accommodating a large number of installed character-cell terminals was not carefully considered. Issues of consistent user concepts for the family of fairly complex applications were not adequately addressed. The direction was finally not successful partly because of the focus on surface user-interface features to the exclusion of the other two areas of concern. In other words, the goals that were the basis for design rationale became unclear, and design thinking drifted off course.

Template 5: Trading Off Expressive Power and Required Skill

There are genuine design trade-offs in human–computer interaction, and working with these trade-offs can be crucial for good design. This template concerns the *expressive power* of the input and the *skill* needed to operate the user interface (Figure 9).

The general idea for this template is that the more range, subtlety of expression, and speed-enhancing shortcuts an input method provides, the more skill is required to take advantage of it. That relationship is expressed by the graph relating expressive power to skill required and by the examples that help to establish the scale.

To use this template, designers should begin by deciding whether the main design driver is maximizing expressive power or minimizing skill

Figure 9. **Expressive power of the user interface and user skill required.**

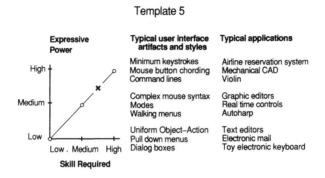

Template 5

required. Then they need to select a point on the *trade-off graph* (marked by the X in Figure 9). Part of this selection is developing a shared understanding of the scale for the graph axes. (See Template 3 for a similar calibration exercise.) Finally, they should design a fallback position in case user testing shows that too much skill is required or insufficient expressive facility is provided in the first design,

Wroblewski (1991) referred to real trade-offs as *design economies*. This is a useful phrase as it suggests that it is possible to buy one feature—expressive power—with the currency of another feature—required skill. There are also false trade-offs that are firmly entrenched among software developers. An example is the claim that a system that is easy to learn will necessarily be clumsy to use. Genuine trade-offs are hard to find, hard to separate from false trade-offs, and hard to express. They are harder still to prove generally true. The power-skill template is a very carefully stated and structured trade-off that I have found useful for organizing discussions in a fairly wide range of practical situations.

For example, a workstation application supported creation of screen objects, arcs connecting the objects, and text associated with objects and arcs. A usability requirement was for creation and manipulation of objects to be fast, even though some of the users might have little practice time. A trade-off of expressive power against required skill was identified:

1. Users would often want to select a single object and perform multiple functions on it by mouse button-clicking on function buttons. This suggested that the object should remain selected after an action was performed on it and until it was deselected.
2. Users would often want to select a single function and perform it on multiple objects by mouse button-clicking on the objects. This suggested that the function button should remain selected after an action was performed on it and until it was deselected.

Both of these methods could be supplied simultaneously at the cost of requiring the user to be aware of both the currently selected function and object—any lapse would result in errors. A popular graphic editor known to the developers proved that skilled users could handle this, and that they liked its versatility and speed. But the same graphic editor proved that unskilled users made many errors.

The designers' difficulty amounted to lack of knowledge of how much required skill the application users would be willing and able to accept. Therefore, they implemented a system slightly more difficult than they believed their users would want but in such a way that it could be quickly simplified without redesign. They were able to do that by carefully looking at the power-skill trade-off. They submitted the first implementation to users for evaluation, confident that they could immediately fall back to a prepared position if appropriate. They converted the test from a success or failure situation into an experiment for tuning a design parameter. There was no resistance to accepting the results of user testing and the experiment led to an excellent design compromise. So, use of the template facilitated an empirically supported rationale for an unusual design.

3.3. Templates for Managing Design Process

There are situations that occur repeatedly in a very similar form during the course of design. I have found it practical and useful to capture these situations and effective actions as templates for process. The three templates in this section illustrate that idea.

Template 6: Exploring the Decision Space

Comparing the kinds of statements made in design meetings with proposed structures for rationales suggests ways to work from partially stated rationales to nearly complete ones (see MacLean, Young, Bellotti, & Moran, 1991 [chapter 3 in this book]).

To use this template, designers apply the following process (Figure 10). For any proposed design *option* (MacLean et al.'s, 1991 [chapter 3 in this book], notation for a candidate feature) ask what *question* (MacLean et al.'s device for grouping options) it answers. Once the question is identified, ask what other options might be included as answers to the question. For any question, ask if there is a bigger, broader, or more inclusive question, which in MacLean's terms converts the original question into a *consequent question*. Finally, for any list of options, ask what the *criteria* are for making a choice.

This is a very simple and effective sequence of moves in a design meeting. In my experience the first query, "If this is the answer, then what is the question?" tends to break log jams and leads to a much expanded explo-

Figure 10. Exploring the decision space with questions, options, criteria, and consequent questions.

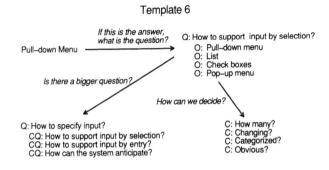

Template 6

Pull–down Menu

If this is the answer,
what is the question?

Q: How to support input by selection?
O: Pull–down menu
O: List
O: Check boxes
O: Pop–up menu

Is there a bigger question?

How can we decide?

Q: How to specify input?
CQ: How to support input by selection?
CQ: How to support input by entry?
CQ: How can the system anticipate?

C: How many?
C: Changing?
C: Categorized?
C: Obvious?

ration of the design space. The query, "How can we decide?" can break the deadlocks that happen when designers commit too quickly to a position and fall into bringing in only supporting evidence for the position.

For example, in a graphic application that involved creating nodes and linking them with lines, the initially designed method for creating nodes was to drag a copy of one of the dozen node types from a set of icons in a control panel. The method was effective and there were no specific complaints. Still, because no other options had been considered, the design team chose to explore the design space further by asking what other ways might be possible for creating nodes. An additional option was entering a node creation mode for a particular node type and creating a node at each mouse button click on any blank area in the workspace. Another was to always create a "nonspecific" node on any mouse button click on a blank area and specify the exact node type later. This led to a new possibility for creating arcs before nodes by letting the application automatically create a "nonspecific" node at the end of an arc that was placed in a blank area. Although the original question concerned creation of nodes, a larger question of general object creation was identified, specifically for lines and labels. This raised new criteria-related issues about consistency. The entire discussion led to considering, as a design criterion, the trade-off between expressive power and required skill. (See Template 5.) By using the process template to focus the discussion, the designers were able to create a convincing rationale for a substantial part of the user interface, starting from one, unexamined design decision.

Template 7: Rules of Evidence

Designers would like to know if the systems we build achieve their goals. Unfortunately, it is seldom possible to determine success directly; rather judgment must depend on indirect evidence such as samples, opinions, and

Figure 11. **Procedure for including data in a rationale.**

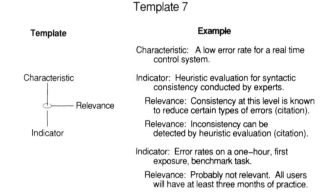

Template 7

Template	Example
	Characteristic: A low error rate for a real time control system.
Characteristic	Indicator: Heuristic evaluation for syntactic consistency conducted by experts.
Relevance	Relevance: Consistency at this level is known to reduce certain types of errors (citation).
Indicator	Relevance: Inconsistency can be detected by heuristic evaluation (citation).
	Indicator: Error rates on a one–hour, first exposure, benchmark task.
	Relevance: Probably not relevant. All users will have at least three months of practice.

performance on benchmark tasks. This template calls such evidence *indicators,* distinguishes indicators from the actual *characteristics* of interest, and suggests the need for a *relevance argument* to connect them (Figure 11).

To use this template, designers need to distinguish between the characteristics of interest and the indicator used as a predictor of that characteristic. For example, ease of use is a characteristic of interest for many interactive systems. Common indicators are tests of a sample of potential users on a few benchmark tasks, or observations of a few users doing work in their own environments, or evaluation of the system heuristically against standards, guidelines, or known effective practice. Besides making the basic distinctions, the designers need to make a case for the relevance of any indicator as a predictor of the characteristic of interest.

Indicators are needed because it is impossible to do exhaustive testing in a real situation; it is impossible to observe every user doing every system-supported task. Yet the relevance of the indicators available for human–computer interaction is always questionable and seems to be the basis of many of the debates in the field. Because the arguments about the relevance of indicators can become heated, cases for or against relevance tend to be a little overstated, and the distinction between desired characteristics and indicators can become a little blurred. This is especially true for developers or designers who are only slightly familiar with the discipline.

I often see two kinds of problems in this area. One is a belief that a favorite indicator should always be used but without any proof of relevance to the characteristics of interest, for example, test results from naive first-time users. The other is a neglect to consider alternative indicators, for example, heuristic evaluation as a less expensive partial replacement for some user testing. This template is designed to keep the distinction between characteristic and indicator in the foreground, and to emphasize

Figure 12. **Postponing, making, and revisiting design decisions.**

Template 8

For a design decision under consideration

If the decision is postponed, what will it cost:
 – In time used to continue consideration of options kept open
 – In designing around constraints that may prove irrelevant after the decision

If the decision is made, what will it cost:
 – In options that are no longer available
 – In risks of losing the benefit of information that may become available later

If a previously made decision is revisited, what will it cost:
 – Directly, in rework
 – Indirectly, because of contingent decisions that will also have to be revisited

that the relevance argument is usually the most difficult part of using indicators and always deserves careful attention.

For example, a prototype system was designed for a hardware maintenance organization to improve tracking of customer contract status, parts inventory, and service technician availability. It was believed that an ability to more easily move information from each of the involved databases to the others would have the characteristic of improving tracking.

The results of two indicators were used to support this belief. The first indicator was an interview in which one worker demonstrated the amount of manual reentry from one database to another. The second indicator was a series of interviews with workers from the same site as they used a workstation with simple cut/paste capability. Both indicators suggested that, for these workers, their effectiveness in the tracking function would be better with support for moving information easily.

The problem for relevance is that there was no proof that these workers were typical in the organization. The best indication was simply that several workers had identical formal job descriptions. But because the planned users were geographically and organizationally dispersed, and procedures varied with location, the argument for relevance was not perfect. Because the rationale was fairly well worked out with the template, the designers were confident in going ahead with the prototype but kept options open for change because of the weakness of the relevance argument.

Template 8: Delay Decisions as Long as Possible But Not Longer

Design is a process of making decisions within a limited time frame. Managing the timing of decisions is a key part of practical design. This process template suggests taking time in design discussions to consider the cost of deciding or not deciding by asking a few key questions (Figure 12).

In any design process, especially during a meeting, there will be moments when it is not clear whether to continue work toward a decision or to postpone the decision—moments when it is not clear whether to revisit a previous decision or let it stand. This template is easy to use. All that is required is to ask the question in the template, that is, to explicitly determine the cost.

The decisions about managing the design process tend to be lost in the clutter of group process. Yet, they are often crucial. As it is not feasible to say in general when to decide, or revisit, or postpone, the best advice is to explicitly bring up the question for team decision on a case-by-case basis.

For example, in an application for storing and editing images optically scanned into computer disk storage, a design debate occurred over choice of a menu word for changing size of the image. The candidate words—each vigorously championed—were *reduce, shrink,* and *scale.* At that point in design there was not yet a decision on whether the function would only decrease size or both decrease and increase size. There was not yet a decision on whether size changes would be in fixed increments or would be continuous. As a complication, it appeared that some team members hoped to drive the functionality decisions in a direction they preferred by setting the choice of menu word at a very early stage of user interface design. After a short discussion, the team agreed that a decision was not useful at that time. In fact, the choice of word was made quite near the end of development with early prototypes using different words. By explicitly examining the process rationale, this team avoided a valueless "rat hole" discussion.

4. CONCLUSION

It is possible to draw a few general conclusions from this summary of a few templates for design in human–computer interaction. Each of the templates bridges a gap between general principles and concrete design. Whereas principles offer general advice, templates require specific decisions or statements. For example, "Be consistent" is a general principle, whereas "The user interface is consistently partitioned along the lines of the user's tasks and subtasks" is an instantiation of Template 2; it is a specific instance of the principle. As a positive result of their specificity, templates are more immediately applicable to practical design. On the negative side, there are probably many more of them than general principles; the eight mentioned here are only a small sample. Also, their use requires the high-skill task of knowing and selecting templates useful in a given design situation. An approach to acquiring that skill is to analyze many examples using templates as an analytical guide. A complementary approach is for skilled designers to use the template format to write down their practical knowledge for others. Templates for design will not take the place of design principles. But they do have potential as a way to actually apply principles and theory to practical design for human–computer interaction.

Besides bridging the gap between principles and practical design, templates enable the construction side of the construct-evaluate loop in a particularly powerful way. Design is a particular kind of making. It is making and evaluating of models. A CHI '90 keynote speaker (Winograd, 1990) posed the question of what equivalent exists in design for human–computer interaction that fills the role of the architect's studio model. This is an important issue, for we must construct models for the construction-evaluation cycle to run its course of producing a series of more and more adequate models until a final design is in hand. Yet, model building in the human–computer interaction domain is very difficult because of the diverse and often intangible nature of the issues. Templates offer a way to capture and make available hard-won wisdom on how to take on that difficult task, to guide construction and to inform evaluation as we design with models.

I find that having a collection of proven templates derived from my practice and that of others does make me a more effective design facilitator by dramatically reducing my cognitive work load and by enhancing my confidence in the results of design choices. Also, the best templates have been quickly understood, adopted, and used independently by development teams that are not very experienced in human–computer interaction. Thus, templates do appear promising as a way of capturing, applying, and transferring knowledge of design.

The results so far have been positive enough to lead me to the following efforts: Create an architectural framework for placing the templates. Collect and validate a fairly complete library of templates. Articulate the relationships among templates. Make the library of templates available without my personal attention. Transfer the process of making templates to others.

From first results, all of these goals appear to be practical, and all fall within the context of a current program for developing techniques and skills in design for human–computer interaction.

NOTES

Acknowledgments. This chapter benefited greatly from the careful reviews and thoughtful suggestions of Tom Moran and Jack Carroll and from editing by Marianne Burnham. The work described would not have been possible without the support of Digital Equipment Corporation. The opinions expressed are solely those of the author.
Author's Current Address. George Casaday, Digital Equipment Corporation, 110 Spitbrook Road, Nashua, NH 03062. Email: casaday@vaxuum.enet.dec.com

REFERENCES

Alexander, C., Ishikawa, S., & Silverstein, M. (1977). *A pattern language*. New York: Oxford University Press.

Card, S., Moran, T., & Newell, A. (1983). *The psychology of human–computer interaction*. Hillsdale, NJ: Lawrence Erlbaum Associates.

Carey, T., McKerlie, D., & Wilson, J. (1995). HCI design rationales as a learning resource. In T. P. Moran & J. M. Carroll (Eds.), *Design rationale: Concepts, techniques, and use*. Hillsdale, NJ: Lawrence Erlbaum Associates. [Chapter 13 in this book.]

Carroll, J. M., & Rosson, M. B. (1991). Deliberated evolution: Stalking the View Matcher in design space. *Human–Computer Interaction, 6*, 281–318. Also in T. P. Moran & J. M. Carroll (Eds.), *Design rationale: Concepts, techniques, and use*. Hillsdale, NJ: Lawrence Erlbaum Associates, 1996. [Chapter 4 in this book.]

Cohill, A. (1991). Information architecture and the design process. In J. Karat (Ed.), *Taking software design seriously* (pp. 95–113). San Diego: Academic Press.

Hewett, T., Baecker, R. M., Card, S. K., Carey, T. T., Gasen, T., Perlman, G., Strong, G., & Verplank, W. (1992). *ACM SIGCHI curricula for human–computer interaction*. New York: ACM.

Laurel, B. (1991). *Computers as theatre*. New York: Academic Press.

MacLean, A., Young, R. M., Bellotti, V. M. E., & Moran, T. P. (1991). Questions, options, and criteria: Elements of design space analysis. *Human–Computer Interaction, 6*, 201–250. Also in T. P. Moran & J. M. Carroll (Eds.), *Design rationale: Concepts, techniques, and use* . Hillsdale, NJ: Lawrence Erlbaum Associates, 1996. [Chapter 4 in this book.]

Norman, D. (1986). Cognitive engineering. In D. Norman & S. Draper (Eds.), *User-centered system design* (pp. 31–62). Hillsdale, NJ: Lawrence Erlbaum Associates.

Rossett, A., & Gautier-Downes, J. (1991). *A handbook of job aids*. San Diego: Pfeiffer.

Schön, D. A. (1983). *The reflective practitioner*. New York: Harper Collins/Basic Books.

Schön, D. A. (1987). *Educating the reflective practitioner*. San Francisco: Jossey-Bass.

Senge, P. (1991). *The fifth discipline*. New York: Doubleday/Currency.

Simon, H. (1969). *The sciences of the artificial*. Cambridge, MA: MIT Press.

Whiteside, J., Bennett, J., & Holtzblatt, K. (1988). Usability engineering: Our experience and evolution. In M. Helander (Ed.). *Handbook of human–computer interaction* (pp. 791–817). Amsterdam, Netherlands: Elsevier Science Publishers.

Winograd, T. (1990). What can we teach about human–computer interaction. *Proceedings of the CHI '90 conference*, 443–449. New York: ACM.

Wroblewski, D. (1991). The construction of human–computer interfaces considered as a craft. In J. Karat (Ed.), *Taking software design seriously* (pp. 1–19). San Diego: Academic Press.

13

HCI Design Rationales as a Learning Resource

Tom Carey
Diane McKerlie
University of Guelph, Ontario, Canada

James Wilson
Eastman Kodak Company, Rochester, NY

ABSTRACT

We report on our experiences applying human–computer interaction (HCI) design rationales as a learning resource for inexperienced designers. The lessons revealed by a series of prototypes are described. A pilot study within an industrial setting developed a framework for access to HCI design rationales, reflecting potential usage scenarios. A prototype system built as an extension to a commercial user-interface toolkit showed that generic design rationale components could guide inexperienced designers, and also revealed aspects of current toolkits that hinder effective access to HCI design rationales. Our current prototype explores the problem of encouraging "use-oriented" design, by addressing representation and rationales of high-level design issues. We discuss how the application of HCI design rationales as a learning resource provides leverage for ongoing design capture and reuse.

Tom Carey is a computer science researcher investigating design methods for user interactions; he is a Professor in the Department of Computing and Information Science at the University of Guelph. **Diane McKerlie** is a usability analyst and a Ph.D. student at the Open University in the U.K. **James Wilson**, a cognitive scientist, is a project research engineer with the Human Factors Group of Eastman Kodak's Electronic Imaging Research Laboratories.

CONTENTS

1. LEARNING FROM HCI DESIGN RATIONALES

Most chapters in this volume deal with the construction and use of design rationales for an artifact during the design—or redesign—process for that artifact. Our focus is different: the reuse of such a design rationale as a learning aid, by others not directly involved in the subject artifact's design or redesign. We are developing a library of exemplar artifacts and design rationales as a learning aid in human–computer interaction (HCI). The library is similar in intent to work reported by Fischer, Lemke, McCall, and Morch (1991 [chapter 9 in this book]); we use a structure for design representations and rationales that is more specific to human–computer-interaction, and employ a "software reuse perspective" to fit our target audience of software engineers. This chapter reports on the lessons learned from our initial efforts, and the directions of our current work.

We have two target contexts for this learning aid: a course-learning situation, such as students studying human–computer interaction, and on-the-job learning situations where software developers are building an interactive software system and need to learn from exemplar artifacts. In both cases, the learners are relatively inexperienced user-interface designers.

In a typical course in HCI design, much of the learning takes place by examining existing user-interface designs and observing their successes and failures. Curriculum recommendations for human–computer interaction courses in college curricula stress this case study approach (SIGCHI Curriculum Development Group report [see Hewett et al., 1992]). The case study approach is particularly appropriate in human–computer interaction, where theory typically follows the appearance of innovative artifacts rather than directing their creation (Carroll & Kellogg, 1989). Thus the artifacts become the "nexus" of an HCI course much as they are for HCI theories.

The on-the-job learning situations are illustrated by the environment within Eastman Kodak Company's Electronic Imaging Division. A wide range of products—including photocopiers, medical imaging equipment, and office workstations—are designed by a multitude of product teams.

Specialized expertise in user-interface design is maintained in a small staff team, the Human Factors Group in the Design Resource Center. This team continually updates its expertise through professional conferences and scientific journals. However, the staff group can provide direct consulting assistance on a limited number of product assignments. Much user-interface design is carried out by product engineers with little formal training in human factors. They learn on the job, mostly by borrowing ideas from previous products and other existing interfaces. One of the motivations for our work was the need for the Human Factors Group to develop mechanisms to make this learning process more effective (and to share more of their expertise outside of resource-intensive direct consulting).

For both learning through courses and on the job, the study of HCI designs is hindered by the lack of explicit rationales to explain how the design achieves its objectives, and why other alternatives were deemed less attractive. In our HCI courses, students conduct a design review of existing artifacts as part of the assigned course work (Carey, 1989). They present the results of that critique to the other class members, so that each student in principle benefits from exposure to critical review of a number of different designs. In practice, many students form shallow judgments of the decisions made in the existing system: either overly subjective ("we didn't like . . .") or not sufficiently critical ("we really liked this system"). In our experience, this holds true even after the students have been instructed in the use of design rationales as a tool for analysis. We return to this problem later.

In learning on the job, much of the learning by inexperienced designers takes place by examining existing user-interface designs, often within a product line or corporate set of products. The existing products serve as a model, good or bad, for the future. Decisions that were necessitated by resource constraints or competitive strategy can be perpetuated; decisions made to address one target market can be taken out of that context and reused in inappropriate ways.

We are currently beginning to build a library of exemplary user-interface designs for use in HCI classes. We are also investigating the feasibility of a corporate exemplar library of HCI designs, for use within Kodak. These design libraries should be more than collections of reusable components. They should also present the way the designer has organized components into a coherent whole, the expected usage of the interface, and the rationale for the designer's decisions.

We have used existing work on HCI design rationales as a base, deriving the content structure for the rationales from QOC (MacLean, Young, Bellotti, & Moran, 1991 [chapter 3 in this book]) and scenario representations (Carroll & Rosson, 1990). The format of the resulting rationales is outlined in Carey, McKerlie, Bubie, and Wilson (1991). We focus here on how this use of HCI design rationales as a learning resource impacts their access structures, content, and capture.

We report in this chapter on what we have learned from this process so far: (a) the questions that would prompt inexperienced interface designers to use the library, and access mechanisms to support them, (b) the role of the HCI design library in comparison with other learning aids, specifically user-interface guidelines, (c) learning needs that inexperienced designers are unlikely to recognize, and how the design library could address them (The resulting content requirements have led us to extend the design library concept beyond design rationales to other design representations.), and (d) the benefits of this use of design rationales as learning aids, on the construction and capture of HCI design rationales.

2. ACCESSING HCI DESIGN RATIONALES AS A LEARNING RESOURCE

We can restate the problem faced by inexperienced designers in a corporate setting as a failure to recoup the investment in design reasoning from past products, because the design was not recorded in a manner that promotes reuse of the reasoning. The problem of reuse is attracting much current interest in software engineering (Prieto-Diaz, 1991). In some cases, this interest has included capturing the reasoning behind software objects (Zavidovique, Serfaty, & Fortunel, 1991). The reuse of interface components is one instance of this more general concern, with the following important difference.

Most attempts to structure software libraries for reuse assume that access will be driven by a functional specification of what the software component is to do. Thus the behavior of the software component is the key ingredient for such systems. In HCI designs on the other hand, the behavior of the component itself (e.g., a "widget" or interactor like a particular type of menu) is much more self-evident; it is the behavior of the user with the component that is much more critical and also less clear. Although widgets can be made accessible through visible palettes of interface objects, this does not mean that their reuse will be appropriate. Design rationales for the widgets could contribute to more informed reuse. (Design rationales would also provide a different structure to the component libraries [Carey & Spall, 1995].)

To accommodate the need to explicitly present information about how people would use a particular interface, we incorporated an explicit scenario component into the QOC (questions, options, and criteria) framework. In preliminary testing with designers, this format allowed more of their reasoning to be expressed (McKerlie, 1991). In using design rationales with student designers, we have also found that explicit scenarios are required to communicate user-interface design expertise. Scenarios are, however, only one component of the content we believe is required for the user-interface design library, a topic we return to later in the discussion of teaching/coaching task-oriented design (see also Casaday, 1995 [chapter 12 in this book] for other design knowledge that could be included for exemplary designs, especially the knowledge about testing in his Template 7).

In addition to requiring different content, a library of reusable HCI designs would require additional access methods beyond specifications of behavior. Toward that end, we developed an access structure to reflect the various ways in which designers would be seeking information from the design library (Carey et al., 1991; McKerlie, 1991). In the simplest example of use, someone facing a decision for a user-interface design would look in an online library to see how the issue had been resolved in a particular previous interface. The library might offer access through an executable version of the exemplar interface, or through a set of screen mock-ups. However, inexperienced designers will often not have the necessary knowledge to determine which exemplars in the library are relevant to their decision. This includes knowledge about the exemplary designs in the library and knowledge of HCI issues.

Based on our experience coaching inexperienced designers, we anticipated that they would frame the concerns motivating their access to the library in the following ways:

1. How should a user accomplish a particular type of function? (e.g., how should users cancel operations?)
2. How should a particular feature be used? (e.g., what functions are suitable for a button array widget?)
3. How can a particular usability criterion be achieved? (e.g., what design decisions have improved "guessability"?)

Our proposed access framework is illustrated in Figure 1. In addition to direct access through links from the exemplar interfaces themselves (or mock-ups), several other views were designed for the library. These views support access directed to specific features, required functional behavior, and impact of particular usability criteria. Multiple views are also valuable for dealing with design rationales of individual projects (MacLean et al., 1991 [chapter 3 in this book]).

The view through features has been implemented by augmenting the palettes in an existing user-interface toolkit, as outlined in the next section. The other views are under development as complementary tables of contents, which we have used in other work to improve access to online information (Carey, Hunt, & Lopez-Suarez, 1990; Hunt & Carey, 1992). The criteria table is based on a published taxonomy (Scapin, 1990). For the function view, we attempted to use an existing taxonomy (Carter, 1986) but found it did not address the variety of issues we needed. We discuss

Figure 1. **Framework for reusing design rationales.**

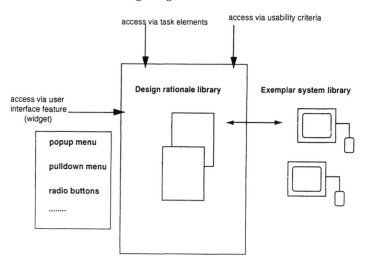

some lessons for providing feature-based and function-based views in the next two sections.

3. HCI DESIGN GUIDELINES AND DESIGN RATIONALES

We have mentioned several methods for communicating design expertise: (a) direct instruction, through courses or textbooks, (b) direct interaction with experts during designs (in either coach/performer or master/apprentice roles), and (c) indirect interaction, through exemplary artifacts and their rationales.

Another mechanism for implicitly communicating user-interface design expertise is a set of guidelines, standards, or approved interface styles. In their least confining form, these are starting points for design—"in the absence of other overriding considerations, do this." In their most confining form, standards can be interpreted as design decisions already made (and sometimes enforced by a constrained construction toolkit).

If we regard all these forms of guidelines in their role as learning resources, we can consider their relationship with other traditional methods for communicating design expertise, and with our proposed new mechanism, the library of exemplary interface designs. It is outside our scope here to compare how guidelines, courses/texts, and direct consulting compare for cost, adaptability, and direct applicability. We will note only that these methods are not, of course, exclusive—a course may contain guidelines, but point out their rationale and range of applicability.

We have experimented with the relationship of user-interface guidelines to the concept of an interface design library. We would expect the guidelines to generalize many specific rationales; indeed, we could lower the cost of creating rationales if designers can refer to guidelines as a default rationale. Exceptions to the guidelines would illustrate the limits on their applicability.

Work to explore how guidelines could fit into the interface design library was begun by graduate student Joseph Crentsil (Carey & Crentsil, 1992). The initial target was provision of design expertise about user-interface widgets. There were two reasons for this choice. First, design expertise was readily available from experienced designers but often not understood by inexperienced designers. Second, the explicitness of widget choice within existing toolkits made it easier to experiment with them than with higher level concepts. This allowed us to provide the feature-based access to rationales described in the previous section.

This research produced a prototype software tool, NIBaid, with the following properties:

1. Usability guidelines and design rationales were integrated into a single framework. Guidelines were treated as generic design rationales; specific rationales could be added as exceptions to the guidelines. The guidelines were obtained by having an expert interface designer generalize her design experience with the toolkit's widget set. This was intended in part to simulate abstracting generic information from an exemplar library. An illustration of this framework is shown in Figure 2. When the generic rationale applies, the designer does not need to include a new rationale.

2. The rationale information, both generic and specific, is accessed through a feature-oriented mechanism—one of the access methods outlined in Section 2. NIBaid is built on top of the NeXT Interface Builder, a contemporary user-interface construction kit. NIB presents designers with a set of widget palettes. Features of widgets can be examined through an Inspector facility. The Inspectors were extended to allow access to design rationale information.

3. The widgets are organized into classes with hierarchical inheritance of methods. Widget classes could be specialized to introduce new methods. One objective of NIBaid was to allow specialization of usability design rationales in a similar fashion: If a widget was being used according to the generic guideline, it could simply be selected from the palette. If a widget was being used in a novel or exceptional way, a new subclass should be created and a new rationale created.

NIBaid was tested with a small number of inexperienced designers, by comparing their interface designs with the existing NIB to those with NIBaid. NIBaid's access to guidelines produced improvements in their interface designs (Carey & Crentsil, 1992). One designer had not originally used the

Figure 2. Sample information structure for generic + specific design rationales (radio buttons).

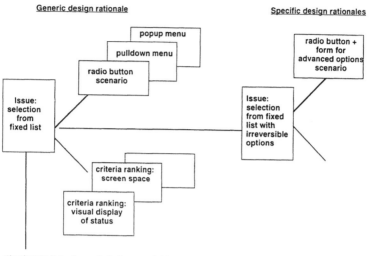

Generic design rationale   Specific design rationales

other issues including radio buttons as viable scenarios

Form widget in the NIB toolkit; as a consequence her screen design was poorly laid out. When the generic rationales were added, the rationale explained that Form widgets were a good choice for organizing a variety of input data elements. Another designer discovered from the generic rationales where to use an array of radio buttons. The generic guidelines served as an online help facility for how and when to use various widgets.

However, the specialization of design rationales within the class hierarchy did not work as well. In particular, we had hoped to make design alternatives apparent through the widget palettes: If the alternatives to a particular widget (Options in the QOC sense) were siblings in the class hierarchy, they would appear on the same palette. This would remove some of the burden of alternative generation from the designer, and possibly permit the inheritance of some design rationale elements. But the class hierarchy was organized around sharing implementation methods. It was not well suited to a design-oriented organization. Design alternatives were often widely separated in the widget palettes, because widgets like buttons and menus were implemented in altogether different ways (despite similarities in facilities offered to users). For example, the Button class appeared in the class hierarchy under the path Object-Responder-View-Control; the Menu class appeared in the hierarchy in the path Object-Responder-Window-Panel. Although menus and buttons might be alternatives in many design situations, the widget hierarchy did not suggest this, and the design rationale information for interaction tasks like "selecting from a set of options" had to be repeated with each widget (instead of being associated with a higher level class).

We are now addressing this problem by developing use-oriented hierarchies of widgets that will complement the existing implementation hierarchies. Such a classification will also serve as a learning resource in its own right.

4. LEARNING "USE-ORIENTED" HCI DESIGN

The NIBaid prototype demonstrated the promise of integrating design rationale facilities with user-interface construction tools. However, it focused on low-level issues such as choice of a particular widget. As Colin Potts points out elsewhere in this volume, a rationale which is explicitly linked to artifact components will typically be subject to this shortcoming (Potts, 1995, [chapter 10 in this book]).

This is particularly troubling when we want to use rationales to communicate design expertise to inexperienced designers, because thinking in high-level task-oriented terms is one of the important components of user-interface expertise.

Kate Ehrlich summarized the situation well in a CHI '90 panel:

> "What we have observed is that software developers are trained to solve technical problems by narrowing the scope of the problem so that there can be an ideal or small set of acceptable solutions. When it comes to user-interface design, this training frequently results in an approach in which the developer takes the functional specification of the product in one hand and the user-interface guidelines in the other and maps, as best as possible, the required functions onto the available widgets. As long as each user visible feature is represented in the interface, the problem can be considered to be "solved." This approach to design frequently results in an interface that is complicated, cumbersome, and provides no clear path through the myriad of features . . .
>
> A user-interface designer, on the other hand, starts from the premise of what task the user is trying to solve and develops a design based on that task . . . This results in a user interface that has some coherence to it, that can be self-explanatory, appears simple and straightforward to use, and does a good job of foregrounding important features and backgrounding optional or advanced features.

Similar problems occur in HCI courses. When we attempted to teach students to apply usability design rationales, they did not connect the role of rationales with the importance of use-oriented design. Rationales were seen as a way of arbitrating among competing concrete, visualizable alternatives for an interface.

An example illustrates how we might go about conveying the significance of design decisions to inexperienced designers in a use-oriented way. The user-interface design library may contain a word processor exemplar. In the word processor, marked text can be moved to a new location in three ways: (a) invoking the *cut* command from a menu, moving the cursor to a new location, and invoking the *paste* command from the menu, (b) following the same process using command keys instead of menu selections, and (c) using the drag-and-drop of the marked text, that is, holding down a mouse button over the marked text and then moving the mouse through the document to the desired new location.

We want to be able to describe these options in a way that communicates why they were chosen, and therefore when other designers should also include them. In a design rationale for these features, we can compare these options on a set of criteria, such as efficiency of operations, consistency with other operations, and so forth. It is easy for the novice designer to imagine the differences in physical movement implied by each option and perhaps the differences in the display style they produce. It is more difficult to see how the visual feedback in the drag-and-drop case can provide a different flow of information back to the user, because the designer trying to relive the user's experience with the options will not share the user's concerns about the text's content and appearance. The ease with which the options can be characterized in terms of physical operations is likely to obscure the fundamental difference in information flows, and make it hard to envision other possibilities that could have been used—for example, including a fish-eye view of the cut text as an additional window on the screen.

Another difficult criterion for inexperienced designers to apply is consistency. If the word processor is a component of a larger system with graphical objects that are frequently moved around the screen display, then this may create the expectation of drag-and-drop for the user when dealing with text. On the other hand, if the work task involves frequent cut-and-paste between documents and drag-and-drop is not provided for that purpose, then offering it for text within a document may have negative implications for consistency.

We want the inexperienced designer to recognize that for different kinds of users doing different kinds of work tasks, the costs and benefits of an additional option for moving text will be different. Our experience with coaching student designers indicates that a full rationale, listing the multitude of relevant criteria and comparing across them, presents a daunting wealth of information that is not easily prioritized. More expertise can be communicated when the decision about how to provide text-moving functions is separated clearly into levels of decisions: about what the users will be trying to accomplish, what information they will need to give and receive to do so, and then how the interaction is implemented. Of course, all of these are interdependent to some degree, so we will also want the decisions to be reexamined from a holistic perspective. But the student designers can only handle a few lessons at a time, and cannot evaluate the whole until they have learned about its constituent parts.

These concerns have led us to conclude that a usability design library must contain more than demonstration software and corresponding design rationale. In order to be an effective learning resource, each system in the user-interface design library must be represented at several levels of semantics, similar to the decision levels in the aforementioned drag-and-drop example. Our original usability design rationales tried to provide additional representation levels by including scenarios of use. However, these are hidden behind particular alternatives for artifact features. They need to be brought out more explicitly.

Our working hypothesis for how to separate out these decisions can be stated strongly: *No rationalization without representation.* The roles of a design representation, which records and structures the design decisions, and a design rationale that justifies/explains them, are complementary but distinct. If we want to present a set of decisions that culminate in a concrete feature, we will need to have explicit representations for the higher levels of the interface so that the visible artifact does not dominate the view of the inexperienced designers.

The most promising mechanism for high-level representation appears to be work scenarios as suggested by Carroll and Rosson (1990). But the leap from work scenarios to concrete interface features is too large for inexperienced developers to make when studying an existing artifact. We believe that the representations in the usability design library must also include an abstract information flow representation. The work task scenarios represent the role of the artifact in the human–computer interaction; the information flow representation is a blueprint for the information "floor plan." We are currently experimenting with a representation that uses work scenarios along with diagrams of information flow states and ERMIA descriptions of information objects (Green, 1991).

An example helps demonstrate how these representations work together. Figure 3 shows part of the initial work scenario for personal computer software to record and manage information about flocks of sheep. Figure 4 shows a sample screen from the existing artifact. Figure 5 shows one of the issues associated with the design for the system, as well as two alternatives to resolve the issue. It is very difficult to identify where this issue affects the screen design, and the evaluation in Figure 6 refers to concerns about how the task is carried out rather than the appearance of the screen. The issue is clarified if we can relate it to another representation of the design, such as the task process representation in Figure 7. Step 2 of this task process raises the issue of redundant entry of the

Figure 3. **Part of work scenario for flock management system.**

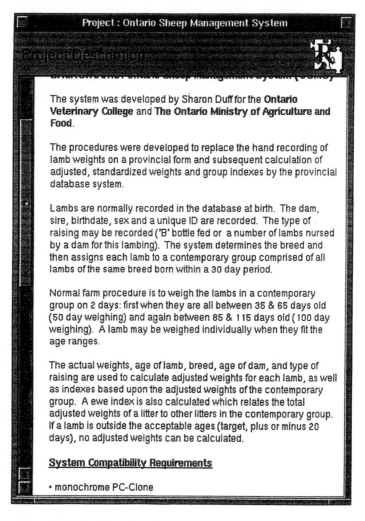

group code, which leads to the alternatives and evaluation in the previous figures. The additional representation helps the inexperienced designer to understand the design rationale, by organizing the design information to separate out the various levels of design issues.

We are developing these design representations for the user-interface design library to augment the learning experience for student designers. This includes preparing a set of descriptions of exemplar systems for use in teaching HCI design. Most of these exemplars are being reverse engineered from existing artifacts. There are many suitable exemplar designs available for this purpose (e.g., Smith, Weiss, & Ferguson, 1987; Stephen-

Figure 4. **Sample screen design for flock management system.**

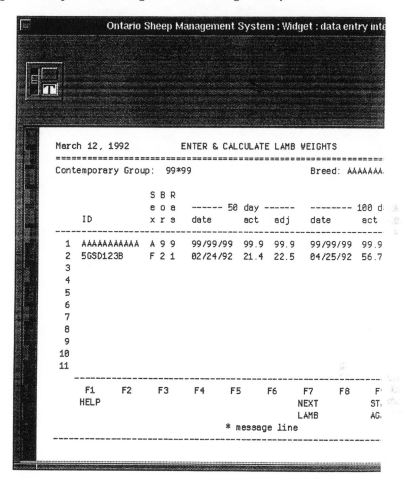

son, 1988) that demonstrate the priority of task design as the driving force behind more concrete user-interface features.

The work with these design representations is still at the first, plan-to-throw-it-away stage. However, it has illustrated some of the relationships between design representation and design rationale. In particular, we suspect that relying on design rationales to help design issues come to the surface may be short-sighted. An effective design representation should be the tool that makes evident what decisions have been made (and which remain to be made). A user-interface toolkit like NIB does this for the visual aspects of the interface, and for the connections between visual widgets and task entities. Higher level representations are still needed for more abstract interface components, like the task types advocated by Brooks (1991). An

Figure 5. Design issue and alternatives for flock management system.

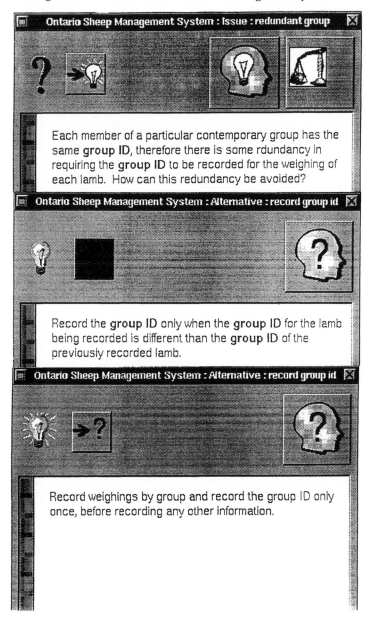

Figure 6. **Evaluation of alternative for flock management system.**

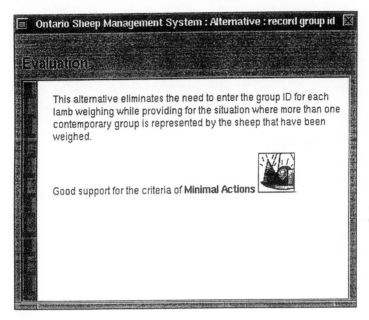

appropriate partnership between design representations and rationales may be able to achieve some of the objectives now set for rationales, such as helping subsidiary issues to surface and tracking dependencies among issues.

5. MAKING HCI DESIGN RATIONALES MORE ATTRACTIVE TO DESIGNERS

To this point we have largely finessed the issue of how design rationales—and the additional design representations with which we are currently experimenting—are generated and recorded for the HCI design library. This is, of course, a more general problem, which has been addressed throughout this volume. However, there are several benefits that arise specifically from our target use of rationales as learning aids.

First, HCI design rationales recorded to serve the needs of ongoing design should be easily accommodated within our design library. We have added complementary views on the design rationales, but these were developed with the intent of incorporating existing design rationales (by automatically constructing the additional links that support learning and reuse). The information for the feature and criteria links is explicit in the QOC rationale format; the task links are explicit in the task-artifact method (Carroll, Kellogg, & Rosson, 1991).

Figure 7. Task process representation for herd system.

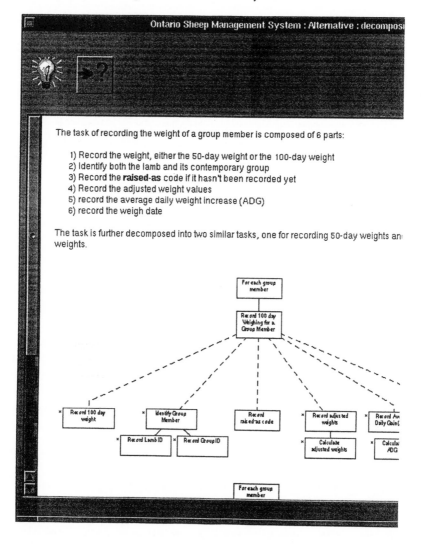

A second implication is that our capture problem is no more imposing than the general problem of capturing and recording design rationales. In some ways it may be easier: It is sufficient for our purposes that the rationale be built after the fact, and be potentially incomplete. Thus in the design library for class-related use, HCI graduate students are preparing design representations and rationales for existing systems. Much of this work consists of reverse engineering, to create a post-hoc representation of the designs at the work task and abstract interface levels.

Although these rationales are based more on heuristic evaluations and empirical reports than on historical design records, they can still accomplish a good deal of what we are after. There is a substantial body of evidence that much design rationale is currently constructed after the fact, often in response to the occurrence of a design review, the need to communicate the design to others outside the original design group (Parnas & Clements, 1986), or the need to accommodate a design to a change in product target. Problems arise whenever subsequent decisions are based on faulty understanding of the original rationale, whether it was recorded at the time of the original decision or later.

In addition, it is sufficient for our purposes that selected key decisions include a design rationale. For purposes of corporate reuse, we have developed the following as a minimal list of situations in which a rationale should be recorded:

1. The designers are uncertain about the decision.
2. The design decisions are critical to product success.
3. A design review raises concerns about the decision.
4. Testing indicates that the decision needs to be revised.
5. The designers are proud of a decision.
6. The designers are unhappy with a decision.

(This last case occurs when project conditions, like schedule constraints, lead to a decision that should be reconsidered when the conditions change. This is particularly noticeable when a product line perpetuates a design decision that was motivated by temporary but critical conditions.)

The reuse of design rationales by nonspecialists has another benefit: It provides a further incentive for specialists to construct and capture design rationales. The mission of specialist groups, like the Human Factors Group at Eastman Kodak Company, includes wide dissemination of their knowledge as well as direct consulting on specific projects. They benefit directly if they can leverage their time to minimize explicit coaching for inexperienced designers. Any additional effort invested in recording HCI design rationales is therefore easier to justify for such a group than for designers who do not have such a mandate.

Such groups may be the major contributors to an interface design library. The library provides them with an opportunity to record how their own expertise has made a difference in the design process during direct consulting assignments. "The absence of a documentation protocol constitutes one of the major impediments to the assessment, and therefore recognition, of . . . ongoing contributions to the design of the artifact" (Fafchamps, 1992).

The richer access structure of the HCI design library may also imply other benefits for the ongoing design process, making design rationale capture more attractive in the long run. For example, options are a key component of QOC design rationales. An initial list of options could be generated automatically from the design library, thus potentially reducing the effort required during ongoing design. We have some work underway to restructure widget palettes and interface object hierarchies to reflect these design options rather than shared implementation methods. Ultimately, effectively identifying and presenting options will depend on richer task semantics in the design representation.

Finally, the growing emphasis on a "reuse culture" within software development should encourage the maintenance of design rationales and representations. For user-interface components, the user-oriented scenarios in the design library serve a function similar to "vignettes," which simulate important uses of other software components (Rosson, Carroll, & Sweeney, 1991). As libraries of reusable software become a more important corporate resource, HCI researchers and practitioners will have to demonstrate that they have analogous but distinct needs.

NOTES

Acknowledgements. We would like to thank the editors and Mak Rusli for comments on an earlier version of this chapter. Steve Portigal suggested the "drag-and-drop" example. Sharon Duff authored the Flock Management example, within the RIDL software built by Michael Ellis. Jack Carroll suggested the term *use-oriented design*, employed in section 4.

Support. This research was sponsored by Eastman Kodak Company and by the Natural Sciences and Engineering Research Council of Canada. Joseph Crentsil was on secondment from the Volta River Authority, Ghana.

Authors' Present Addresses. Tom Carey, Director, Learning Technologies Teaching Support Services, University of Guelph, Guelph, Ontario, Canada N1G 2W1. Email: tcarey@ snowhite.cis.uoguelph.ca; Diane McKerlie, Department of Computing & Information Science, University of Guelph, Guelph, Ontario, Canada N1G 2W1. Email: mckerlie@snowhite.cis.uoguelph.ca; James Wilson, Electronic Imaging Research Laboratories, Eastman Kodak Company, Rochester, NY.

REFERENCES

Brooks, R. (1991). Comparative task analysis: An alternative direction for human–computer interaction. In J. M. Carroll (Ed.), *Designing interaction* (pp. 50–59). New York: Cambridge University Press.

Carey, T. T. (1989). A design-oriented HCI course. *SIGCHI Bulletin, 21*(1), 21–22.

Carey, T. T., & Crentsil, J. (1992). Integrating widget design knowledge with user interface toolkits. *Proceedings of the IEEE Workshop on Computer-Assisted Software Engineering*, 204–212. Los Alamitos, CA: IEEE Computer Society Press.

Carey, T. T., Hunt, W. T., & Lopez-Suarez, A. (1990). Tables of contents as hypertext overviews. *Proceedings of the INTERACT '90 Conference*, 581–592. Amsterdam, Netherlands: Elsevier Science Publishers.

Carey, T. T., McKerlie, D., Bubie, W., & Wilson, J. (1991). Communicating human factors expertise through usability design rationales. *Proceedings of the HCI '91 Conference on People and Computers VI*, 117–130. New York: Cambridge University Press.

Carey, T. T., & Spall, R. (1993). Supporting design rationales in user interface toolkits. In R. Hartson & D. Hix (Eds.), *Advances in HCI*. Norwood, NJ: Ablex.

Carroll, J. M., & Kellogg, W. (1989). Artifact as theory nexus: Hermeneutics meets theory-based design. *Proceedings of the CHI '89 Conference on Human Factors in Computing Systems*, 7–14. New York: ACM.

Carroll, J. M., Kellogg, W. A., & Rosson, M. B. (1991). The task-artifact cycle. In J. M. Carroll (Ed.), *Designing interaction: Psychology at the human–computer interface* (pp. 74–102). New York: Cambridge University Press.

Carroll, J. M., & Rosson, M. B. (1990). Human–computer interaction scenarios as a design representation. *Proceedings of the 23rd Annual Hawaii International Conference on Systems Sciences*, 555–561. Los Alamitos, CA: IEEE Computer Society Press.

Carter, J. A. (1986). A taxonomy of user-oriented functions. *International Journal of Man–Machine Studies, 24*, 195–292.

Casaday, G. (1995). Rationale in practice: Templates for capturing and applying design experience. In T. P. Moran & J. M. Carroll (Eds.), *Design rationale: Concepts, techniques, and use*. Hillsdale, NJ: Lawrence Erlbaum Associates. [Chapter 12 in this book.]

Fafchamps, D. (1992). Cooperative processes to facilitate the design process. *SIGCHI Bulletin, 24*(1), 22–27.

Fischer, G., Lemke, A. C., McCall, R., & Morch, A. I. (1991). Making argumentation serve design. *Human–Computer Interaction, 6*, 393–419. Also in T. P. Moran & J. M. Carroll (Eds.), *Design rationale: Concepts, techniques, and use*. Hillsdale, NJ: Lawrence Erlbaum Associates, 1996. [Chapter 9 in this book.]

Green, T. R. G. (1991). Describing information artifacts with cognitive dimensions and structure maps. In D. Diaper & N. Hammond (Eds.), *Proceedings of the HCI '91 Conference on People and Computers VI* (pp. 297–316). New York: Cambridge University Press.

Hewett, T., Baecker, R. M., Card, S. K., Carey, T. T., Gasen, T., Perlman, G., Strong, G., & Verplank, W. (1992). *ACM SIGCHI curricula for human–computer interaction*. New York: ACM.

Hunt, W. T., & Carey, T. T. (1992). *Accessing online documentation with complementary tables of contents* (Tech. Rep. No. CIS92-009). Ontario: University of Guelph.

MacLean, A., Young, R. M., Bellotti, V. M. E., & Moran, T. P. (1991). Questions options and criteria: Elements of a design space analysis. *Human–Computer Interaction, 6*, 201–250. Also in T. P. Moran & J. M. Carroll (Eds.), *Design rationale: Concepts, techniques, and use*. Hillsdale, NJ: Lawrence Erlbaum Associates, 1996. [Chapter 3 in this book.]

McKerlie, D. L. (1991). *Exploring the use of design rationales and scenarios as an approach for communicating human factors expertise*. Unpublished master's thesis, University of Guelph, Ontario.

Parnas, D. L., & Clements, P. C. (1986). A rational design process: How and why to fake it. *IEFE Transactions on Software Engineering, 12*(2), 251–257.

Potts, C. (1995). Supporting software design: Integrating design methods and design rationale. In T. P. Moran & J. M. Carroll (Eds.), *Design rationale: Concepts, techniques, and use.* Hillsdale, NJ: Lawrence Erlbaum Associates. [Chapter 10 in this book.]

Prieto-Diaz, R. (1991). Implementing faceted classification for software re-use. *CACM 34*(5), 88–97.

Rosson, M. B., Carroll, J. M., & Sweeney, C. (1991). A View Matcher for re-using Smalltalk classes. In S. R. Robertson et al. (Eds.), *Proceedings of the CHI '91 Conference on Human Factors in Computing Systems,* 277–283. New York: ACM.

Scapin, D. L. (1990). Organizing human factors knowledge for the evaluation and design of interfaces. *International Journal of Human–Computer Interaction, 2*(3), 203–229.

Smith, J. B., Weiss, S. F., & Ferguson, G. J. (1987). A hypertext writing environment and its cognitive basis. *Proceedings of ACM Hypertext '87,* pp. 195–214.

Stephenson, G. A. (1988). Knowledge browsing: Front ends to statistical databases. *Proceedings of the IV International Working Conference on Statistical and Scientific Database Management, 2,* 55–65.

Zavidovique, B., Serfaty, V., & Fortunel, C. (1991). Mechanism to capture and communicate image-processing expertise. *IEEE Software 8*(6), 37–50.

DESIGN RATIONALE IN ORGANIZATIONAL CONTEXT

14

A Process-Oriented Approach to
Design Rationale

E. Jeffrey Conklin
Microelectronics and Computer Technology Corporation (MCC)

KC Burgess-Yakemovic
NCR

ABSTRACT

We propose an approach to design rationale (DR) that emphasizes supporting the design process in such a way that a trace of the rationale is captured with little disruption of the normal process. We describe a rhetorical method for design dialogue called IBIS (meaning "issue-based information systems") and two implementations of this rhetorical method: a graphical hypertext tool for conducting IBIS discussions called gIBIS and a simple indented text notation. We describe a field trial in an industrial setting in which the "low-tech" indented text IBIS was used to capture more than 2,300 requirements and design decisions. We also explore the implications of this experience for the design of computer tools that, like gIBIS, seek to capture DR nonintrusively.

Jeff Conklin is a computer and cognitive scientist with an interest in workgroup computing, hypertext, and linguistic models; he is Chief Scientist at Corporate Memory Systems. **KC Burgess-Yakemovic** is a systems designers interested in applying technology to helping people work together; she is currently Managing Partner and Chief Information Officer at Group Performance Systems, Inc., in Atlanta.

CONTENTS

1. INTRODUCTION

Due to the increasing complexity of computer systems and demands for higher reliability, there is growing interest in the notion of capturing the design rationale (DR) of these systems. Yet, because of the wide range of kinds of artifacts and kinds of design, there is not yet a broad consensus about the form and function of DR or about the best way to extract it from the heads of designers.

Although he never used the term *rationale,* Naur (1985) proposed that "the primary aim of programming is, not to produce programs, but to have the programmers build theories of the manner in which the problems at hand are solved" (p.). Naur felt that these theories were so complex and rich as to be only transferable through direct personal interaction and that this is why it is so important for a system maintenance team to include at least one member of the development team. The theory behind a system provides the conceptual background for the specific features and structures of the system design, and it provides a coherence and intelligibility to the huge number of otherwise isolated design elements. For Naur, this theory filled the information gap between developers and maintainers, the gap that so often makes maintenance a matter of guessing what the original designers and implementers had in mind.

Simon (1969) described the design process as being akin to search in artificial intelligence (AI)—in which one is searching the design space. The design decisions in this search process correspond to the branch points in AI search, and the central functions of search—the *successor function,* which generates the alternatives for a given decision, and the *evaluation function,* which determines which of these alternatives to pursue first—are performed by the designer. The notion of searching the design space continues to be a useful metaphor, because the DR can be considered to be the path of decisions and selected alternatives that join the initial state (in which no decisions have been made) to the final state (in which all design decisions have been resolved). Mostow (1985) described an approach to design in which AI techniques are used to capture and structure the design rationale.

Capturing design history may also enhance the design process. Recent evidence suggests how poorly human beings manage exploration of the space of design decisions, especially on large, complex development projects. One study of a team design effort that lasted over many months (Walz, 1989) showed that key design decisions were discussed and explored by the group only to be taken up again weeks or months later as if the decision had not already been explored. Often, the issue had even been resolved when discussed earlier. Another study suggested that the number of issues raised and the proportion remaining unresolved may be the most valuable leading indicator available to management of a system's development progress (Curtis, Krasner, Shen, & Iscoe, 1987).

In this chapter, we use *DR* to mean information that explains why an artifact is structured the way that it is and has the behavior that it has. Rationale provides a dimension of description that is usually missing, by augmenting the "what" of the artifact's structure and function with the "why" behind its design. It is thus a kind of communication from the creator of an artifact to those who later must use or understand the artifact. Therefore, like any communication, the successful use of DR depends on the writer and reader of the rationale document having some degree of shared background.

1.1. DR Capture Is Not New

The notion of DR capture is not entirely new to engineering practice. System design teams have for years tried to document some account of the key decisions and trade-offs they were making in the course of the development process, if only as notes, white papers, minutes of design meetings, and memos.

Indeed, some standards definition groups have made a concerted effort to document every standards decision, including the alternatives and pros and cons (see Figure 1). Defining a standard is, after all, a kind of high-level design, the output of which is a standard: a very abstract design on which all development efforts are expected to build. By recording their DR, these standards groups are able to accumulate a very rich and semiformal docu-

Figure 1. **A sample page from the (unpublished) issue log of the American National Standards Institute graphics standards group X3H31.**

X3H31/82-51   **X3H31 ISSUES LIBRARY**   02/07/83

TG-ID: **2-20-010**

KEYWORDS:
  attribute
  polygon hatch table
  hatch
STATUS: Resolved by TG

ISSUE: **Should PHIGS define a mechanism for defining the hatch table?**

DESCRIPTION:

GKS 7.2 defines a mechanism for selecting a hatch style from a hatch table. GKS defines a function which defines elements of the pattern table,, but it does not define a mechanism for defining the elements of the hatch table. Should PHIGS define this function?

ALTERNATIVES:
1. PHIGS *will* specify a function for defining the hatch table.
2. PHIGS *will not* specify a function for defining the hatch table.

ARGUMENTS:

a. Pro 1, Con 2: This function is useful in many application environments. When devices do not support definition of fill area patterns, but the workstation could support a wide variety of hatch patterns in hardware or software; this functionality would be very useful.
b. Pro 2: Strict GKS 7.2 conformity.
c. Contra b: This is really just an extension to GKS 7.2.
d. Pro 1: Consistency in functionality between raster and calligraphic devices (compare to pattern table).

RESOLUTION: Alternative 1: PHIGS will specify a function for defining the hatch table.

REFERENCES:

ORIGINATOR INFO: X3H31

ORIGINATOR ID: NA

TASK GROUP HISTORY:

This issue was identified by a subtask group at the Berkeley meeting in October 1982.

  October 82 letter ballot vote results:

    8 persons did not vote.
    Alternative 1 - 12
    Alternative 2 - 1

Discussed at the Silver Spring meeting in Dec., 1982. Vote was: 13-0-0 with 2 abstentions.

---

mentation of the learning and exploration process that took place over years. In general, when new members joined the standards committee, they were required to "come up to speed" by reviewing this rationale (R. McNall, personal communication, June 1988). Similarly, a recent programming language book (Steele, 1990) includes an appendix in which all of the decisions that went into the language standard are presented.

1.2. Why DR Capture Is Not Common Practice

Despite the theoretical and practical evidence that a record of DR would be of value, most efforts to capture it eventually fail, for many reasons. Writing down a DR is just a kind of documentation, resulting in a description of the design, a description that is not executable or even implementable, and is thus expendable when deadlines loom. In some kinds of design, such as architecture, which deal with the design of concrete and visible artifacts, documenting design decisions as somehow separate from their realization in the artifact itself can even be disruptive and can prevent the smooth flow of reflective exploration (Fischer, Lemke, McCall, & Morch, 1991 [chapter 9 in this book]).

As documentation, a set of design decisions is inherently unstable, especially if it is being written during an exploratory process. Designers are reluctant to document carefully a wrong turn in the exploration of the design space once it has become apparent what the right turn is.

Moreover, a DR document, in order to be effective, cannot be a "write-only" trace of the exploratory process. It must be kept up to date, earlier decisions that have been undecided or redecided must have the documentation updated to reflect this, and this record of decision evolution must be ongoingly referred to in order to avoid unnecessary rehashing of decisions. Thus, as a development project matures, the DR document can grow into an unwieldy amount of loosely organized textual information. In addition, the chunks of this textual information tend to be highly interrelated, making awkward any linear storage medium, such as paper or computer files.

If, as is often the case, there are many people involved in the development process, there is also a high likelihood that the same issues will come up repeatedly, in different groups and even within the same group over longer periods of time. These repeated occurrences of an issue will usually be worded and even conceived of differently. If potential duplicate decision instances are not detected and eliminated, then the DR document will grow to contain inconsistent information in the form of multiple instances of an issue with differing rationales and resolutions.

One of the most difficult problems for those researchers who would create the DR documentation during, as opposed to after, the design process is that design is often marked by breakthroughs of understanding, so-called "Ah ha!" events. Previous decisions and assumptions may become incorrect, misleading, or irrelevant after such conceptual restructuring. When this happens, much of the existing DR must be updated to reflect the new understanding.

The current interest in DR may be stimulated in part by the emergence of technologies such as hypertext, computer-supported cooperative work tools, and large text database technologies. It now appears possible that these and other technologies may be able eventually to make the capture, analysis, and reuse of DR practical on the large development projects for which this information would be so useful.

It is worth noting that the value of recording rationale extends beyond design to any activity for which there is a need to be able to look back and ask, "Why did we do it that way?" This includes activities similar to design, such as strategic planning and the creation of legislation. In these activities, a complex intellectual effort results in the production of some artifact; however, even in organizational operations, where there is no specific artifact except "to keep things running smoothly," there is often a legitimate need to know why things were done the way they were. Further study will be needed to determine how the current research on DR applies to these other domains. In the mean time, fundamental problems in the design of software systems appear to be driving much of DR research.

This chapter has five parts. The next section provides some background for our approach to capturing DR, the third section describes a field trial that demonstrates this approach, the fourth section presents some of the lessons we learned from the field study, and the final section draws implications for future research.

2. A PROCESS-ORIENTED APPROACH TO RATIONALE

In this section, we describe briefly the events that lead to our current understanding of the design process and the nature of rationale and its capture.

2.1. A Brief History of IBIS and gIBIS

IBIS (issue-based information systems) is a rhetorical model for design and planning dialogues. It specifies a simple set of conversational elements and moves: An IBIS discussion always starts with a root issue, which specifies the main question or problem at hand. One or more positions respond to the issue with potential resolutions, and arguments either support or object to these positions. Secondary issues question, challenge, and generally expand on the discussion of the root issue, and each of these new issues may be explored with positions, arguments, subsequent issues, and so on.

Figure 2. The gIBIS screen. The left half of the screen contains a graphical browser that displays the local structure of the hypertext network (this is especially effective in color). Inset into the lower right corner of the browser is a subwindow that shows the "view from 10,000 feet" of the entire network. The right of the screen contains, from top to bottom: an index window for manipulating linear lists of nodes or links from the network, a control panel for overall gIBIS functions, and a view window that displays the contents of individual nodes from the browser. In this instance, the view window is split in two, and the lower subwindow is editing the contents of a node.

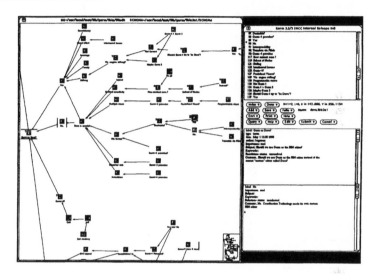

IBIS was developed during the 1970s by Horst Rittel and others (see Kunz & Rittel, 1970) to provide a structure for planning dialogues in which a wide range of stakeholders addressed complex problems (see Fischer et al., 1991 [chapter 9 in this book], for a more detailed history of IBIS). Various extensions to IBIS have emerged over the years, for example, Lee and Lai (1991 [chapter 2 in this book] and Potts (1995 [chapter 10 in this book]).

gIBIS (pronounced "gibbous") is a graphical hypertext software tool for building IBIS networks (Conklin & Begeman, 1988, 1989). gIBIS runs on Unix workstations (see Figure 2) and has a limited "groupware" capability, allowing several users on a local area network to contribute to an IBIS discussion at the same time. gIBIS was developed as a prototype tool for research on using hypertext, groupware, and rhetorical models (e.g., IBIS) to facilitate and capture software system design decisions and their rationale. It evolved substantially during 3 years of active experimentation and field testing at MCC and its member companies.

During much of this time, gIBIS was also being used by researchers at MCC and some of its member companies as a practical tool for design and planning. However, gIBIS exposed a paradox of technology transfer. The investment to use gIBIS paid off best when it was used for a project that had specific goals, an extended time frame, and a clear process that was used by all project members. These are characteristics of commercial development groups much more than of research groups. However, development groups generally are not equipped as well as research groups, and in the MCC member companies development groups almost never had networked Sun workstations for each member of their project teams. As a consequence, the engineering groups that wanted to use gIBIS did not have the necessary equipment for it, whereas the researchers, who had the equipment, only used it in a casual way.

Nonetheless, gIBIS served as a productive vehicle for research on DR. The first author used it to explore a variety of possible extensions to the IBIS method for representing some of the subtle and complex phenomena that arise in the logic of DR (Conklin, 1989b; see also Lee & Lai, 1991 [chapter 2 in this book]), some of which are described in Section 4.

2.2. A Lesson in Technology Transfer

Several efforts were undertaken in the MCC member companies to develop versions of gIBIS that would run on the existing engineering computer environments, including PCs, Apple Macintoshes, and mainframes with terminals. Most of these reimplementations never reached "industrial strength" and were never used on engineering projects. One, however, was incorporated successfully into an engineering project (at NCR). This successful implementation of IBIS is very simple and is really little more than a notation. Called "itIBIS" (for indented text IBIS), it uses indentation to represent the hierarchical relationships among the nodes. Issues are labeled with I, positions with P, supporting arguments with AS, and objecting arguments with AO. Issue resolution is marked using * to show that an issue has been resolved or that a position has been selected, ? to show that no decision has been made, and – to show rejected positions. The example in Figure 3 shows what a simple issue deliberation might look like. Team members could use an outline processor, their favorite word processor, or even pencil and paper to construct itIBIS documents. Perhaps the most salient feature of itIBIS is that it was developed by the people who were to use it (the ultimate in user-centered design) and that it solved an immediate problem that they had: having effective documentation of a long series of design meetings. The itIBIS notation turned out to be especially useful when used to capture notes during meetings or to restructure them afterward.

The NCR project, which is described in detail in Section 3, consisted of five people working on one component of a rather complex software system.

Figure 3. **An example itIBIS discussion. the root issue (I) is marked as resolved (*I) and is addressed by three positions (P), the second of which is selected (*P). The positions have various supporting (AS) and objecting (AO) arguments. The argument objecting to the third position has a subissue that seeks verification of the argument's assertion.**

```
*I:  Which processor should be used?
  ?P:    A.
         AS:  Fast.
  *P:    Processor B.
         AS:  Already in use, thus cheaper.
  –P:    Processor C.
         AO:  Will not be available in time.
                 *I:  Can it be delivered sooner?
                     *P:  No.
                     AS: Design will not be completed until next year.
```

The members of the team ranged in experience from 2 years to 12 years. They had not used any form of IBIS before this project. The importance of a "champion" in technology transfer is widely recognized (Bouldin, 1989), and in this case the champion for the use of itIBIS was the second author. She convinced her management to support the use of itIBIS as a low-tech experiment in capturing DR. Then she convinced her colleagues on the project team to try using it to structure their meeting records and some of their most critical documents (i.e., requirements document). At the end of 18 months, the project team had generated more than 8,000 itIBIS entries documenting more than 2,300 requirements and design decisions.

There were other, somewhat smaller, uses of itIBIS. For example, one member company engineering group used itIBIS as the focus of its documentation during an intensive 6-month period in which it met for 4 hours per day every day. During this time, this eight-person team generated 246 issues with 400 positions and 482 arguments. The team ran its meetings according to the IBIS structure; that is, as each person spoke, he or she used the IBIS model to label comments. At the end of 6 months, the planning was done and the team stopped using itIBIS and started writing the document that would become the official output of this project.

From a research standpoint, it was striking to find the users for whom gIBIS had been designed (e.g., designers and engineers) eschewing use of this multifeatured hypertext system, preferring on real projects to use the itIBIS notation for which there was virtually no machine support! There are many reasons this was so, but we believe that a critical component was that, even though the capture of DR was a new idea for the team members, they found in itIBIS a way to capture DR that required minimal changes to their existing work practices. This experience, among others, lead us to pursue the capture of DR from the process-oriented perspective of seeking to build technologies that matched as closely as possible the existing tools, methods, and practices of engineers and designers. The remainder of Section 2 outlines this process-oriented approach.

2.3. The Fundamental Trade-Off: Cost Versus Payoff

There are clearly some costs to making and/or using DR. In practice, either designers must document the DR as the design process unfolds, or the maintainers must reconstruct the DR after the fact, using whatever documentation (or personal contact with the designers) is available. Although current practice leaves all the cost for later (i.e., to the system maintainers), it is possible that overall life-cycle costs can be reduced dramatically if the DR is treated as a first-class document. But the trade-off between the cost of creating a DR and the cost of using it later is a fundamental one. At one extreme is the use of videotape to capture the design process. This has minimal cost to the members of the design team (unless you count their loss of privacy), but, because no abstraction or structuring of the process record has occurred, there is a very high cost to maintainers who try to recover rationale from this record. At the other extreme is construction of an expert system to design a class of applications automatically; in this case, the DR is so highly abstracted that it has largely disappeared into the rules of the expert system. All approaches to DR capture that tend toward automated or semiautomated design are at this end of the cost–payoff spectrum. Obviously, this has very high cost to the designers (at least initially) because the design space must be analyzed completely and structured for reuse, but system maintainers or later implementors merely need to make changes to the functional specification to make any needed modification to the application system.

Any compromise between these two extremes assigns some cost to the designers (to structure and summarize their reasoning process) and some cost to the maintainers (to read and understand the DR documentation). Most current DR research has moved away from the extremes of videotape and expert systems and is looking for tools, methods, and notations that lower the cost to one of these groups without appreciably raising it for the other or equivalently, that raise the payoff of DR capture enough to offset the cost; see also Grudin (1995 [chapter 16 in this book]).

2.4. Two Approaches to the Trade-Off

The range of design situations and artifact types makes it necessary to focus one's DR research efforts. Current efforts can be divided broadly into two categories. The first approach studies the design of relatively stable and well-understood design domains, in which there is a relatively high degree of standardization. This approach emphasizes the careful construction of DR as a map of the design space and focuses on a rigorous and logical representation of the rationale. This approach maximizes the payoff of DR through its reuse and/or through lowering the cost of system maintenance and is represented by the work of MacLean, Young, Bellotti, and Moran (1991, [chapter 3 in this book]).

The second approach to DR studies more dynamic design domains in which the problems are vague (or "wicked"), the solution technology is poorly understood, or both, and in which there is, therefore, little or no standardization of designed artifacts. This approach emphasizes minimizing the cost and intrusion to the designers by making the DR capture as transparent as possible. In this approach, any intrusion into the pure design process must be offset by some process benefit that makes up for the DR capture disruption. This approach is represented by the work of Conklin, Burgess-Yakemovic and Begeman (see Conklin, 1989a; Conklin & Begeman, 1989; Burgess-Yakemovic & Conklin, 1990).

Let us call these the "structure-oriented" and "process-oriented" approaches, respectively, because the former emphasizes the DR as a logical structure (a knowledge representation of the design space), whereas the latter emphasizes the DR as a history of the design process. In the structure-oriented approach, the DR is prescriptive, because it summarizes the design decisions and trade-offs so that others will reuse the reasoning. In the process-oriented approach, the DR is merely descriptive; its reusability is incidental.[1]

Any approach to DR capture results in the production of a new artifact—the DR itself. For the structure-oriented approach, this is precisely the point: One *designs* the DR (MacLean et al., 1991 [chapter 3 in this book]), and the quality of the final product depends in part on the quality of the DR. For the process-oriented approach, however, the fact that DR is itself an artifact is problematic, because one cannot simply instrument the pure process and reify its design decisions without having *altered* the process. (It thus falls to the process-oriented approach to show that it does not on the whole damage the design process.) We discuss this irony later.

Another feature of the process-oriented approach is that, by keeping and indeed emphasizing the temporal unfolding of the artifact and its DR, the process-oriented DR provides a dimension of narrative. This narrative, or story, about the evolution of the artifact includes not only the wrong

---

1. A hybrid approach has been taken by Fischer, McCall, and Morch (1989) and by Fischer et al. (1991 [chapter 9 in this book]), in which a minimally intrusive IBIS system is coupled with the use of a domain-specific knowledge base. In essence, the human and machine have a dialogue in which the human side is process oriented and the machine side is structure oriented.

turns and rejected alternatives in the design of the artifact but also the order in which they were taken, and to some extent this temporal sequence can help in painting a compelling picture about the artifact's rationale. Thus, another distinction between the process- and structure-oriented approaches is that process DR focuses on the rationale behind the design of a specific artifact, whereas the structure DR focuses on mapping the space of design issues for a family of related artifacts (i.e., a generic artifact).

The process-oriented approach reflects an assumption that the DR dwells implicitly within the process of design and that to capture DR is to reify the decisional elements within the process. A closely related assumption is that the technology of this reification must disrupt the process as little as possible, because revolutions in design process are rare, whereas the process of design naturally evolves through small, incremental steps.

The use of IBIS is thus an important part of our approach. The IBIS model is very simple—indeed, it is a minimal model of design decisions. Because of our process-oriented approach, however, the IBIS model is more than a description of DR structure. It is also the rhetorical model within which "legal" moves are made in the IBIS design conversation. Part of the cost of producing the DR artifact in this approach is thus subsidized by supporting communication and coordination efforts that would have to occur in any case, to the extent that the IBIS model is easy to learn and/or natural to use. In the next section, we present evidence that it is both.

3. AN INDUSTRIAL-STRENGTH FIELD TRIAL OF THE METHOD

As a result of efforts to transfer gIBIS technology to the MCC shareholders, it became apparent that there is a large gap between the environments in which the current version of gIBIS can, and would be, used and those in which the capture of DR is economically important: Commercial development groups with scarce resources, hard deadlines, and a central accountability to produce industrial-strength products at a profit. The difference in the environments includes both available computer platforms and organizational practices and attitudes.

3.1. Use of the IBIS Method

The second author had many years of experience as a systems analyst in the software development environment at NCR and, during this time, collected information about the decisions made during software development projects, primarily by keeping detailed handwritten notes on most meetings she attended. Moreover, she acquired a reputation for being good at retrieving this information and other informal documents from the three-ring binders in which she stored them, thus helping the projects she was on to avoid wasting time and losing information. Having been exposed to the IBIS method and the gIBIS tool, when she began the requirements analysis for a new commercial software development project, she decided to see if the IBIS method could be used to help structure the information she normally collected.

The project at NCR involved the development of hardware and software for a special purpose workstation controller, known as the KDS controller, which was being developed to fit into an existing system. NCR distributes responsibility for various aspects of product development to different organizations; for this project, marketing, product management, hardware development, and the four organizations that were responsible for four different parts of the software system (referred to as KDS, POS, MWS, and MAINTAIN) were involved. The KDS team, of which the second author was a member, was thus one of several teams working on the same system. The KDS software had both real-time and distributed system requirements. Because the development environment for the KDS project did not allow the use of gIBIS, itIBIS was used.

Initial Use of the IBIS Method. As an approach to starting the requirements analysis process, the second author prepared an itIBIS document that captured the DR found in the initial requirements documents, as well as (interactively) from the product management group. Although not explicitly stated in the requirements document, the root issue the document addressed was, "What is required for this product?" This root issue was specialized into subissues, such as, "What is required for the product hardware?" and "What is required for the product software?" The initial requirements document contained mostly positions (again, not explicitly stated as such) and some arguments supporting these positions. The itIBIS document became the focus for discussions between the design team and the author of the requirements document; additional supporting arguments were supplied as this itIBIS document was reviewed. Following these discussions, a new version of the requirements document was prepared by the product management group, which incorporated much of the rationale discovered through this review process.

KDS Team's Training and Use of itIBIS. The KDS team was exposed to the IBIS method during the requirements document analysis described before, and they became convinced that there might be some benefit to using the method for capturing other project-related information. Because of the many people and organizations involved in this project, a significant amount of time was spent in meetings where the KDS team discussed not only hardware, software, and user requirements but also design approaches, interfaces with other systems and other departments, and so on. These meetings traditionally had resulted in the generation of prose meeting minutes. The KDS team chose to use itIBIS to replace prose as the format for these minutes. In these project meetings, there was usually a central issue that was the reason for the meeting, such as "What do we need to supply?", "What does the hardware need to do?", "How should we solve this design problem?", "What information is needed for a particular interface?", or "Who should perform system integration testing?" Discussion during the meeting contributed positions and related arguments and raised further issues.

In addition to the introduction to IBIS given to the team through watching its use for requirements analysis, the KDS team was given a more formal introduction to the IBIS method and the itIBIS conventions in a 30-minute session. Further training on how to prepare itIBIS meeting minutes was done by having both a trained and an untrained scribe participate[2] in the same meeting with each recording minutes. Following the meeting, the trainer and trainee compared the notes they had independently derived; after a couple of meetings, the form and content captured by both the more and less experienced person became nearly identical. During the training period, two of the five team members acted in the trainer role. A hard-copy version of the file containing the issues, positions, and arguments recorded during a particular meeting served as a replacement for the customary prose meeting minutes for these meetings.

The rest of the KDS team began using itIBIS approximately 3 months after the initial itIBIS document was created. Team members elected to use either an outline processor or a text editor to create itIBIS documents, depending on which they were more comfortable using. Regardless of the tool used, the result was an ASCII text file, which was stored on the team's network server disk. An itIBIS document (file) contained a group of related issues, positions, and arguments: for example, the DR derived from the hardware portion of the product requirements document, the DR resulting from a series of meetings held to verify the design against user events for the cases when the user was a supervisor, or the DR based on issues relating to the software requirements from a meeting with the marketing organization. Although the primary use of itIBIS was for requirements analysis and meeting minute preparation, two of the developers also made use of the method for a variety of individual analysis tasks, including analysis of design options for particular modules within the application. During the 15 months that the entire team used itIBIS, the KDS project was involved in requirements analysis, high-level design, and detailed design.

---

2. The team member responsible for note taking was also involved actively in the meeting as a participant. The team was too small and resources too scarce to have nonparticipants act as scribes.

Figure 4. Distribution of IBIS nodes in the four project categories. Platform discussions dealt with selection of the enabling substrate, whereas requirements and design discussions dealt with the KDS system itself. The process discussions were about project procedures and policies.

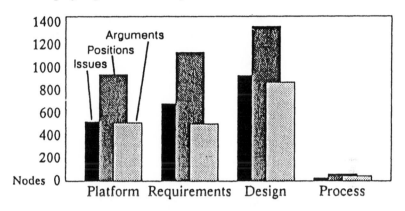

Sixty-six files containing IBIS DR were created using itIBIS; these contained approximately 16,000 lines of text in total and required 780 KB of disc space. As a high-level organizational aid, each file was placed in one of four categories: (a) *platform:* information about the hardware and operating system platform; (b) *requirements:* total system requirements; (c) *design:* application design; and (d) *process:* development process. Figure 4 shows the total nodes by category. The entire set of itIBIS files contained approximately 8,000 nodes and addressed 2,260 issues. By the time the project moved into its second year of development, it had become very difficult for any team member to determine which file contained what information, either for retrieval or for extension.

Conversion From itIBIS to gIBIS. Although the team members continued to contribute to the creation of IBIS information, the difficulty in finding a file that contained a desired issue, position, or argument limited the team's participation in the use and maintenance of the electronic version of the information. Because we thought that a tool that specifically supported IBIS could help with this management problem, a Sun workstation was at last acquired and the indented text files were converted to gIBIS format, through a semiautomated process.

There were two major manual processes required for conversion: one to create the labels used in the gIBIS graphical browser and the other to clean up the results of the automated layout that placed the nodes in the graphical view. While working with itIBIS, the users often added new nodes or updated the status of existing nodes on the printed copy of the DR but did not update the electronic version. In addition to the updates required by conversion, the resolution status of issues was examined and updated as needed, and additional rationale was also added to the gIBIS networks. This additional rationale included information that individual team members had added to their printed copy of the DR, as well as such things as adding to a position the arguments that were discussed but never recorded.

Because of the amount of time required to perform this conversion and DR update, we decided not to try to convert the entire IBIS corpus but, rather, to concentrate on the design category. As a result, only the DR for the application design, which represented 35% of the total, was reviewed in detail. The decision to convert and review only this category of DR was based on two factors: Many design issues were still open, and it seemed that a review of this category would provide the most support to the project activities at that time. In addition, there had been a complete change in direction with regard to the product platform, which meant that the platform DR category could be archived. Of the 66 itIBIS files, 25 of the design files and 4 of the requirements files were converted to gIBIS format and reviewed and updated. Within this set, the resolution status of the issues was updated from 36% open to 12% open. Approximately 800 nodes, including new issues, were added during the update process. Although gIBIS was used for several months at the end of this study, the great majority of the experience reported here is with itIBIS.

3.2. Observations and Results

Two empirical issues were explored during the use of the IBIS method at NCR: (a) whether the IBIS method could be used, over an extended period of time, in a commercial environment, with minimal disruption of the existing practices of the trial group, and (b) whether the method could improve a process of capturing DR that was already in place and extend the use of the process beyond a single individual.

Because this field study was conducted in a commercial product development environment, rather than in a laboratory under controlled conditions, the results of this study are qualitative and open to interpretation. Overall, the results were encouraging for DR capture research and tool development: There were several advantages seen when using IBIS to structure note taking, and there were some obvious disadvantages from the lack of real technical support for the method. We present our observations in terms of two broad categories: impact on access to project rationale and impacts on the development process.

Impact on Access to Project Rationale. At several points in the nearly 2 years covered by this study, there were changes in personnel, and the captured informal information was very useful at these times. In one case, the design expert who was centrally involved in shaping the overall system structure left the company before the detailed design was complete. Because not only his design but also his design decisions and much of the reasoning behind them had been recorded, the team was able to finish the detailed design without difficulty. Indeed, during the project, whenever a "Why did we do such a such?" type of question was asked, there was often an answer in the IBIS record of the decision and its rationale. Perhaps, because of the frailty of human memory, system designers need their own rationale records during development as much as the maintainers who work on the system later.

Impact on Requirements Analysis Process. The use of itIBIS had an impact on several aspects of the development process. The method was found to be very helpful in the design group's analysis of the requirements document, which came from another group, exposing faulty assumptions and weak or missing rationale for the requirements. For example, in the requirements document, a particular feature was requested (a position on the issue, "What is required?"), with no supporting argument. The document's author assumed that everyone knew that the feature was required because there had been a number of customer complaints with a previous system that failed to support this capability. Actually, most of the development team was unfamiliar with the previous product and, therefore, did not know why the feature was required.

Impact on Design Process. Several software-engineering techniques were used by the KDS team to expose problems with their design prior to implementation, including involving the entire team in both high- and low-level design walkthroughs. However, during the process of inspecting and updating the IBIS DR following the conversion to gIBIS, 11 additional problems were found with the existing design; without this examination, it is most likely that 7 of these problems would not have been discovered until the related code was written and 4 not until the integration test was occurring. The 11 errors discovered during this process are outlined in Figure 5. Based on industry average figures,[3] the early detection of problems resulted in a substantial savings for the project. The savings was between three and six times greater than the cost of using gIBIS, calculated from

---

3. The industry figures we used were derived from material quoted within NCR for quality assurance purposes. This material (attributed to Albertson) appears on page 21 of the course material accompanying Computer-Aided Software Engineering Environment USC Institutional Television Network, Dr. Ellis Horowitz, Professor.

the actual number of hours spent working with the tool reviewing the itIBIS information and performing updates.[4]

The information captured in an IBIS provides a different view of the software design from that presented by usual design documentation, so reviewing the issue base allows the design to be reviewed from a different angle, exposing different problems from traditional design reviews. The second author has found, from previous work with DR in an unstructured form, that the act of searching for specific information often reveals problems similar to those found during this study. The process of browsing the DR and looking for a specific entry seems to serendipitously trigger detection of problems or inconsistencies in the data. Further studies will be needed to clarify the connection between review of DR and detection of design errors. We recommend three factors to observe in such studies: (a) whether the DR review is concentrated or spread out over time, (b) whether the reviewer is also an author of the DR, and (c) how the goal of the review (e.g., searching for a specific item) affects the detection of errors.

The detection of design errors on this project is a preliminary result, but it points to the possibility that comparing and reviewing DR might actually reduce the cost of the design process by making it more rigorous and error free. Moreover, we speculate that any tools for browsing and reviewing the DR that make review activities on large projects less formidable will probably facilitate this mode of error detection.

Impact on Project Team Communication. Learning about and using the IBIS method had a direct impact on the communication processes among the members of the group, which also affected the development process.

---

4. Two numbers were calculated for the comparison of actual DR capture costs, C_{DR}, versus potential error cost, C_{ERR}. C_{ERR}, which represents the potential cost had the errors not been found, was calculated as $E_C A_C + E_I A_I$, where E_C is the number of errors that would most likely have been found during coding, A_C is the industry average dollars per error found during coding, E_I is the number of errors likely found during integration, and A_I is dollars per error found in integration testing. C_{DR}, which came out roughly three to six times smaller than C_{ERR}, is the cost of finding the errors in the DR (the cost of reviewing and updating the IBIS DR) and is simply the number of hours multiplied by dollars per hour. We did not include the cost of initial entry of data into the IBIS because, for the most part, the itIBIS was built by creating text files in IBIS format that replaced documentation (e.g., meeting minutes) that would have been produced anyway. The actual number of hours spent reviewing the information and performing updates was tracked fairly closely, for reasons external to this field study. Unfortunately, the proprietary nature of the information prevents us from revealing either the number of hours or the dollars per hour directly. Because the review/update time included time that was "overhead" (e.g., learning to use the gIBIS tool) and because the industry average figures were used rather than NCR specific figures, we have given the difference as between three and six times.

Figure 5. **Summary of the 11 design errors identified during conversion from itIBIS to gIBIS. Each is classified as to when it would have been detected without the IBIS conversion process, which determines how much it would have cost. Key: *I* means "not likely to have been found before integration test," and *C* means "probably would have been found during coding or unit test."**

1. I: Design failed to implement a requirement correctly (an object of a certain class should have been created for all systems, but the designer thought it was needed only under certain conditions).
2. C: Failed to create two objects before their addresses were used.
3. C: A class of objects was used but instances were never created.
4. C: A single data field used for multiple and conflicting purposes.
5. I: Design failed to implement a requirement correctly (handling of multiple cursors was in error under certain conditions).
6. C: One part of the application system status was not initialized.
7. C: Addresses for a group of objects were needed but no method of obtaining them was defined.
8. C: Change in subsystem functionality not reflected in the design.
9. C: Change in subsystem functionality not reflected in the design.
10. I: Design failed to implement a requirement correctly (a timer was not started when it should have been).
11. I: Design failed to implement a requirement correctly (user was not notified of a system action when he or she should have been).

Once the entire KDS team was trained in the IBIS method, it seemed that team meetings were more productive. We propose that a knowledge of the underlying form of the dialogue allowed the team to determine when the discussion had wandered from the intended topic. By recording the discussion, team members were able to "save" any open issues raised during tangential discussions—so that they could be addressed later—and were able to return to the original topic quickly, by reviewing what had been recorded prior to the tangent. (It should be noted that while making use of IBIS during meetings, the team members spoke normally and did not preface their comments with, "My issue/position/argument is . . .") Open issues were tracked, and other important informal information (often in the form of positions or arguments) was more readily referred to by virtue of being logically tied to some design issue. Thus, the itIBIS documentation is perhaps more properly termed *project rationale* than just *design rationale.*

The IBIS method was also used for setting meeting agendas: The attendees were provided with a set of issues and, if the topic had been discussed before, any existing positions and arguments. This proved particularly useful when the issues being discussed were interrelated in ways that were not obvious. In one particularly notable case, where the IBIS was used for meeting preparation, there were a set of seven design choices, all of which depended on one or more of the others for resolution, and each of which had thus evaded independent resolution more than 6 months. Figure 6

Figure 6. **The seven interconnected KDS system design issues. The full rationale of these issues is proprietary—they are presented here to provide a sense of the kind of issues that were involved.**

1. Where should the subsystem's current state data be kept?
2. What needs to be maintained to identify the current state?
3. How should the current state be maintained?
4. How does the subsystem determine the current state after a load?
5. When is the subsystem's parameter file built?
6. How are parameter changes made by external systems to be handled?
7. What changes can external systems make?

Figure 7. **Interissue connections among the seven system design issues. For example, Issues 1 and 2 were interdependent, in the sense of having a constraint on their resolutions (see Section 4).**

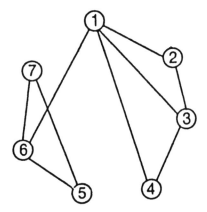

outlines these issues, and Figure 7 shows which issues were related to which other issues. One reason for the difficulty in achieving resolution was that five different groups had a vested interest in the resolution of the issues, and no one group was responsible for resolving all of them. Figure 8 gives an indication of the stakeholder interests in each of the issues.

The set of seven issues constituted a very complex design decision, which no one designer, including the leader, had completely grasped. Review of the IBIS by this head individual, in preparation for a meeting that was originally planned to address two of the issues, allowed the precise iden-tification of the seven issues and their proposed resolutions, as well as identifying which development groups were stakeholders in this issue network. Having then determined what needed to be discussed (the seven decisions) and who needed to be at the meeting, an agenda was prepared and distributed to the proposed attendees, listing the set of issues, with all previously discussed positions and arguments (but not in the itIBIS

Figure 8. Interests of the five development groups in the seven system design issues. For example, the KDS team had a vested interest in the resolution of Issues 4 and 6 and was dependent on the resolution of Issue 2 to take further action.

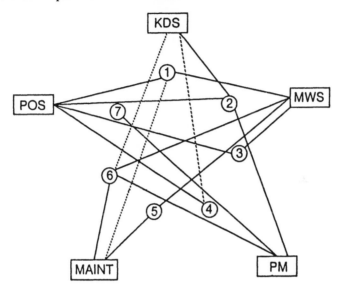

Link Key:

┄┄┄┄┄ = Stakeholder action is dependant on the issue resolution
───── = Stakeholder has vested interest in the issue resolution

Group Key:

KDS   = team using IBIS
MWS   = one software development team
POS   = another software development team
MAINT   = a third software development team
PM   = product management (representing the customer)

format—to be discussed). The relationships between the issues (as shown graphically in Figure 7) were also described in the agenda. This allowed those attending the meeting to come prepared to discuss any issues in which they were personally involved, with an awareness of the related issues and the prior discussions. The global perspective provided by this IBIS agenda allowed the groups to devise a new, previously unconsidered solution that elegantly resolved all seven issues while incorporating the arguments from the prior discussions.

Other Development Process Observations. Two other observations can be made about the development process. First, the KDS team had little success in persuading other groups with which they had to work closely (e.g., marketing) to use the itIBIS method for capturing the discussion that

Figure 9. **Example of an indented text IBIS discussion and a prose text version of the same discussion.**

---

itIBIS form:
I: How should we handle situation X?
    *P: Using Y
        AS: simple
        AS: some of this work needs to be done anyway
        AS: may be able to borrow from an existing implementation
    –P: Using Z
        AO: <etc.>
Prose form:
    The first issue we addressed was how to handle situation X. The first solution we discussed, and the one we finally selected, was using Y. This approach has the advantage of being simple. Also some of this work needs to be done anyway. In addition, we may be able to borrow from an existing implementation. Another potential solution discussed was using Z, which was rejected because <etc.>

---

occurred during their joint meetings. In one case, with the product management group that supplied the requirements, the two groups were able to use IBIS to facilitate the intergroup communication, as described earlier. In the other cases, where interorganizational communication was required, itIBIS was considered too awkward or technical. When queried about not having read the itIBIS minutes published following a meeting with the KDS team, a member of one group commented that the itIBIS minutes were "too technical." It was our impression that the "technicality" was related to having to translate the key (I:, P:, AS:, or AO:) into issue, position, supporting argument, and objecting argument. Following this conversation, when a meeting was held with a non-KDS group, itIBIS minutes were produced for KDS team use, and these minutes were manually converted to a prose form for the other attendees. Figure 9 shows an example of the type of conversion performed. The goal was to take the same information and put it into a more familiar format.

Although the content was identical, the other groups accepted these minutes with no complaints. However, converting itIBIS DR into prose was time consuming and seemed to have limited value. Therefore, after a couple of attempts, the KDS team decided not to use the itIBIS format for minutes when meetings involved non-KDS team members. Intergroup communications with these other groups were troubled by the traditional breakdowns, such as lack of clarity about the resolution status of important requirements issues, that had often been a problem in the past. We now feel that, if the use of itIBIS had been continued as a method of documenting the communication between the KDS team and the other groups, some of these problems could have been avoided or identified sooner.

The second observation that can be made about the development process is that tracking the open issues on the project made it clear that another important aspect of teamwork (i.e., action items) was missing from the group's tracking system. The connection between IBIS and project management in general is discussed in the next section; however, we note that the primary tracking problem was with action items that required action by someone who was not a member of the KDS team. This indicates that action item handling is, at least in part, an interorganizational communication problem.

4. IMPLICATIONS FOR gIBIS

There are two central themes in our research on DR. One is an emphasis on supporting the design process so that we can unobtrusively "catch" important informal information such as design decisions as they are made. To do this, we study what is happening or trying to happen in situated design practice and build tools to give it expression and support.

The other theme is an emphasis on industrial acceptance and technology transfer. In other words, much of our research on DR has been shaped by the need (encouraged no doubt by our funding source!) to show some measure of commercial acceptability in our research ideas and to move our prototypes into field testing as soon as practical. Indeed, we have come to regard technology transfer as an intrinsic and important part of our research. One strategy for easing this transfer process is to replace existing practices and/or technologies, as opposed to introducing new ones (Bouldin, 1989). This is illustrated in the use of itIBIS in the NCR field trial. Similarly, technology transfer is facilitated by involving the users of the new technology in its design and development.

The themes of supporting the design process and facilitating technology transfer result in three major issues (*issue* here in the sense of a class of requirements) for functionality of gIBIS and the family of tools that have evolved from it. The first is support for the processes other than design deliberation that are a part of the larger development process. The second is the need for the DR capture technology to merge as transparently as possible into existing practices and tools. The third deals with computing over the structure of the DR network database.

4.1. Design in the Larger Context of Work

Design is always embedded in a larger context of work, and design decision making is especially sensitive to the processes and practices of getting the work done—however that is organizationally defined. These processes include project management (i.e., people and resource management and planning); communication and coordination with other groups and organizations; developing and maintaining relationships, meetings, and presentations; document preparation; evolution of the understanding of the design problem to be solved; and, of course, production of the actual design or implementation itself. In that these activities all overlap in time, people, and materials, any tool that supports exploration and reasoning should integrate smoothly with the tools for the other activities.

One might object that the activity of design deliberation can be reasonably segregated and supported with a stand-alone tool, and this may be true in certain instances. Indeed, in the research environment, gIBIS has been moderately successful because it is possible to segregate design deliberation. But our field experiences have shown that, in general, design deliberation is not a distinct activity separated from the rest of the process; it is part of formal and informal meetings, phone calls, chance encounters, lunch discussions, and so on, and its progress is intimately bound to the progress of these other aspects of work.

A simple example from the field trial was the desire expressed by the NCR design team's manager to be able to generate a report showing both the open issues and the action items connected with each issue. Once the design team had an explicit representation of open issues, it was natural to want to track the action items that, in many cases, were the stepping stones to getting the information needed to resolve the issues. (Action items were partly supported in gIBIS, with due dates but no reminder mechanism when the due date arrived, but were not added to the itIBIS notation.) This experience suggests that if a DR capture tool also supported tracking of the action items associated with issue resolution, then use of DR would be that much more integrated into daily practices. This makes perfect sense from the simple intuitive perspective that any kind of work, including design, consists of thinking (reflection, deliberation, exploration) and doing (construction, action).

As the action item example illustrates, one of the most important attributes of the development process is that it happens in time. The temporal dimension of DR may be of secondary importance when one is taking a structure-oriented approach, but the process-oriented approach requires that one provide such services as versioning and configuration management of the DR "hyperdocument" (hypertext document) itself. Users of gIBIS and itIBIS reported that they frequently needed to amend either the text within some node or the structure of the network but needed (presumably because of the tentative nature of the exploration process) to be able to "save state" in case they had to review later what the previous node contents or network structure had been.

In any case, this very demanding class of requirements opens the door to integrating support for DR capture with full-scale project management tools. However, much of the project management functionality that is needed is still only available, if at all, in research tools.

4.2. Minimizing the Net Cost of Adopting DR Capture

There is a nearly universal intuitive impression that DR would be a good thing to have in the long term but that the immediate cost of capturing DR discourages the practice of DR capture. In order to make DR capture attractive to designers, this impression will have to be offset by perceptions that DR capture actually has a short-term payoff and that the cost of DR capture has been minimized. Thus, a second class of requirements, stemming from our orientation to technology transfer, deals with ways of minimizing the net short-term cost or maximizing the net short-term payoff of adopting the tools and practices of DR capture in industrial system design.

There are several ways in which designers can be shown short-term payoffs from the practice of DR capture. The NCR field trial demonstrated several such payoffs: for example, increased rigor and clarity of thinking, better communication among team members and stakeholders, and improved meetings.

Also, as Guindon's empirical studies have shown, the process of design is, at a cognitive level, exploratory and opportunistic in nature (see Guindon, 1990). That is, in solving a design problem of real-world complexity, a designer's thought does not proceed in a linear or top-down fashion but, rather, jumps among a range of levels, open issues, and partial solutions. By using a DR capture tool, the designer gains direct support for the memory-intensive process of keeping many issues open while exploring the design space (Simon, 1969).

> The [design] environment should support rapid access and shifts between tools to represent and manipulate different kinds of objects . . . Some of these objects are informal requirements, information about the problem domain, issues and criteria about the system and the design process, design decisions expressed in a formal or semiformal notation, and design process goal management. (Guindon, 1990, p. 341)

Thus, there should be a perceptible payoff simply from augmenting the designer's memory with a DR tool.

At the same time the payoff from capturing DR is increased, the direct costs must be minimized. Some of the cost is simply the effort of writing out the DR and maintaining the accuracy and consistency of the DR hyperdocument. These are mechanical problems that, although nontrivial, may be reduced by engineering solutions such as voice input and limited natural language analysis (i.e., using statistical or neural net technologies).

Some of the other immediate costs may be less tangible, coming for example in the form of organizational inertia. The very concept of DR as primary documentation is quite different from existing practices, and, in any case, producing documentation of any sort is often only weakly an existing practice. For DR capture to become common practice, then, will require a shift in the culture of design toward more of a process and quality orientation. One technique for altering organizational culture is to introduce minor modifications to existing practices. In the field trial, for example, the NCR design team, which already had the practice of writing meeting minutes for design meetings and sending them to their management, found that switching to an IBIS notation for their minutes was an acceptably small shift in practice.

Another way to minimize disruption of the design process is to provide the designers with a clear model of precisely when it is important to capture rationale. If DR must be regarded as an ephemeral substance for which the engineering staff members must be ever vigilant, then they will soon grow tired and resigned about is capture. But if a process model specifies the specific occasions and events when DR is to be captured, then the cost of DR capture is limited to those occasions, and the engineering staff members can be confident that they are producing an appropriate level of DR documentation. For example, many development processes call for periodic reviews (e.g., requirements, design, and implementation reviews); such review meetings might be augmented to include recording the key issues and DR about the artifact under review.

Another idea for reducing the cost of introducing a DR capture tool came from another MCC participant company, one that makes heavy use of email and bulletin boards for its design project communications. This company suggested that an email version of gIBIS would be the most effective tool for their DR capture. Because most design deliberation is already conducted using email and a bulletin board tool, DR might be more readily captured if the email and DR capture tools were integrated.

One of the biggest problems reported by the NCR design team was that the diagrams and texts of its evolving design could only be represented outside of itIBIS or gIBIS. This was clearly a problem, because the activity of design is normally devoted to construction of the artifact called *the design,* and most organizations are already stocked with and accustomed to editors and other design-aid tools (e.g., Computer Aided Design and Computer Aided Engineering tools) for constructing this (primary) artifact.

Thus, for DR capture to occur at minimum additional cost, DR capture tools will need to be closely wed to the editors and design-aid tools being used to shape the design artifact: In the DR community, this problem is being best addressed by the work of Fischer et al. (1991 [chapter 9 in this book]), in which the construction and argumentation tools are tightly integrated with each other. However, hypertext is the only practical medium for representing the highly interlinked chunks of DR, and the great challenge facing the hypertext community is precisely this one: how to bring integration and interoperability into the hypertext picture. At the moment, and for the near future at least, hypertext links will join only those documents whose editors have been extended carefully to support a shared link traversal protocol. Failing this, DR support tools will have to: (a) be embedded into existing editors and design tools, (b) supply powerful text- and graphics-editing capabilities of their own, or (c) take advantage of emerging software integration platforms such as Portable Common Tool Environment (Thomas, 1989).

4.3. Computing on Rationale

The third class of requirements for gIBIS deals with computing over the structure of the DR representation. Our process orientation has focused our research efforts on minimizing the cost and intrusion of DR capture to the designers. Once captured, the DR can be valuable to the designers not only as a record of the design process but also as the basis for analytic tools to add value to the design process. For this short-term value to be added, however, our process-oriented DR representation must be such that useful computation can be done on it, and this need conflicts with the need for the capture process to be natural and unobtrusive (as discussed in Section 2.3 as the fundamental trade-off of DR capture).

In fact, the issue of the trade-off between the expressive power of the DR notation and its ease of use is one of the central problems of the DR field. Lee and Lai (1991 [chapter 2 in this book]) show that certain extensions to the IBIS method, such as goal nodes, allow more detailed trade-off analysis among an issue's positions, which in turn supports a quantitative decision support capability that gIBIS lacks. Decision matrices were also proposed by MacLean, Young, and Moran (1989) as a way of having a computable trade-off analysis for a decision. Our work undoubtedly represents the low-tech, high-informality corner of the field, which we have generally chosen in favor of lower expressive power and greater ease of learning and use.

However, there are a few ways of computing on the issue base that tend to require minimum knowledge engineering effort: multiple *views* of the DR representation, issuing *reminders* about time-sensitive elements, *syntactic checks* on the DR structure, and *semantic checks* on the DR contents.

Views. One straightforward step is to provide customized reports and views on the issue base. Instead of working with the hypertext network itself, the design team and its management can have, for example, a view or report that shows the open issues, ranked in order of decreasing priority, with any open action items that are linked to them. Similarly, customized lists of all decisions made (issues closed) within a specified period could be sent to the various stakeholders in those decisions (e.g., management, marketing, cooperating design teams, etc.). We expect the important views in a given organization to be dependent on organizational practices and goals and to evolve over time.

Two primary views are those of the issue base from the process and structure perspectives. In the process view, time is a major axis, and the temporal unfolding of the issue base—wrong turns, retracted arguments, and all—is the salient property. This is partly what gIBIS shows now. In the structured view, time has been factored out so that the user sees only the current logical state of the design and its rationale. gIBIS also shows this to some extent, depending on how it is used. Sorting out how to separate—and then smoothly integrate—these two views is a major challenge for our research. In particular, the process view is problematic because some aspects of the temporal unfolding of DR are quite useful in making the DR clear to later readers, providing a dimension of narrative that allows the reader to appreciate more easily why certain avenues were explored and others ignored. Conversely, some aspects of the temporal evolution of the DR hyperdocument (e.g., corrections of spelling errors) are trivial and distracting. It remains to be seen if the salient features of the exploration process can be extracted automatically or if the creation of a compelling history remains a matter of human ingenuity and discrimination.

Reminders. A process-oriented DR naturally contains a variety of time-sensitive items, such as due dates on action items, deadlines on issue resolution, and arguments whose validity might be undone by time (e.g., an argument objecting to using processor C because it will not be available until a date 3 months away should signal for reevaluation after that date has passed). A process-oriented DR capture tool will thus offer mechanisms for notifying its authors and other stakeholders whenever a temporal dependency has come due. It is possible that receiving reminders from DR tools will further orient users to thinking of and relying on these tools consistently.

Syntax Checks. Another kind of simple computation is to perform certain well-formedness checks on the syntax and structure of the issue base. A pattern-recognition engine could, for example, detect if a position were selected, even though its only arguments were all objections, and warn the user, or it could warn the user who tries to mark an issue "resolved" without selecting any positions. Although one must be careful about when and how to intrude with such warnings, we are experimenting with a version of gIBIS that is coupled to Prolog and that detects such structural conditions in the network (Hashim, 1990). Because most of these syntactic conditions are defined in terms of network (node/link/attribute) structure, experience with this kind of checking has been limited by the lack of a "structural query" (Halasz, 1988) capability in gIBIS (or any other hypertext system).

Semantic Checks. The most challenging computation on issue networks is to monitor the logical and semantic consistency of the issue base and to aid a user (e.g., a moderator) in maintaining the hygiene of the issue base. We mention here four kinds of semantic structures: repairs, decision support, strategies, and constraints.

1. *Repairs:* One of the most common patterns of design reasoning is to observe a problem with a proposed design component and to devise directly a "repair" that revises the component and removes the problem. In gIBIS, this takes the form of an argument that objects to a solution presented in a position, followed by an issue that challenges the argument, saying, in effect, "Isn't there some way that this position can be repaired that eliminates or addresses the objection of the argument?" Such structures could be detected and monitored to ensure consistency between the selection of the repair and deliberation of the originally faulty solution.

2. *Decision support:* When working with IBIS argumentation, users sometimes feel that there is more going on in the resolution of issues than qualitative judgments. Some issues are more important than others, some positions are clearer or more cogent than others, arguments have differing levels of importance and validity, and it seems natural to represent these more quantitative factors as part of the IBIS network. However, allowing these measures into the DR places it on the slippery slope of decision support systems and group decision support systems (for a good review, see Zachary, 1986). One of the great advantages of IBIS for upstream design is precisely that its qualitative approach avoids the complexities of multiattribute utility theory (see Keeney & Raiffa, 1976) and the quagmires of arbitrary and hair-splitting quantitative judgments. One way to avoid this pitfall is to regard these measures as being a quick and convenient way of recording and communicating extremes of these measures to other people. For example, gIBIS currently supports the attribute *importance* on issues and arguments, and the values of the attribute can be "high," "medium" (the default), and "low." If a node's importance is either high or low, a small subicon is displayed next to the node in the browser to draw attention to it.

One version of gIBIS, called *fuzzy gIBIS,* uses a truth maintenance engine to propagate automatically measures of strength and belief around an IBIS network using fuzzy logic (Lubars, 1989). As might be expected, however, such detailed modeling of the logic of a set of interacting issues requires vastly greater amounts of time and care than simply creating a traditional

IBIS network covering the same issues. When in general it will be appropriate to use these much stronger computational models is an open research issue, but our process orientation and our field trial experience make us cautious about expecting designers in commercial settings to adopt tools requiring such large input efforts—even if large payoffs can be promised.

3. *Strategies:* These are sets of positions from different issues that are all consistent with each other and that reinforce each other. Strategies are used to capture the intuition that in complex problem solving one is rarely resolving individual issues singly or independently. When a strategy is selected, all of its member positions become selected, and all of its issues become resolved. This structure was developed to help represent the process by which the field trial team resolved the set of seven interacting issues described in Section 3.2.4.

4. *Constraints:* It is a commonplace of system design that two or more decisions interact such that certain patterns of resolution among them do not work well, or at all, and must be prohibited. For example, suppose that Position P1.1 on Issue I1 is selected but that this makes selection of Position P2.1 on Issue I2 invalid because the two positions happen to interact in an undesirable or impossible way. We call this phenomenon an "interissue constraint," but we have had difficulty representing constraints using only minor extensions to the IBIS model. The decision representation language shows one way of representing such dependencies (see Lee and Lai, 1991 [chapter 2 in this book]).

Whatever mechanism for constraints is finally developed, however, we can be certain that it, like the other semantic-processing schemes proposed earlier and in the literature, will have the effect of shifting the status of an IBIS network from being a vehicle for structured communication between people to being a formal or semiformal knowledge representation. Our experience with trying to formalize the relationships among the seven interacting issues from the field trial, however, taught us an important lesson: What was operationally important in coping with the seven issues situation depicted in Figure 6 was simply observing that some kind of dependency existed between the issues and not in identifying precisely the logic of those relations. In this case, the power of the IBIS DR was that it simply made apparent to its users what the interacting issues were and that it facilitated devising an entirely new solution that observed the unformalized constraints.

5. CONCLUSIONS

For DR capture to become accepted practice (from the process-oriented perspective) the DR documentation must become "living"—it must take on the same qualities of dynamism that weekly sales charts have in the back room of an automobile dealership. Meeting minutes, rather than static notes that are dutifully written up and filed away after every meeting, must be merged into a single, growing meeting minutes hyperdocument. Issues raised repeatedly must be documented as a single issue's discussion, growing and evolving, and not as a series of unrelated incidents.

For the documentation to be living, it must possess three qualities: ease of input, effective views, and activeness. We have said much about ease of input in this article. This comes down to minimizing the increase in documentation efforts of the people who understand the DR and can be done by advanced input technologies, use of scribes, or capture of DR from existing forms of documentation.

The problem of use of and retrieval from the living hyperdocument presents a greater challenge, because the scalability problem will never be permanently solved. No matter how effective one's technologies and practices are in providing effective access to a DR hyperdocument of size S, the steady growth of the hyperdocument through accretion of new knowledge will always overwhelm those technologies and practices when it reaches $2S$, or $10S$. The natural approach from a database perspective is to develop views and filters that show the user only that portion of the DR needed for the task at hand. But views and filters must always trade-off between hiding potentially useful information and overwhelming the user with an unmanageably large chunk of the database. Thus, efforts at managing large DR libraries will always be with us, in terms of both continued basic research on the scalability problem and continued need for skillful librarians.

The third property of a living document is that it is active. It has the ability to detect important events and conditions that develop within it and to initiate action, from simply notifying a user or librarian to starting substantial computations. A simple example would be the capability to signal that an important issue had remained unresolved or unaddressed for some threshold period of time. A better example would be that a summary of the state of the DR is automatically issued on a weekly basis (e.g., percentage of issues open, percentage revolved within the last week, and list of the top 10 unresolved issues) and this report is used as a management tool.

Our orientation to the introduction of DR capture clearly has been one of minimizing disruption to the existing design practices and tools of both individuals and groups, while seeking to offer as much long- and short-term improvement to the development process as possible. Nonetheless, it is inevitable that many forms of resistance and inertia will be brought out by the introduction of something so new and different. We believe that our field trial experience has taught us some lessons about how to avoid these pitfalls when introducing DR capture technology, principally:

- the importance of a local champion (see also Bouldin, 1989);
- the importance of data showing the cost of not capturing DR (One of the most critical shortcomings in the field of DR research is any

study or estimate of the cost of the failure to capture DR, in any area of design and for any type of artifact.);

- that the imposition of even minimal structure on the informal information that DR represents is generally immediately beneficial; and
- that the practice of DR capture will be easier to accept if it can be limited to specific events or occasions, such as design meetings or reviews.

However, if we are successful and DR capture really catches on in an organization, the new practices and the existence of the DR are very likely to have significant impact on that organization. It is difficult to say from present experience what that impact will be. If it turns out to be possible for the DR to be maintained as living documentation, then that organization will have acquired a new capacity in the evolution of organizations: organizational memory. That is, rather than depending on the memory and permanence of individuals, organizations will have a capacity to form a record of key events and decisions and to make effective use of that history in building its future.

On the other hand, there are some practical and ethical questions that are raised by the existence of DR documentation. After all, on the face of it, DR capture is one of those things, like travel restrictions and layoffs, that is good for the organization but not necessarily good for the individual. Some questions that future research in the field should address are:

- Who does the routine work of DR capture and DR maintenance, and why?
- How is politically or personally sensitive information handled in the DR hyperdocument?
- How are the people who were honest enough to document their wrong turns and bad ideas rewarded? How are they protected?
- What prevents the DR from being used against the designers in litigation over the damages stemming from a design error? If the O-ring fiasco had been documented clearly in the Challenger DR and people were fired on the basis of the DR, why would the shuttle staff members continue to document their DR?
- How can sincerity of the DR authors be ensured? Can it be ensured?
- Will the DR become a medium of persuasion? If so, what keeps it from becoming a tool of organizational politics?

These are questions that perhaps cannot be answered but must be addressed. How DR capture, whether structure or process oriented, takes its place within organizational dynamics and politics will depend very much on the kinds of organizations adopting it and on their sophistication in the use of other high-technology communication media. But DR, like any technology, is not impervious to being misused. Imagine an organization, for example, in which individual performance reviews were based on statistics on one's issue base participation during the year and that bonuses were based on total number of issue nodes created and resolved during the year.

On the other hand, an enormous amount of chronic organizational waste and irrationality might be alleviated by the appropriate use of organizational rationale. Potential benefits of this research go back at least as far as Doug Engelbart (see Engelbart, 1963), who articulated the notion that individual and group work could be augmented powerfully by appropriate technologies. Engelbart's efforts in this direction over the last 30 years have led him to a unique research approach, in which the developers of augmenting technologies "bootstrap" themselves into their own tools and test their ideas, tools, and methods in doing their own daily work. When we start hearing news on DR tools for which the DR has been captured and is ongoingly used, we will know that the field of DR has passed a critical watershed.

NOTES

Background. This chapter first appeared in the *Human–Computer Interaction* Special Issue on Design Rationale in 1991.

Acknowledgments. Many thanks go to the people who have contributed to this work over the years, including Peter Marks, David Creemer, Mitch Lubars, and Colin Potts. The first author did all of the work reported here while in the Software Technology Program at MCC in Austin, Texas, and wishes especially to acknowledge Les Belady and the wonderful research environment that he created there. The second author is particularly grateful for the efforts of Allison Kemp, Tom Klempay, David Pynne, and David Witherspoon at NCR, who were trying to get their real work done while this study was taking place, and for the NCR management support provided by Jim Brown, Hildegarde Gray, Ed Krall, and Roy Kuntz. Finally, we thank Tom Moran and Jonathan Grudin for their guidance in refining previous drafts of this chapter.

Authors' Present Addresses. E. Jeffrey Conklin, Corporate Memory Systems, Inc., 11824 Jollyville Road, Austin, TX 78759. Email: conklin@cmsi.com. KC Burgess-Yakemovic, 4776 Village North Court, Atlanta, GA 30338. Email: Kcby@gpsi.com

REFERENCES

Bouldin, B. (1989). *Agents of change: Managing the introduction of automated tools.* Englewood Cliffs, NJ: Yourdon.

Burgess-Yakemovic, KC, & Conklin, J. (1990). Report on a development project use of an issue-based information system. *Proceedings of CSCW '90,* 105–118. New York: ACM.

Conklin, E. J. (1989a). Design rationale and maintainability. *Proceedings of the 22nd International Conference on System Sciences* (Vol. 2, pp. 533–539). Los Alamitos, CA: IEEE Computer Society Press.

Conklin, E. J. (1989b). *Interissue dependencies in gIBIS* (Tech. Rep. No. STP-091-89). Austin, TX: MCC.

Conklin, E. J., & Begeman, M. L. (1988). gIBIS: A hypertext tool for exploratory policy discussion. *Transactions on Office Information Systems, 6,* 303–331.

Conklin, E. J., & Begeman, M. L. (1989). gIBIS: A tool for all reasons. *Journal of the American Society for Information Science, 40,* 200–213.

Curtis, B., Krasner, H., Shen, V., & Iscoe, N. (1987). On building software process models under the lamppost. *Proceedings of the 9th International Conference on Software Engineering,* 96–103. Los Angeles: IEEE.

Engelbart, D. C. (1963). A conceptual framework for the augmentation of man's intellect. In Howerton & Weeks (Eds.), *Vistas in information handling* (pp. 1–29). Washington, DC: Spartan.

Fischer, G., Lemke, A. C., McCall, R., & Morch, A. I. (1991). Making argumentation serve design. *Human–Computer Interaction, 6,* 393–419. Also in T. P. Moran & J. M. Carroll (Eds.), *Design rationale: Concepts, techniques, and use.* Hillsdale, NJ: Lawrence Erlbaum Associates, 1996. [Chapter 9 in this book.]

Fischer, G., McCall, R., & Morch, A. (1989). JANUS: Integrating hypertext with a knowledge-based design. *Proceedings of Hypertext '89,* 105–117. New York: ACM.

Grudin, J. (1995). Evaluating opportunities for design capture. In T. P. Moran & J. M. Carroll (Eds.), *Design rationale: Concepts, techniques, and use.* Hillsdale, NJ: Lawrence Erlbaum Associates. [Chapter 16 in this book.]

Guindon, R. (1990). Designing the design process: Exploiting opportunistic thoughts. *Human–Computer Interaction, 5,* 305–344.

Halasz, F. (1988). Seven issues for the next generation of hypermedia systems. *Communications of the ACM, 31,* 836–852.

Hashim, S. (1990). AiGerm: A logic programming front-end for Germ. *Proceedings of the SEPEC Conference on Hypermedia and Information Reconstruction: Aerospace Applications and Research Directions.* Houston: University of Houston–Clear Lake, Software Engineering Professional Education Center.

Keeney, R. L., & Raiffa, H. (1976). *Decisions with multiple objectives: Preferences and value trade-offs.* New York: Wiley.

Kunz, W., & Rittel, H. (1970). *Issues as elements of information systems* (Working Paper No. 131). Berkeley: University of California, Berkeley, Institute of Urban and Regional Development.

Lee, J., & Lai, K.-Y. (1991). What's in design rationale? *Human–Computer Interaction, 6,* 251–280. Also in T. P. Moran & J. M. Carroll (Eds.), *Design rationale: Concepts, techniques, and use.* Hillsdale, NJ: Lawrence Erlbaum Associates, 1996. [Chapter 2 in this book.]

Lubars, M. (1989). *Representing design dependencies in the issue-based information system style* (Tech. Rep. No. STP-426-889). Austin, TX: MCC.

MacLean, A., Young, R. M., Bellotti, V. M. E., & Moran, T. P. (1991). Questions, options and criteria: Elements of design space analysis. *Human–Computer Interaction, 6,* 201–250. Also in T. P. Moran & J. M. Carroll (Eds.), *Design rationale: Concepts, techniques, and use.* Hillsdale, NJ: Lawrence Erlbaum Associates, 1996. [Chapter 3 in this book.]

MacLean, A., Young, R., & Moran, T. (1989). Design rationale: The argument behind the artifact. *Proceedings of the CHI '89 Conference on Human Factors in Computing Systems,* 247–252. New York: ACM.

Mostow, J. (1985). Toward better models of the design process. *AI Magazine, 6,* 44–57.

Naur, P. (1985). Programming as theory building. *Microprocessing and Microprogramming, 15,* 253–261.

Potts, C. (1995). Supporting software design: Integrating design methods and design rationale. In T. P. Moran & J. M. Carroll (Eds.), *Design rationale: Concepts, techniques, and use.* Hillsdale, NJ: Lawrence Erlbaum Associates. [Chapter 10 in this book.]

Simon, H. (1969). *The sciences of the artificial.* Cambridge, MA: MIT Press.

Steele, G. L. (1990). *Common LISP: The language* (2nd ed.). Burlington, MA: Digital Press.

Thomas, I. (1989). PCTE interfaces: Supporting tools in software-engineering environments. *IEEE Software, 6,* 15–23.

Walz, D. (1989). *A longitudinal study of group design of computer systems* (Tech. Rep. No. STP-110-89). Austin, TX: MCC.

Zachary, W. (1986). A cognitively based functional taxonomy of decision support techniques. *Human–Computer Interaction, 2,* 25–63.

15

Organizational Innovation and the Articulation of the Design Space

Wes Sharrock
University of Manchester

Bob Anderson
Rank Xerox Research Centre, Cambridge

ABSTRACT

In this report of a case study of the work of a design team we describe some of the organizational contingencies that provided the conditions for the team's work and the ways in which the teams responded to these. We focus on the strategies by which the design team managed their circumstances in innovatory ways and manipulated the design space within which they were operating.

Constraints run through the whole (design) process from beginning to end. Constraints though are not given a priori but must be fleshed out, filled in, fabricated, even thrown away or simply forgotten in the process of dialogue and negotiation. Constraints, in a sense, are to be designed. (Bucciarelli & Schön, 1991, p. 19–20)

Wes Sharrock is Professor of Sociology at the University of Manchester; his main interests are in ethnomethodology and in the relation of philosophy to social science, and he is currently involved in field studies. **Bob Anderson** is a social scientist with an interest in the intersection of technological, organizational, and business processes; he is currently Director of Rank Xerox Research Centre, Cambridge (formerly EuroPARC).

CONTENTS

1. INTRODUCTION: ARTICULATING THE DESIGN SPACE

When we first began working on our contribution to this volume, we found it hard to encapsulate what was meant by "a" or "the" *design rationale.* As the term was being used in the sources we consulted (Carroll & Rosson, 1990; Conklin & Burgess-Yakemovic, 1991, [chapter 14 in this book]; MacLean, Young, & Moran, 1989), the concept seemed many sided and by no means uniformly employed (Lee & Lai, 1991 [chapter 2 in this book]). As we proceeded though, it became clear that there was a common thread between other studies of design rationale-related topics and the discussion we wished to present of our materials. This thread was a preoccupation with delineating and exploring what we have come to call the unfolding of the design train of thought.

As we discuss more fully in a moment, our materials were gathered in the course of a field study of a large development organization. As a consequence, they reflect somewhat different emphases and objectives to more standard investigation of design processes. At the cost of oversimplification, we would suggest the more traditional studies are:

1. Primarily interested in the deliberations and decisions of individual designers. Even where the focus is on team design, less attention is paid to the role of the group as a group. Though this has been satisfactory for many purposes, as we indicate later, we detect a movement away from this tradition, with design rationale studies themselves being part of this movement, routinely concerned with design as part of group work. The emerging approach places more central emphasis on designers working in teams and

on projects in organizational settings. Several studies have sought to tie the unfolding of design decisions back to the group and organizational processes within which they are embedded. Olson and Olson (1991), for example, used the notion of distributed cognition to make this connection, whereas Conklin and Burgess-Yakemovic (1991, [chapter 14 in this book]) and Grudin (1995, [chapter 16 in this book]) stressed the important issues of usability of design rationale tools in commercial design and development environments. Grudin (1995) in particular brought out some of the features of large-scale organization that, by implication, will constrain opportunities to deploy design rationale tools.

2. Prone to be "design-centric" in outlook. Issues and possibilities are reviewed almost exclusively from the point of view of the design, whereas we have less to say about the way their design skills are involved in delivering a well-designed product and more about the way in which their skills are deployed in keeping the project "on the road" in "working the system" and "stretching resources."

3. Concerned with explicating design-in-principle; that is, with attempting to isolate general principles and methods of design practice rather than with design-as-work, with this latter emphasis turning attention much more toward the environments within which design projects must be carried through and within which design methodologies, tools, and so on, must be implemented.

We focus on the organizational setting of design. Through the analysis of a single case, we analyze the ways in which organizational exigencies affected design activities and the *design train of thought.* We intend the notion of design train of thought to capture the sense in which the work of the design team comprised a succession of tasks in hand and of problems to be solved, activities and problems that were unified by their perceived place within the process of working out and implementing a design. It contrasts with the selective tracing of design-relevant considerations found in, for example, QOC constructions through its emphasis on the sheer multiplicity of considerations that must be contended with in working through a design. Later we describe a tendency that we detect in studies of design to pay more explicit attention to the "practicalities" of design, which include increasing recognition of the collective nature of design work and the practical character of tool use in actual working environments as well as in research and experimental situations. The tendency, which is manifest in other studies in this volume, is to broaden their focus beyond narrowly conceived design issues and to consider any of those aspects of work and its organization that bear on the carrying out of design. In particular, we want to bring out the ways in which, as part of their carrying through the design, the designers on Centaur (an "add-on" high-capacity photocopier feeder) creatively sought to manage both the design and the unclear and changing circumstances within which they found themselves.

Many people will ask if anything valuable can really be learned from a single case. For ourselves, we are wary of rushing to generalization from any set of materials and certainly would not claim any easy universalization of our limited experience. The issue, we suspect, has a more methodological frame to it. It could be heard as an objection to case-based analyses as an entirety. There are at least two possible responses to this. One is simply to refer to the body of literature outlining and defending the use of case studies in the social and related sciences (Dilman, 1973; Mitchell, 1983; Schwartz & Jacobs, 1979). The other is to argue that it seems likely that the phenomena we are interested in understanding are particularly suited to investigation on a case-by-case basis, for such things as engineering projects are likely to vary significantly one to another. This is especially so when one of those phenomena is the designers' own conception of their situation. For them, each new project may well be a mixture of the old and the new, the recognizably familiar and the largely unknown, the standard and the unprecedented. Because our concern is with the character of the designers' day-to-day work, then we automatically notice the extent to which designers have to be adaptive to whatever circumstances they find themselves in. These circumstances are often less than ideal.[1] The very emphasis of our description is that of designers managing their work under problematic conditions.

We have chosen to use the notion of a *design space* as the fulcrum for our discussion. In some discussions on design rationale, the idea of a design space is used deliberately to bring out the fact that an artifact emerges from the design process as the result of decisions and selections. Design problems (being typically "ill-structured" problems; Simon, 1984) rarely have unique solutions. The solution implemented is selected from among alternatives. The identification of possible solutions and the process of selection involve the elaboration of the features the solution might require, the arrangements that will provide these, and the criteria to be applied for discriminating among these solutions and arrangements. The configuration of decisions that, from a design point of view, makes up the artifact is, then, drawn from a "space" of possibilities. We do not want to take the expression design space too literally and most particularly we want to avoid reifying the idea. In our view, the notion that the expression seeks to capture is more widely employed than the expression itself. The coinage design space is a useful shorthand for the fact that there is a range of candidate solutions to any design problem. In turn, this range is constrained by features of the design task and its environment that delimit

---

1. We are sure that we deliver no news to many working designers when we remark about the "unsatisfactory" character of the circumstances under which they are called on to achieve adequate results. For another description of designers contending with less-than-ideal conditions drawn from our researches see Button and Sharrock (in press).

what can be included among serious candidates for the solution of a given problem.

As it is used in design rationale, the notion of space is an analytical conception. The space being created and explored is a conceptual one: namely the repertoire of questions, options, and criteria (MacLean, Young, Bellotti, & Moran, 1991 [chapter 3 in this book]), the sequence of decision points mapped (Conklin & Burgess-Yakemovic, 1991 [chapter 14 in this book]), or in terms of arguments (Fischer, Lemke, McCall, & Morch, 1991 [chapter 9 in this book]). What is attractive about this way of speaking, though, is that it allows analysts to play off the senses in which the space is both cleared and opened up for design. For designers, design is, at once, a process of survey and systematization as well as exploration and discovery.

We think of formation of the design space for a project as its *articulation*. We use this term in a double sense to mean both "spelled out" and "arranged in a structured way." By viewing it this way, we try to emphasize how the boundaries of specific design activities are explicitly drawn and the opportunities that the room for maneuver thus created makes available are expressly identified and deliberated on. At the same time, we are also concerned with the ways in which the activities of drawing out, working within, and deliberating on the design space are themselves interrelated. Obviously, design rationale techniques seek to ensure that the space will be more—and more reflectively—articulated in both senses. As a result, design possibilities will be more consciously spelled out and more methodically ordered. It follows that the focus of design rationale techniques is on capturing, preserving, and supporting the flow of designer reasoning that can be seen at the level of the technical issues surrounding the design of a particular artifact. The design rationale is not solely a history of the project, but a record of its unfolding "train of thought"—the course of reasoning about the design questions posed and the answers given.

Just as the development of an artifact through the design process is an exigent outcome of decisions within the design space, so too is the development of the design train of thought. In many respects, the eventual dimensions of the design space are also the outcome of other processes relating to the organization and conduct of work that are not usually brought within the explicit remit of the design rationale. Those engaged in design rationale work are, of course, very aware of the presence and influence of these processes, as are designers themselves. They feature for designers as things with which they must, practically, come to terms if they are to do their work. Hence, their presence is and can be extensively taken for granted by those seeking to explicate the technical aspects of design rationale. However, because design rationale approaches aim to have an effect on the way that actual design teams in actual settings go about their work, it will be important at some point to adumbrate the interconnectedness of the technical and the organizational constraints on the dimensions of the design space more fully.

The materials we discuss are drawn from fieldwork carried out on a project to design and develop a high-capacity feeder for a photocopier. One of us (Sharrock) spent from early February to mid-April 1990 working alongside the Centaur team, attending their meetings, following design activities, sitting with designers. Many of these sessions were audiotape recorded and all of them summarized in field notes. In addition we accumulated copies of diagrams, meeting agendas, summaries of actions to be taken, report forms, and all the other multifarious bric-a-brac ethnographic field-workers accumulate. Our motivation for following the day-to-day, hour-by-hour life of the Centaur project was to try to come to some understanding about design in practice and in real-world circumstances. We were also interested in the temporal features of designing and hence in the arrangement and synchronization of design activities in real time. Although, after the event, it is always possible for analysts (or anyone else) to see why things had to turn out the way they did, at the time and in the middle of it all, 20/20 hindsight is not available. Decisions have to be made in the light of whatever information can be got to hand and in the certain knowledge that things will almost always turn out in ways one could not have foreseen. Following designers through their daily round is, then, the best way we can think of to get some grasp of the real-time realities of their work, of what it is they are trying to do, and of the local culture that informs their understandings of what is going on around them. Although snapshot visits and knowledge-capturing interviews will always provide some insights, in our view if you want to get close to and understand the day-by-day, week-by-week rhythm and pacing of project activities, the project's local history, and its own unique character, there is no alternative to being there.

Our description of these materials brings out some of the exigencies that shaped the evolution of the design space. In particular, we want to concentrate on (a) aspects of the way in which the design task is defined, structured, and redefined and restructured, and (b) the way in which the design task is carried out under those definitions and structures. In other, plainer words: We look at an organizational context and its consequences for a project.

2. CENTAUR AND ITS PROBLEMS

Centaur was a comparatively small project (involving between 8 and 18 people at any one time in an organization where projects would routinely involve up to 200 participants) delivering a high-capacity feeder for a model of photocopier that had already been launched, but that was not selling as well as had been hoped. There was a pressing need to increase the sales of the model and so reduce the inventory. Centaur was a possible solution to this problem. There was a potential market niche created by the copier's particular suitability for use in (college) libraries. The only drawback was its lack of a high-capacity feeder. In addition, it appeared there was a narrowly defined window of opportunity, framed by the buying cycles of the targeted organizations and the imminent launch by a rival firm of a closely comparable machine. These combined to set a time limit to the project. The whole thing would have to be carried through, from inception to launch, in a little over a year.

It is this context that determined the predominant feature of the Centaur project. The team was "up against it" right from the start. Everyone recognized there were serious, possibly insuperable problems to be confronted. Those on the project were unable to tell whether they were going to be able to overcome them. In any event, it was clear that even if they did, it would, to quote the Duke of Wellington, be "a damn close run thing."

Here are the most prominent of these problems and the implications that they had for the project's definition of its design objectives.

2.1. Keeping to Schedule Was the Paramount Concern

The aim of the project was not just to produce a viable and marketable product, but to devise an add-on that would transform the sales prospects for an existing model and thereby to contribute to the solution of pressing budgetary problems. But this opportunity was rigidly confined. If a machine was not available for display at the relevant equipment exhibitions that were themselves timed relative to the annual buying cycles of the targeted library organizations, then the sales would be foregone for a further year, perhaps forever. Then there was the launch of the rival machine to be taken into account. Any schedule slippage would be fatal to the plan.

Once the project got underway there was even more pressure to expedite things. A demand was received for the early provision of a version to be located for trial use at a prominent user site. This could provide advance advertising of the machine's availability and demonstration of its particular suitability. In addition, because the project was always specifically rationalized as a solution to a problem, there was the perennial possibility that other solutions to that problem might be found elsewhere. Significant price discounting was repeatedly rumored to be another, very viable alternative option being considered.

This urgency meant that the scheduling was worked out on the explicit assumption that "everything goes right first time." There was no room for iteration, even at the prototyping stage. Further, the scheduling was done against the background of an unresolved and major practical difficulty resulting from a discrepancy between the requirements of the formal product development procedures and the lead-time requirements of manufacture.

2.2. Standard Decision Procedures Had to Be Circumvented

The site at which the project was initiated had recently undergone significant staff reductions, the character of which had been dictated by the quality of redundancy entitlements and the prospect of employment elsewhere. Reductions in the size of departments had been uneven. Some were virtually depopulated. One so affected was responsible for the site's marketing activities. The formal product development procedures (collectively known as the Product Development Process or PDP) require the provision of a business case to show that the product could be an economically viable. In the absence of anyone formally designated to do this for Centaur, the working out of exactly what opportunity the project might take advantage of had proceeded relatively informally.

In addition, the formal processes for product development require explicit phasing, with reviews to evaluate success at the end of each phase and to make commitments to subsequent phases. The steps and requirements involved in these phases effectively set an irreducible time line for the project as a whole and for individual phases. Even with the assumption that everything that needed to be done before a review could be held would be completed in optimum time, there was still a limit as to how soon the review can take place and hence a decision to proceed be made. Of course, such limits and phases are negotiable (and more negotiable for some than for others—indeed, the situation under consideration was itself subject to negotiations). However, invoking the standard procedures would inevitably have meant that the earliest possible point at which a review could authorize movement from the concept and design development to manufacturing, was later than the date at which lead times required that manufacturing operations be initiated. In other words, if the serial review procedure was adhered to, expenditure on manufacturing preparations could not be authorized early enough to allow manufacturing targets to be achieved.

2.3. Flux in the Environment Had Its Effects

Contraction in the site's staffing impacted the team structure. The staff reduction process had been driven by the need to achieve downsizing rather than by attempts to retain a balance of skills of those who would remain. This meant that project teams had to draw on whoever was available or call on the assistance of other sites. For Centaur, this meant (a) because the site was now "top heavy" with relatively senior staff, some four out of the eight main participants were of equivalent seniority, including the project manager; (b) the skills available to the project did not always match the project's precise requirements; in several respects the team was operating with "the next best thing" to what they needed, and sometimes less;

and (c) for one important role, they had to begin the project with stand-ins while attempts were made to find someone suitable.

2.4. UMC: A Problem That Wouldn't Go Away

By no means least among the problems that Centaur was up against from the start was that of the unit manufacturing cost (UMC). The project was targeted at a particular market niche. The estimated size of that niche indicated a production run of 4,000 high-capacity feeders. The high costs of production tooling are typically defrayed over a long production run. In this case the run was not long enough. the target unit manufacturing cost was, of course, set with an eye to the pricing of the product. Given the tooling problems, a realistic estimation of the tooling cost meant that the UMC would inevitably be much greater (almost double) than the target cost.

3. MANAGING THE DESIGN SPACE

The problems we have just outlined might appear to be intractable. This is certainly a view that was at times shared by the project team. Nonetheless, they proceeded to look for ways to make the intractable, if not tractable exactly, then at least less of an obstacle. In effect, what they did was to look for innovatory strategies for managing their design space. What they innovated upon and around were the prescribed procedures for design and product delivery. Some of the unorthodox approaches they used are institutionalized in this site, and presumably elsewhere. Others were unique responses to the particularities of their case. As each was put in motion, it generated its own complexity and complication that had, in turn, to be handled within the ongoing project. In their own ways, each one had its effect upon the course of design thinking that was being explored. As the design team saw it, they were faced with two sets of problems. One was to design a specific device, the feeder. The other was to find ways of reducing production costs to a level that would be acceptable and to initiate manufacturing early enough to allow targets to be met. The first problem appeared to be relatively straightforward, although not without its own wrinkles. The second looked to be a killer. Given that this was the case, the problems readily prioritized themselves, with almost all the innovative energy being directed to finding ways of managing the cost and time constraints.

For the purposes of this discussion, we concentrate on just three of the innovative strategies that members of the Centaur project used. These are: (a) improvising on the formal procedures, (b) working around the normal work practices, and (c) revising the design task from within the design.

3.1. Improvising on the Formal Procedures

Within the design organization that we studied, standard procedures and protocols are used for every stage in the product delivery process (PDP). The purpose of these procedures is, of course, to ensure quality control, comparability of cases, and standardization of decision making. As we have seen, if Centaur were to have followed these procedures to the letter, it would be impossible for them to have delivered on their targets in time. The team's response was not to disregard the formal steps and processes. Rather, they sought to fulfill them while at the same time reducing the constraints they imposed on the team's room for maneuver. Here are some of the ways this was done.

Informalizing the Review. One of the major problems, it should be remembered, was that of bringing the project to a point at which a formal review was possible. This meant both achieving the conditions required for review and organizing the allocation of time to be taken from other project work in order to prepare for such a review. One way of dealing with this might be to arrange a relatively early and informal review. In particular, this offered the possibility of a timely commitment to manufacturing spending. Extensive telephone and email negotiations were undertaken to seek official approval for such a proposal and to bargain for the staff time required. Eventually, they were successful. By being able to hold an informal review, the team was able to get "in principle" permission to proceed before they were actually ready to ask for definitive permission. This meant that they could begin to think about addressing the manufacturing problems, for instance, before they actually reached the phase in the project where that activity would normally have been scheduled.

Opportunistically Exploiting the Black Economy. The problem of unit cost was constantly with them. Though it seemed they could not make more than marginal differences to the discrepancy between the target and projected cost, matters of detailed costings were nonetheless carefully attended to. Much satisfaction was derived from anything that was judged low cost. Complaints were registered about and close scrutiny given to anything that made a "hit" against costs. Endless negotiations between the project team and the managers at the proposed manufacturing sites continued to see what could be done about the cost and schedule problems.

A chance conversation between managers at the local site revealed that, as one of the by-products of restructuring mentioned earlier, the Procurements Department was short of work, and was casting about for things to do. Through its knowledge of and relations with suppliers, a shadow operation was set up by which the Procurements Department compiled an alternative costing and scheduling of parts production to that being provided by the official manufacturing operation. The aim was to see if lead times (and costs) could be significantly cut by contracting out. In turn, such possibilities might be used to force a relaxation of the UMC and time constraints being insisted upon by the manufacturing groups. Achieving the latter might relieve the pressure of getting everything right first time and offer the chance of at least some limited iterative prototyping.

Massaging the UMC. The UMC was affected by two principal things: the cost of developing tooling for the manufacture of nonstandard parts and the size of the production run. Because these were the only two things that could be varied, ways of varying them had to be considered.

The extension of the production run was one possibility. Lifting the run from 4,000 to 10,000 units would have brought the UMC considerably closer to target. It would still be high, but by a "reasonable" amount. A search for someone within the organization who might be able to advise on the source of a significant demand for these units was, therefore, set in motion. If such a demand could be discovered, it could be factored into the UMC calculations.

The cost of tooling was usually looked at in two parts: the development of "soft tooling," that is, tools made out of inferior materials for producing the parts required for the prototype; and the development of "hard tools" for use in the production run. Soft tools are intended only to produce a small number of prototype parts, and are, in any event, likely to require redesign for the production run. Because this project had only one shot at the prototype and because the production run was short, the possibility of using soft tools in production was considered. Negotiations with parts suppliers on this possibility were initiated.

Adopting a Deflationary Approach to Problems. The organization uses a "management by problem solving" approach to the conduct of projects. This provides mechanisms whereby project monitoring is carried out. These include a weekly project meeting for the identification and recording of problems according to certain formal processes and classifications. Part of this means that problems are classified by seriousness. There are three levels: ordinary, major, and critical. The ordinary are effectively minor problems: They have not been solved, but only routine measures are required for their resolution. Major problems are ones whose solution will have consequences for the cost, quality, or delivery of the product. Critical ones are identified as difficult to solve and may even be insoluble, and to which attention must be given. The existence of critical problems calls the continuation of the project into question.

Given the position in which it found itself, Centaur adopted a "go ahead anyway" approach even to critical problems. They were known to exist and strategies for dealing with them were underway. But they were not as yet resolved. In the meantime, it was necessary to get on with the project's other tasks. In effect, the formal classification system was applied and then recalibrated, with critical problems being treated as if they were major. In addition, Centaur's schedule was, as we have mentioned, constructed on a "right first time" basis. When things, as they inevitably did, failed to go right first time, it was impossible to halt work on dependent problems because, then, the schedule would slip irrevocably. The net result was that the formal requirements of problem solving were circumvented from the start. There was no time for serial problem solving, iteration, trial and error, and refinement.

3.2. Working With and Around the Normal Work Practices

The culture of design in this organization is, as with any work group, composed of the patterns of normative activity and the value systems espoused by those who identify with it. These are what any designer in the organization knows about how things are to be done. In the course of actual designing, the making of decisions and the solving of problems, this knowledge is deployed not as procedural rules or even as rules of thumb, but as ways of making design sense of the issues on hand, and therefore deciding just what to do. Knowing, then, how long some activity should take to complete or what quality of output from some process one should expect is determined *in media res*. As the design goes along, and as the design tasks are encountered, the configuration of this knowledge changes. In turn, this dynamism resonates back onto the ordering of design decisions, the possibilities explored, and the route to be taken. On the Centaur project, this reciprocal fitting of work practice and workplace knowledge to the design tasks in hand could be seen in a number of ways. We detail just a few.

Cutting Corners and Watching for Potholes Became a Way of Design Life. Schedules are usually compiled more in hope than expectation. The interlocking of steps in sequences means that exceeding the estimated time always has knock-on effects that have to be either anticipated or actively managed. One case that occurred on Centaur involved working out the detailed features of the design and the production of the technical drawings. The concern was to get decisions to the point at which technical drawing could begin while at the same time attempting both to truncate the process of producing usable drawings and to prevent any slippage at that point. To do this, a policy was adopted of using less-than-finished drawings wherever possible. Where it was thought that the supplier of a part was well enough known to be relied on to understand and implement a rough drawing, it was agreed these could be issued at an early stage. Even so, the quantity of drawings to be produced was under the demanding scheduling, recognized to be such as to require longer than was previously allowed for and thus to threaten delays.

In addition, the production of technical drawings was recognized to be at risk from the 90% finished problem. Compared to the total number of drawings to be produced, it may seem that most are done and there is not much left to do. However, the relatively small number remaining may be the ones that are going to take a long time and may even be left to the last because they are the difficult ones to do. On Centaur, the lead designer sought to avoid the 90% finished problem by making his own calculation of the relative difficulty of particular tasks and therefore of the proportions of the available time particular jobs should take. Projecting the rate of work actually done against the rate required to meet the schedule meant that deviations from the necessary rate would show up early. Corrective steps could then be taken and the work got back on schedule before it was too late.

Problems Were Traded Off Against One Another. We mentioned that the composition of the Centaur team was affected by the contraction of the site's work force. The team was made up of individuals, many of whom were senior and long serving, and who regarded themselves as equally experienced in project leading as their project manager. These individuals treated their mutual relations as delicate. There was a ready possibility for misreading motives, and especially for disagreement to be construed as personal criticism. Objections to aspects of the project's organization particularly might have been (mis)understood as expressions of sour grapes. Any attempt to force disagreements to conclusions in design meetings might well have been taken as attempts to "show up" the project manager in front of colleagues. This does not mean disagreements did not occur, but they were muted and were not pressed particularly far. If dissent on some point revealed that the project manager had a strong preference, then this was deferred to. This display of restraint as a way of handling this issue had one crucial consequence for the development of the project.

Timing diagrams are a particularly important tool for the design of photocopiers. They involve working out the precise timings for the movement of paper sheets through the machine. One important aim in photocopier design is to achieve maximum possible speed in the copying of sheets and, thus, to keep the sheets moving through the paper path as close together as possible, but without leading to overlaps or conflicts and hence misfeeds or paper jams. Some of these separations are timed in milliseconds. The timing diagrams provide a working out by the millisecond how the designed system will move paper. Whether or not the design will work (at all, or without significant revision) cannot be a matter of intuition. It has to be a matter of calculation. The difference between a working design and one with a serious problem may be a matter of a few hundred milliseconds. The production of the timing diagrams for Centaur was, then, a matter of some importance. However, there was disagreement on just how urgent it was. The project manager appreciated the task was important, but did not feel it was quite as critical or urgent as did some of his colleagues.

From the project manager's point of view, although working out the timings would be a difficult task, the team had no one who was experienced or appropriately skilled in it. On the other hand, the feeder was to be compatible with a machine that had already been built, and for which there would or ought to be extant timing diagrams. He also knew there was someone on the site who was experienced in the work. Furthermore, she had done the very timing diagrams for the relevant copier. The project manager set about tracking down these diagrams and tried to borrow the relevant skilled person from the project to which she was currently attached. In his view, the diagrams were in hand. He had other more pressing issues to resolve. The structure of problem classification no doubt played a compounding role in his prioritization here. Naturally attention is likely to focus on the "critical" matters, especially on a time scale as short as Centaur's. The provision of the timing diagrams, although being seen as something important to do, was not even worth classifying as a problem. Anyway, it was in hand.

Other members of the team did not agree. They regarded the timing diagrams as critically important, and thought that they should be produced as soon as possible. When it became clear that the project manager did not share their view, they acceded to his argument, without accepting it for one moment. They foresaw problems resulting from delay in getting the timing diagrams out. However, their choice was between two kinds of trouble on the project: that which would result from late availability of the diagrams, and that which would result from creating personal animosities. Furthermore, those animosities would have been created without necessarily getting the result they wanted anyway. It seemed better to deal with the problems created by postponing working out the timings when they arose than to create conflicts over the running of the project.

Necessity Was the Mother of Redeployment.   Relative to the project's life, the search for the copier's timing diagrams and the negotiations with the other project for assistance in working them out was protracted. The necessary diagrams were eventually found and then only by happy coincidence. The negotiations to borrow the skilled person were not successful, and so the work had to be done—now belatedly—by someone within the team. Even here, the one who had the most relevant skills was not the one who did the work. He already had his hands full with other, equally critical tasks, such as designing the printed wiring boards (PWB). These tasks were not further deferrable, so working out of the timings had to be assigned to someone who had just enough skill to do them effectively, albeit with difficulty, and who had other tasks that could be deferred. It was not that delay in the performance of these tasks might not eventually create problems of its own, simply that it would make no immediate, tangible difference.

3.3. Revising the Design Requirements From Within the Design

Clearly, one of the ways to find an achievable route through Centaur's design space, would be to relax some of the constraints encapsulated in the design requirements. If this could be done, then the problems being faced might reorder themselves sufficiently to enable some relatively obvious design solutions to be offered. A several points a number of attempts to relax these constraints were made.

Amending the Customer Requirements. A key feature of the project was that the attachment of the Centaur feeder should not require any communications with the central processor of the host machine. This meant that the design was constrained to use the communications facilities already designed into the host. This gave rise to one of the tricky design problems, namely the use of the host's sensors to communicate the various states of the paper trays. Much time was devoted to working out the possible arrangements of trays, sensors, and papers. It became apparent that all possible combinations could not be accommodated, and so one of the immediate responses was to see if the range of possibilities could be constricted in some way.

The issue could only be resolved at the level of marketing strategy. The design had been developed on the assumption that the machine was to be produced in the routine way to meet the requirements of various markets of an international company. It would have to operate in different climatic conditions, with different sizes and qualities of paper, with instructions in different languages, and so on. The difficulties in accommodating all these possibilities without independently communicating with the central processor led the team to ask if this was realistic. The project was, as everyone was well aware, designed to solve a specific problem by exploiting a specific market niche, one most prominently based in the United States. How many actual markets, therefore, was the machine to be designed for? How much variability in paper sizes and climatic conditions should be involved? What was the actual pattern of paper use within the main projected market? Answers to these questions might allow a drastic reduction in the range of alternatives to be designed for. And that might enable them to design the communications systems according to specification.

This example illustrates the ways in which the various strategies the team engaged in interacted with each other and how problems can move in and out of the foreground of the design team's attention. We pointed out earlier that the project had been initiated relatively informally. This informality now began to have consequences, in that there was no formal mechanism to clarify the actual requirements for the design. There were informal contacts with relevant parties in the local marketing function, but these were rendered problematic by changes in personnel. In the event, an agreement was made with someone in the main marketing organization to collate information on the actual pattern of paper types, sizes, and so forth, in the target market. The need to obtain clarification of the marketing policy and to establish whether the feeder could be launched in one or two countries only was also recognized. Launching the machine for specific markets offered the possibility not only of restricting the combinations to be designed for—and thus simplifying the task—but also of being able to dispense with the need to design packaging and arrange translations for all the different markets. This promised cost savings. However, though this information was badly needed, it could not be speedily obtained. There appeared to be no easy way to clarify market potential. Personnel in the international marketing arm were changing positions, and the steps involved in moving informal support for Centaur to formal decisions about its launch policy had not been taken. They had gone as far as deciding that Centaur should be aimed at the North American college market, but no further.

Assuming the Best Solution Will Be Available. Obviously decisions are interdependent. Designers seek to line them up so that one decision can determine the character of others. On Centaur one issue was how much power it was going to need. This would depend on the size of motor required which, in its turn, would fix the amount of space available for the installation of the PWBs. Engineering lore lays down that the final size of a motor is always greater than that envisaged or wanted. Hence, the size of the motor could not be determined prior to decisions about the power supply or about the architecture of the boards. The investigation of the available motors, the decision about the power supply, and the design of the PWBs all had to proceed simultaneously, with decisions about each involving a certain amount of risk. Allowing for the worst possible outcome with respect to the size of the motor was not possible because there was a constraint on the space into which the motor and PWBs had to go. The risk was in judging what one would get away with in terms of motor size and then hoping that this estimation would be fulfilled.

It is this kind of problem that puts designers in "no choice" situations. There are things they would like to know and decisions they would like to have settled, but the speed of the formal "decision loops" and the time scale of design tasks mean that important considerations cannot be resolved. The designers have no choice save to go on working out the design—even its detailed features—while awaiting the resolution of issues on which the effectiveness of their decisions depend. These are not decisions with defined contingencies: The designers do not know what they are going to do if and when the assumptions they have made prove false. That is something to be faced if and when it comes to it. They have no choice, in the circumstances, but to assume a favorable outcome and continue designing on its basis.

Adopting an Alexandrian Approach to Requirements. The final production of the timing diagrams revealed serious problems in the timing and tracking of sheet movements. There were, too, still unresolved questions about the capacity to communicate between the feeder and the host machine. These were both "life threatening" problems for the project. To enter a review (even an informal one) with two critical problems of this order would almost certainly be fatal. Time was running out. It was becoming clear that a drastic solution was needed. As various possibilities were examined to solve the paper-feeding problems, it became increasingly apparent that the main obstacle was the requirement about communication with the central processor. If that requirement was not in place, then there was a simple solution. All that would be needed to solve the timing problems successfully was a single wire between the feeder and the central processor. The installation of such a wire would make only marginal difference to the installation work. (Minimizing the installation task was a prime reason for the prohibition on communication with the central processor.) But if communication with the central processor was available then other outstanding problems would also be solved. Signaling the state of paper and trays could also be done through that connection (and with minimal alteration to the software). The only way forward was to go back on what had hitherto been treated as an immutable constraint, and cut the gordian knot it had created. Once it was accepted that this must be done, the outstanding problems evaporated and the project entered the review in sufficiently good condition to have its mechanical and electrical designs approved.

4. CONCLUSIONS

Our account has instantiated the ways in which the development of the design train of thought is intimately and elaborately intertwined with organizational contingencies bearing on the project and with which the project team has to cope. It is this emphasis that marks our materials off from those of QOC and similar analyses, especially insofar as those draw on experimentally contrived situations, for these usually enable the participants to disregard anything other than the explicit features of the design concept. For our designers, the elaboration of the design concept was the very least of their problems. Their troubles began once the design concept had been defined. In bringing this chapter to a conclusion, we consider some implications of our materials for the development of tools to support design rationale methodologies. To keep the discussion within bounds, we focus on just two aspects. In both cases, let us hasten to add, we do not see the issues we raise as invalidating attempts to build tools. Rather, we offer our comments in the spirit of furthering this as joint endeavor. As was recently noticed in another but closely related context, ". . . maintaining a user-centered perspective during design must be done in concert with engineering realities of function to be provided, schedules to be met, and development costs to be managed" (Karat & Bennett, 1991, p. 270). We see our contribution as the elaboration of some of these realities from the point of view of the organization of design. What we seek to provide is a set of sensitivities to the ways in which the tools being used might interact with the organizational setting of design.

4.1. Some Features for Tools

The decision what to include within the specification of design rationale tools obviously depends on the envisaged uses to which they are to be put. We can imagine two such different functions and would like to make the following observations about them.

An Account or Record of Design. If design rationale tools provide a record for and of the design project, then it seems clear that they will also have to find ways of marking and tracing through a number of organizationally generated contingencies. Again, to be succinct, we nominate just three:

1. Project team membership is generally changing. Any tool will have to mark the variability of skills available and courses of action decided on as a response to the variability of membership.
2. Every member of a project team is simultaneously a member of other project teams. Each of these requires a commitment of time and effort. The prioritizations of activities are not always in the hands of the designers concerned and can be affected by a multitude of exogenous forces. Tools should adaptive to the consequences for design scheduling of potential discontinuities in the management of the prioritization process.
3. Organizationally defined procedures such as the PDP are constantly under review and subject to upgrade, improvement, and/or reconfiguration at almost any level of granularity and at any stage. Design rationale tools should indicate just how such changes impact design decisions and those reformulations of the formal procedures.

A Forcing Device Within Design Decision Making. One commonly agreed use of design rationale tools is as a device for deliberately expanding the design space beyond what might be intuitively obvious. In principle such an exploratory strategy could refuse to be bound by any of the presuppositions apparently built into a set of specifications. However, from the material we have gathered, two forms of resistance are likely to prevent this radicalisation of design from occurring.

The possibilities that are laid out for any design and the constraints that operate on them are not all of the same order. Designers have a sense of their relative weightings and the interactions between them. Choices are made in the context of these weightings. One of the most important elements in the weighting system is the family of solutions embedded within a particular development organization. These *standard solutions to standard problems* have what we might think of as an *interactional efficiency* in design. The overhead of discussion, explanation, and plausibility construction is greatly reduced by their use. This being the case, "blank sheet design" is by and large a myth.[2] Forcing explicit review of a wider range of design possibilities may merely provide in the eyes of designers a more roundabout route to conclusions they would have reached in any case. Like the skilled chess player, it is part of their competence to cut through possibilities, eliminating many without consideration. They may, therefore, treat the explication of the design space as a mere formality, one more requirement to be satisfied only because it is formally necessary and because someone else wants it done.

4.2. Context of Use

No tool can be introduced into any set of working practices without some consequential effects. Indeed, the purpose of introducing the tool is often to bring about certain kinds of effects. We would recommend the following be borne in mind when thinking through how it is envisaged that any particular design rationale tool might be used:

1. Records are organizationally pertinent documents. They have particular orders of usage, lifetimes, and necessary components. Given their

---

2. It has been suggested to us that our talk of resort to standard solutions carries a pessimistic implication with respect to innovation in design, but it is not intended to do so. First of all, we do not mean to suggest that designers were thoughtlessly bound to such solutions, merely that the availability of these could facilitate many aspects of their work and that exploration purely for exploration's sake or purely for the record would seem to them superfluous to the needs of their work in hand. On the Centaur project itself, an exercise in "value engineering" was scheduled to provide a testing reexamination of the prototype, particularly with respect to cost.

consequential nature, their production is often formatted in particular ways—as with the minutes of meetings—or ritualized—as with roll calls of attendees. In organizational settings, organizational consequences are known to need management. In Centaur, to get the documentation kept, assurances sometimes had to be given that there would be "no witch hunts," that is, that records would not be used to attribute blame for any problems they revealed. In view of this, the presumption of goodwill on the part of all current and subsequent parties to the record may well be misplaced. This, indeed, may feature quite intrusively in situations where what the record shows may be held to be consequential for the allocation of responsibility for outcomes or a matter of controversy. In such cases, the work of constructing the record can tend to displace that which it is supposed to record.

2. The point of constructing design rationale is to preserve an account of design decisions for later reengineering or maintenance. There seems to be an inherent uncertainty in this. What specifically will later users of the rationale wish to have available? How much is enough? Who will want to know what? It is not until the information is needed that its utility can be estimated. In addition, there is the problem of reworking at some later date, precisely why some record has the shape and contents it does.

3. To the aforementioned considerations, we have to add the mobility of design. In our study, important design decisions were made on the phone, in the corridor, in the car park, as well as in formally constituted meetings. Indeed, often they only appeared in such meetings as confirmations or even consecrations (such as formal sign-offs). Any tool that was to be of real practicable value would have to be usable in all these contexts. It is in this respect that the practical fact of life that information is not a free good becomes visible. If a design rationale tool imposes an overhead of information management, then it is unlikely to be seen as of value unless the obligation to provide this information clearly, concisely, and intelligibly expressed, often for the benefit of unknown others, conformity to the demands of the tool may be seen as adding another bureaucratic task to the work load of individuals who already feel bureaucratically imposed on by the range of form filling and record keeping they currently manage.

4. Even where design decisions are made in formal and informal meetings, our experience is that this is often a delicate balance between being persuasive and seeking control, allowing adequate turn management and giving everyone sufficient air time, managing the meeting and managing the activity the meeting is in support of. If we then introduce all the impedimenta of meeting support likely to be required for any effective design rationale tool, do we know what the impact on the character of such decisions will be? Do we know enough about any kind of meeting structure to predict effects, let along those between people from different background and management groups and with vastly different interest?

As we said at the beginning of this chapter, ours is a less design-centric approach to design than is more usually encountered in design rationale studies and our emphasis has been on the ways in which the development had to be adapted to the practicalities of project work and to the unclear and shifting circumstances of the project. This difference is on view in the range of issues we have outlined. It can also be seen in the conclusion that we draw from our experience and that we found to be somewhat ruefully shared by many of the designers we met. In the end, the fate of a project and hence the success or failure of the design team is only partially under the control of or even the business of the designers themselves. A reminder of that fact alone may be the most important contribution our approach can make to design rationale.

5. POSTSCRIPT

Several readers of earlier drafts have wondered about the fate of Centaur. Did it ever get into production? Or did it go the way of premature cancellation many predicted for it? Some time after the review we described previously, the project was terminated. At the meeting at which this was announced, some of the team insinuated that they had been let down by the project managers. This was met by the response that experienced members of this environment know that cancellation is a common fate of development work. The team was quickly redeployed although some effort was made to assemble and retain whatever remnants of the project could be preserved. Those who did this did so on the grounds that they would not be surprised if the project was resurrected at some future point. Some 6 months later there was a brief attempt to resuscitate Centaur and a few months after that there was speculation it was about to be revived. On a visit to the United States, the project's manager discovered a firm that specialized in customized items. This firm was already producing a high-capacity feeder. Not long after, he found one of his own company's other sites was manufacturing a comparable add-on. And that really was the end of Centaur.

NOTES

Acknowledgments. We would like to thank Tom Moran, Victoria Bellotti, and Allan MacLean for extensive discussions and comments that have enabled us to see more clearly the potential interrelationships between our field-based studies of design and current approaches to design rationale. We would also like to thank Jack Carroll, Jonathan Grudin, and an anonymous reviewer for many helpful comments.
Authors' Present Addresses. Wes Sharrock, University of Manchester, Department of Sociology, Oxford Road, Manchester MI3 9PL, UK. Robert Anderson, Rank Xerox Research Centre, Cambridge Laboratory, Ravenscroft House, 61 Regent Street,

Cambridge CB2 1AB, UK. Email: msrssws@cms.mcc.ac.uk and anderson.
europarc@xerox.com.

REFERENCES

Bucciarelli, L., & Schon, D. (1991). *Generic design process in architecture and engineering: a dialogue concerning at least two design worlds.* Unpublished manuscript.

Button, G., & Sharrock, W. (in press). Occasioned practices in the work of implementing development methodologies, environments and languages. In J. Goguen, M. Jirotka, & M. Bickerton (Eds.), *Requirements engineering.* New York: Academic Press.

Carroll, J., & Rosson, M. (1990). Human–computer interaction scenarios as a design representation. *Proceedings of the 23rd Annual Hawaii International Conference on System Science,* 555–561. Los Alamitos, CA: IEEE Computer Society Press.

Conklin, E. J., & Burgess-Yakemovic, KC. (1991). A process-oriented approach to design rationale. *Human–Computer Interaction, 6,* 357–392. Also in T. P. Moran & J. M. Carroll (Eds.), *Design rationale: Concepts, techniques, and use.* Hillsdale, NJ: Lawrence Erlbaum Associates, 1996. [Chapter 14 in this book.]

Dilman, I. (1973). *Induction and deduction.* Oxford, England: Blackwell.

Fischer, G., Lemke, A. C., McCall, R., & Morch, A. I. (1991). Making argumentation serve design. *Human–Computer Interaction, 6,* 393–419. Also in T. P. Moran & J. M. Carroll (Eds.), *Design rationale: Concepts, techniques, and use.* Hillsdale, NJ: Lawrence Erlbaum Associates, 1996. [Chapter 9 in this book.]

Grudin, J. (1995). Evaluating opportunities for design capture. In T. P. Moran & J. M. Carroll (Eds.), *Design rationale: Concepts, techniques, and use.* Hillsdale, NJ: Lawrence Erlbaum Associates. [Chapter 16 in this book.]

Karat, J., & Bennett, J. L. (1991). Working with the design process: Supporting effective and efficient design. In J. Carroll (Ed.), *Designing interaction* (pp. 269–285). Cambridge, England: Cambridge University Press.

Lee, J., & Lai, K-Y. (1991). What's in design rationale? *Human–Computer Interaction, 6,* 251–280. Also in T. P. Moran & J. M. Carroll (Eds.), *Design rationale: Concepts, techniques, and use.* Hillsdale, NJ: Lawrence Erlbaum Associates, 1996. [Chapter 2 in this book.]

MacLean, A., Young, R., Bellotti, V., & Moran, T. (1991). Questions, options and criteria: Elements of design space analysis. *Human–Computer Interaction, 6,* 201–250. Also in T. P. Moran & J. M. Carroll (Eds.), *Design rationale: Concepts, techniques, and use.* Hillsdale, NJ: Lawrence Erlbaum Associates, 1996. [Chapter 3 in this book.]

MacLean, A., Young, R., & Moran, T. (1989). Design rationale: The argument behind the artifact. *Proceedings of the CHI '89 Conference on Human Factors in Computing Systems,* 247–252. New York: ACM.

Mitchell, J. C. (1983). Case and situation analysis. *Sociological Review* (New Series), *31*(3), 187–211.

Olson, G., & Olson, J. (1991). User–centered design of collaborative technology. *Journal of Organizational Computing, 1,* 61–84.

Schwartz, H., & Jacobs, J. (1979). *Qualitative Sociology: Method in the madness*. New York: Free Press.

Simon, H. A. (1984). The structure of ill-structured problems. In N. Cross (Ed.), *Developments in design methodology* (pp. 145–166). Chichester, England: Wiley.

16

Evaluating Opportunities
for Design Capture

Jonathan Grudin
University of California, Irvine

ABSTRACT

A key motive for developing ways to structure and preserve records of decision making in software development is the belief that the high cost of system maintenance could be reduced by providing access to these records. Development projects vary widely. Tools and methods can be appropriate for one development environment but not for another. In this chapter, I describe four kinds of software development environments and outline the opportunities and obstacles for using design rationale in each. Some development projects have good reason to avoid investing resources early in development, even when great benefits may accrue later. Many off-the-shelf product development efforts are arguably of this nature. In other projects, such as customized software development, the value of upstream investment seems more compelling. I reflect on the difficulty of learning from failed projects and on the need for researchers to distinguish carefully between the requirements of science and those of engineering.

Jonathan Grudin has worked as a software developer and as a researcher in HCI and CSCW, where one of his interests has been organizational constraints on interactive systems design and development; he is currently an Associate Professor of Information and Computer Science in the Computers, Organizations, Policy and Society (CORPS) group at the University of California at Irvine.

453

CONTENTS

Interviewer: *What percentage of development projects are*
                                     terminated before they're completed or before
                                     the product gets out the door?
Successful development manager
at a major software company: *Ninety percent.*
Interviewer: *(laughs) Seriously, what would you estimate?*
Development manager: *Ninety percent.*

1. INTRODUCTION

The recording of design rationale is an investment of resources early in the development process ("upstream"). It is expected to provide significant savings downstream during design change, enhancement, and other forms of maintenance. Maintenance is often cited as consuming up to 60% or even 90% of total software development costs (e.g., Boehm, 1981; Computer Science Technology Board, 1990; Fischer et al., 1992; Jarczyk, Loffler, & Shipman, 1992; MacLean, Young, Bellotti, & Moran, 1991 [chapter 3 in this book]). This provides a powerful argument for upstream investment, but a crucial datum is always overlooked: Many development projects have little or no downstream activity that will benefit from such an investment—*because they are not completed.* Even when a project is completed, a different organization is often responsible for maintenance, in which case an upstream investment would not benefit the development organization. And any project, by diverting resources to capture design rationale, may reduce its likelihood of surviving or succeeding. The challenge for those who promote design capture is thus twofold: first, to identify projects that have a relatively high probability of surviving long enough to require maintenance; then, to assure that a favorable balance of costs and benefits is realized by those participants who contribute to recording design rationale.

Consider the situation described in the opening interview. Assume that the manager is correct and assume that nothing is preserved when a project is terminated (these assumptions are examined later). Ninety percent of the time, investment in recording design rationale for the benefit of corrective, adaptive, perfective, or preventive maintenance is lost outright due to project termination. Furthermore, a project that has a chance to survive, that could join the successful 10%, might, by diverting resources to record design rationale, slow its progress, fail to reach a critical checkpoint, and be terminated. When the upstream waters are infested with piranhas, your best strategy can be to get downstream as quickly as possible, even if you arrive less than fully prepared for all possible contingencies. (We meet some piranhas later.)

In commercial product development, project completion rates are low. (The case study by Sharrock & Anderson, 1995 [chapter 15 in this book] is a nice description of a terminated project.) But in other development contexts, they are higher. For example, expensive systems built under contract are usually delivered. All else equal, a tool to capture design rationale for facilitating maintenance will be more valuable where project completion rates are high. However, completion rate is only one factor in assessing the appropriateness of design capture. Another factor is the continuity of personnel. Maintenance may be handled by the original developers, by different groups within the same organization, or be the responsibility of an independent organization. Clearly, this also affects the value to the developers and the development organization of an upstream investment to facilitate subsequent maintenance.

The message for those developing systems and methodologies to support the recording of design rationale is this: Give careful thought to your prospective users. Many design rationale efforts are directed toward an audience that may not offer the best conditions for success, the developers of commercial off-the-shelf software. Design rationale research will be more successfully applied if it is directed to appropriate audiences.

Of course, one can always work to alter adverse conditions. Organizational change could improve the conditions for design capture—clear away the upstream piranhas. But organizations change slowly. Someone is fond of the piranhas or they would not be there, and design capture does not yet have a track record that gives weight to arguments for major change.

In the sections that follow, I first present the perspective on design rationale taken in this chapter. Then I suggest that many analyses have been limited by being restricted to the examination of successful projects. The utility of capturing design rationale is discussed for four development contexts: off-the-shelf product development, internal or in-house system development, large competitively bid development contracts, and smaller noncompetitive consulting or third-party development projects. Each context presents unique opportunities and obstacles for utilizing design capture techniques, but they are far from equal in promise. The possible value of information preserved in the course of unsuccessful projects is then considered. Finally, I examine the fit between approaches based on the methods of scientific research and the requirements of engineering and development practice.

2. THE USE OF DESIGN RATIONALE IN SYSTEMS DEVELOPMENT ORGANIZATIONS

In this section, I focus on considerations affecting the capture and use of design rationale in systems development organizations, primarily software development. This is a natural focus for the developers of software tools to support design capture, although Sharrock and Anderson (1995 [chapter 15 in this book]) and Gruber and Russell (1995 [chapter 11 in this book]) present examples from other design domains. Consideration of other design domains could only add to the variation in design environments and strengthen the case made in this chapter for considering the conditions that exist in a given setting for capturing rationale.

As noted by Conklin and Burgess-Yakemovic (1991 [chapter 14 in this book]), design rationale is captured and used routinely in development organizations. It is captured in the memories of individual developers, in meeting minutes recorded by secretaries, in dictaphone tapes, and in memos. Written specifications may include some rationale. Development organizations sometimes preserve video records of design meetings and presentations. Engineering notebooks are often used conscientiously to record rationale. When working as a developer, I used all of these techniques at one time or another. Each had its uses and its weaknesses. Among the weaknesses are tendencies to be incomplete, unstructured, and/or dispersed throughout an organization; thus records are difficult to retrieve and interpret later, and can be impossible for third parties to access.

The current interest in this topic derives from the observation that technology could address these weaknesses. It can help with retrieval and access. By supporting formal or semiformal systems for representing information, technology might assist in building more interpretable and complete accounts of the design process. This chapter focuses on such computer-based systems, although some of the conclusions will apply to other forms of design recording.

What is the rationale for design rationale research? The most common reason given for capturing design rationale is to use it for subsequent consultation, for use in redesign or maintenance. Another reason is to produce improved designs through the discipline or focus of attention required to derive or capture design rationale (e.g., Carroll & Rosson, 1991 [chapter 4 in this book]). Several authors mention both as potential benefits;

for example, MacLean et al. (1991 [chapter 3 in this book]) present an extensive case for the benefits of design rationale in system maintenance, but also speculate that designers might be rewarded through immediate design improvements. A third reason for recording rationale is for communicating progress to other project members.

This chapter focuses on the first rationale noted previously: the benefits to subsequent development or maintenance. This has tremendous intuitive appeal. It is less clear that formal or semiformal design rationale representations will be useful in communicating design progress; the little available evidence is negative (Conklin & Burgess-Yakemovic, 1991 [chapter 14 in this book]). Similarly, although outstanding researchers have improved designs when using these methods, trials with less brilliant users have not fared well (e.g., Fischer et al., 1992). One impressive design improvement occurred not when the design rationale was being recorded by developers, but when researchers translated it from one representation format to another (Conklin & Burgess-Yakemovic, 1991 [chapter 14 in this book]).

My analysis is comparative. Conklin and Yakemovic discuss the trade-off between costs and payoffs for design capture. Some cost is assumed: If a design tool cost nothing to use, it could be used without harm in every development context. Similarly, a limited benefit is assumed: If a design tool provided virtually infinite benefit, it would be used in every development context. Existing tools lie in between: Mastering and using them requires additional effort on the part of developers, and it is hoped that a net benefit will result. The additional effort is concentrated, along with design activity, at the outset of a project, consistent with the evidence that design errors are less costly if caught early. My central question is: Under what conditions will the early, extra effort most likely be beneficial? Having answered that, we can dig further for insight into improving the tools.

The same analysis holds for tools or methods that improve design directly. The utility depends on the cost of the effort and the benefit it provides. If an immediate benefit outweighs the cost, it is a winner. If realization of the benefit is delayed, then more or less favorable conditions probably exist for applying it.

These cost–benefit analyses are one of three major considerations in applying design capture in specific settings. The other two are not addressed further in this chapter but must also be evaluated in specific development contexts:

1. Will social, political, and motivational concerns prevent the explicit statement of the real reasons underlying design choices? This can occur even in engineering settings. The rationale might contain strategic information of use to competitors; a certain group might not be deemed capable of a tricky implementation, but it is not politic to state this explicitly; the introduction of the design capture tool could shift decision-making power

to skillful tool users, who could game the new system; fears of these things could affect tool adoption. And so on. Examples can be found in Sharrock and Anderson (1995 [chapter 15 in this book]) and a discussion of some of these issues concludes Conklin and Burgess-Yakemovic, 1991 [chapter 14 in this book]).

2. Accurate interpretation of captured design rationale could require more knowledge of the context that existed at the time of capture than can possibly be recorded. This issue is addressed by Gruber and Russell (1995 [chapter 11 in this book]).

3. "REVERSE-ENGINEERING" SUCCESSFUL PROJECTS: PROBLEMS WITH A SYSTEM PERSPECTIVE

As more large software programs are written and as they age with varying degrees of gracefulness, challenges in software maintenance become more pressing. Boehm (1988) argued for considering maintenance to be inseparable from development, for freeing it from its present "second-class" status in software engineering. In "Scaling Up: A Research Agenda for Software Engineering," the Computer Science Technology Board (CSTB) also focused on maintenance. They wrote "System maintenance may constitute up to 75% of a system's cost over its lifetime. The high cost of maintenance makes designing for a system's entire life cycle imperative, but such design is rarely if ever achieved." (CSTB, 1990, p. 282)

Given the apparent benefit to be gained through upstream design investment that anticipates maintenance needs, why is such design "rarely if ever achieved"? Why do we neglect an apparent opportunity to benefit economically? Why, as recounted by Conway (1968), is there "never enough time to do something right, but always enough time to do it over" (p. 29)?

The answers are not hard to find. In the next section, obstacles to upstream investment in design are identified in each of several different systems development contexts. They have not been recognized because of assumptions that have biased analyses of systems development projects:

1. *Analyses are system centered.* In considering the development process, our focus is on the system being developed. This diverts attention from the specific individuals involved in the process. Focusing on development phases reduces our awareness that different people are generally involved at different times. Focusing on the system being developed also obscures aspects of the organizations in which development occurs, including "tangential" organizational factors that can act to terminate projects prior to completion.

2. *Analyses are based on large systems.* The study of large systems, particularly in the context of government contracts, gave rise to the field of software engineering. The challenges and risks are greater for large proj-

ects; nevertheless, it is a mistake to apply data or models derived from their analysis to contexts in which smaller projects are the rule.

3. *Analyses of development projects are retrospective and aggregate.* This leads to a model of a complete development process, with no consideration given to early termination. One rarely sees descriptions of the waterfall model in which the water fails to fall to the bottom. Only recent variations such as Boehm's (1988) spiral model provide explicit provisions for exiting when a risk analysis produces an unfavorable assessment.

This idealized project perspective obscures the fact that many projects never incur maintenance costs because they are canceled. Statements such as "maintenance . . . can occupy as much as 90% of the effort in the software life cycle" (MacLean et al., 1991 [chapter 3 in this book]) are clearly based on successfully completed projects. One does not read "in many contexts, maintenance occupies 0% of a typical system's cost," although this is equally true. The system-centered perspective obscures the fact that design, development, and maintenance engineers are typically different individuals, which can reduce or eliminate the former's incentive to design for maintenance. Even when an incentive exists, lack of communication among the engineers reduces the awareness of how useful design rationale might be. Our attention to relatively large projects, often contract or internal development projects, neglects important development contexts in which different conditions prevail.

A corrective is to broaden our analytic focus, considering not individual projects but instead the organizational contexts in which development occurs. A decision to embrace design capture, like any approach to design and development, is taken on the basis of evolving conditions within a development organization. Examination of these conditions enables us to assess the obstacles and opportunities for innovation in upstream design activities.

Four organizational contexts for systems development are examined: product development, internal development, competitively bid contract development, and customized software development. Although not an exhaustive set, they provide a range of opportunities for and obstacles to recording and using design rationale. Design capture tools or techniques may require adaptation for use in specific environments. They do not seem equally promising in all cases.

4. PROSPECTS FOR DESIGN CAPTURE IN SEVERAL DEVELOPMENT CONTEXTS

Substantial design rationale research is carried out by employees of computer and software vendor companies and by academic researchers sharing their concerns. In the late 1970s, large markets for word processors, spreadsheets, graphics, and other interactive applications caught the attention of computer companies and spawned software development companies. In the early 1980s, conferences and journals concerned with creating usable software products appeared with names involving permutations of the words "human," "computer," and "interaction." Exploration of design rationale is found in these settings, much of it carried out by cognitive psychologists with an enduring interest in issues in problem solving, memory, and other related activities. Much of the work in this book was first outlined in a workshop devoted to design rationale at a Computer and Human Interaction Conference (CHI '91) and published in the journal *Human–Computer Interaction*. This book's nonacademic contributors were from Digital, IBM, Kodak, MCC, NCR, and Xerox—all commercial product developers and marketers.

Important as commercial, off-the-shelf product development is, more resources are devoted to in-house software development, carried out by information systems groups in hospitals, banks, insurance companies, government agencies, and so forth. Variously called data processing (DP), information systems (IS), management information systems (MIS), or information technology (IT), this community has conferences and journals that focus primarily on large systems that serve organizations.

A third influential development context is competitively bid contract development. The U.S. government, the largest computer user, sponsored most of the major software development projects of the 1950s. As a result, many software engineering methods in use today evolved in the context of contract development. These large systems encounter significant maintenance issues, and interest in design rationale is growing in this community.

Many smaller organizations that have recently computerized cannot afford full-time software developers, yet want software customized or adapted to their individual needs. A growing number of consulting and third-party development companies specialize in customized software development. Addressing specific vertical markets or application domains, these companies often seek to reuse software in moving from customer to customer. With less formal or restrictive procurement, the conditions surrounding development are different than for large, competitively bid contracts.

When we think in terms of "the computer industry" we overlook how very different these and other branches of systems development have become. Different issues, approaches, and challenges govern each (Grudin, 1991a). Developing and positioning methods and tools to support design argumentation require careful consideration of the contexts in which the methods and tools are to be applied.

4.1. "Off-the-Shelf" Product Development

Product developers can see the potential benefits in recording design rationale. Proceeding under time pressure, projects add new personnel, including programmers and members of support groups involved in documentation, training, quality assurance, performance analysis, and marketing. The education of new arrivals can be extremely time consuming for existing team members; it can be very frustrating when new arrivals challenge design decisions that were made earlier for reasons lost from memory. Major productivity gains could plausibly result if a record of design rationale handled much of the education, freeing existing team members to continue working as developers. The design rationale could subsequently assist in the development of new versions or releases.

However, commercial products are generally failure-prone. According to Business Week, "estimates of new products' failure rate vary hugely, but that rate could be anywhere from 66% to almost 90%. In a 1991 survey, marketers expected 86% of their new products to fail, up from 80% in 1984" ("Will it sell?", 1992). Reliable figures for commercial software projects are difficult to find, but in an informal poll of several experienced development managers from an array of computer companies, completion rate estimates ranging from 10% to 50%. Experienced developers reported having never seen a project through to completion and a well-known consultant recalled involvement in only one successful project.

Projects are terminated for different reasons. A company might initiate competing projects with the intention of terminating one (see Kidder, 1982, for a well-known example). Companies change strategic direction or decide that a window of opportunity has closed. Schedule or budget overruns shake management's confidence in a development team. Early field testing reveals that a product will not recoup the cost of launching it. The list goes on. Note that the first three reasons provide clear incentives for developers to push ahead quickly: to beat internal competitors to milestones, to get something out and capture market share before a window closes, to meet project milestones and reassure management. These are upstream piranhas.

In this climate, it seems sensible to limit upstream activity to proofs of concept and to push toward completion. Resistance to expanding upstream activity may reflect the awareness of project managers of the precariousness of their enterprise.

The capture of design rationale in product environments is further hindered by the distributed nature of design decision making. Managers, programmers, domestic and international marketers, human factors engineers, technical writers, industrial design engineers, mechanical engineers, training developers, performance analysts, and others are involved. Contributors to the design often work at different sites on different platforms. Conklin and Burgess-Yakemovic, 1991 [chapter 14 in this book]) encountered this problem when marketing and other groups would not adopt the development group's design rationale representation format. Not only would others not use the system, the development group found it necessary to manually convert their design rationale into prose before meetings. This should be unsettling to those who envision design rationale as *contributing* to communication on a project!

A third obstacle is that the product developers who have to record their reasoning are unlikely to be the principal users of the rationale. Buyers of off-the-shelf products are largely responsible for adapting them. Vendor companies provide limited assistance, but it is delivered through customer support or field service groups—not by the initial development team. True, a company might derive an overall benefit by providing customer service with the development rationale, but development and customer service budgets are distinct and such "altruism" is often absent (Grudin, 1991b). A chronic cause of failure in group-work situations (Grudin, 1988, 1994) occurs when one group incurs additional work (in this case, the developers) and another group benefits (in this case, customer service).

Similarly, the initial development team often does not revise a successful product. New releases can be seen as less challenging, a form of maintenance. Product enhancements are often turned over to a subset of the team or to a different group as key designers are given new assignments.

Thus, the advocate of design rationale must ask a developer to undertake an effort that will slow down the project, be discarded unused in the likely event that the project is terminated, and in the best outcome will benefit someone else. And when a project is canceled, it will be natural to wonder: Might we have progressed more quickly and escaped the ax if we avoided that side effort? Convincing examples of successes with design rationale are needed to counter these concerns. Until then a developer has little to go on.

Some assumptions underlie this bleak picture. Perhaps design capture could provide immediate benefits to the developer. Perhaps the effort required to explicitly structure design arguments could be minimized. Perhaps benefit could be obtained from a record of design rationale left by a terminated project. So far, immediate benefits have not been shown and existing systems require substantial effort to capture design rationale. In a unique case in which builders of a design rationale tool captured their own design rationale, described in Fischer, Lemke, McCall, and Morch (1991 [chapter 9 in this book]), the group ended up pessimistic about the approach. Chapters of this book reporting substantial trials (Buckingham Shum, 1995 [chapter 6]; Potts, 1995 [chapter 10]) report difficulties in using existing methods to record rationale. The unlikelihood of benefiting from rationale captured in the course of a failed project—of using the rationale to learn from failure—is described in Section 5.

In sum, much design rationale research is done in the product development context and is presented to audiences of product developers. The concept of design capture appeals to them when they focus on the positive possibilities. Rational approaches to problem solving appeal to engineers. It is attractive to individuals who envision themselves as particularly effective users of such tools. Unfortunately, product development is also marked by tight time constraints, frequent progress monitoring, low project completion rates, and developer ignorance of postsale use. These render product development a relatively unpromising context for exploiting design capture.

4.2. "In-House" or Internal Development

When an information systems group within a large organization, such as a bank, hospital, or automobile manufacturer, undertakes a development project, conditions are unlike those in product development. The system or application is usually large and focused on organizational support rather than individual use. Although management usually initiates a project with some specific functional goals, an internal development project often has a broader charter to address both functionality and interface issues than does a product development team formed after most functionality has been defined.

Are internal development project completion rates higher than in product development? Statistics are difficult to find, but there are disquieting indications. One survey reported that 75% of in-house projects are terminated prior to completion or produced software that was never used, the in-house equivalent of an unmarketed product (Gladden, 1982). In such a setting, a general policy of upstream investment to facilitate maintenance could be difficult to justify.

Positive factors exist for design capture in internal development. In-house projects are notorious for running over schedule, but their relative duration motivates the recording of design information as a hedge against losses of individual and corporate memory. As in other large projects, participants join a project as it progresses, requiring education about the rationale for past decisions. Also, the same organization and perhaps the same individuals will be responsible for system maintenance, and a system or application is likely to be operated and maintained for a particularly long time. Large organizations often use systems for decades and are less motivated than product developers to make major changes that would require retraining users.

It may be difficult to introduce design capture tools and methods in this context. Tools available to in-house developers are usually older. Innovation is not attributed to such environments; innovation in the use of design capture is unlikely to be a widespread exception. Nevertheless, opportunities might present themselves to vigilant advocates of design capture.

4.3. Competitively Bid Contract Development

Large, contracted development projects have high completion rates, because both parties have strong incentives to avoid a cancellation. Delivered systems are often unusable without substantial further work (Martin,

1988), but corrective maintenance argues *for* supporting downstream activities. In addition, not only must the contractor educate new project members as development proceeds, but the delivered product is often maintained by people who have no access to the initial developers. Attributions of a high percentage of project costs to maintenance come from contract projects, where cost accounting is more routine. The high completion rate of contract development projects explains the inattention to project mortality considerations in the literature.

Also conducive to design capture in competitively bid contract development is the detailed project documentation that already is required for the written specifications. The trend in government procurement is toward requiring a better record of design decisions and rationale: Concurrent engineering (CE) and computer-aided acquisition and logistic support (CALS) are new approaches that aim to integrate development and to allow the tracing of design features to specific requirement specifications. This perspective was shared by several participants in the AAAI '92 Workshop on Design Rationale Capture and Use.

However, there is a major disincentive to focus on upstream investment in large contract projects: The design is often carried out under one contract, the development of the resulting design is awarded in a second contract, and the particularly lucrative system maintenance can be carried out under a third. Each contract is let independently, potentially to a different organization. And even if the same company wins successive contracts, different people inevitably are assigned to complete them. This fragmentation of development effort can eliminate the incentive for those engaged in the upstream activities to allocate resources for supporting downstream activity. The contracting agency can try to mandate such support, but it is an open question whether useful compliance will result, especially given the experimental status of tools and approaches to capturing design information.

4.4. Customized Software Development

Smaller contract development efforts in specific market niches are particularly promising arenas for design capture, as well as being increasingly important centers of system development in general. As before, with a contract involved, albeit less formal and with less detailed specifications, these projects are relatively unlikely to be canceled. But in addition, those developing systems for specific customers in particular markets are more likely to work with a customer after implementation, participating in various forms of maintenance (corrective, enhancement, etc.). (In contrast, close, long-term relationships with customers are often impossible in competitive procurement, where barriers are erected to prevent subsequent favoritism to one development company and where adherence to the letter of the contract can take precedence over the usability of the system.)

Another positive factor is that a consulting company that targets a specific application domain will seek similar projects that allow the reuse of software and domain expertise. A record that captures the initial rationale plus exceptions on subsequent projects could be very useful in a new project, as well as in maintaining existing systems.

A major hurdle to applying design capture techniques in this context is that many of these are smaller companies with fewer resources to invest in using and developing new or experimental tools. However, there are large consulting firms, and companies operating in specific markets are growing in size, number, and profitability.

Fischer et al. (1991 [chapter 9 in this book]) explore the creation of design environments that support reuse in a manner conducive to customized software development, although they have thus far developed tools to support design activities outside of software (network administration, kitchen design, etc.).

5. WHAT CAN BE GLEANED FROM A TERMINATED PROJECT?

It has been argued that "the best prototype is a failed project" (Curtis, Krasner, & Iscoe, 1988). Project members learn from their experiences, whatever the outcome. Could design capture be exploited if employed in a prematurely terminated project? Key decisions affecting success could be reexamined and progress could be exploited. Design possibilities that missed a "window of opportunity" on a project could be more reliably or accessibly preserved for use in subsequent projects than occurs through the memory of individual participants.

This seems worth exploring in situations where a similar project is likely to follow a failure. But some hard questions have to be asked. Would people make the effort to record design rationale with this purpose in mind? Would such records be used? If created and used, would such records be more useful than project members' memories? Having experienced numerous project terminations, I am skeptical. The emotional aftermath of project terminations is not conducive to the use of project records. A dead project is about as inviting to linger over as the body of a victim of a highly contagious disease.

Projects are not terminated casually. A project can undergo major shifts in direction, schedule, budget, and even management without being terminated. Many factors may contribute to a termination decision. Precise reasons for a cancellation may not be announced, but it is fair to assume that the product design was not brilliant. Any decision might have contributed to the failure in some way. A clear determination of responsibility may be difficult or impossible to reach; if reached, it may not be announced; if announced, it may not be accepted or believed. A pall descends over a terminated project. If a similar effort is initiated, the new management is keen to differentiate the new project, to put its own stamp on things. The desire to avoid overt associations with the decisions of the past can be superstitious (yet no less potent for that reason), but it can also be quite rational. Both symbolically and objectively, the new is dissociated from the tried and apparently untrue.

At the level of the individual contributors, disappointment and uncertainty follow a cancellation. Management works to move quickly past this state and to avoid extended recriminations. Public postmortems are rare. Less is learned from the experience as a result, but this cost is outweighed by the benefits of quickly confining skeletons to closets. The paucity of detailed analyses of failure testifies to this dynamic.

Individuals who were involved in a terminated project retain the ideas that particularly impressed them. Failed projects contribute when dispersed project members reintroduce certain points on their merits. Documentation from old projects is rarely explored systematically for ideas. Quite the contrary, in my experience the tangible traces of failed projects are discarded remarkably quickly.

This is true in research as well as development environments. At the research consortium MCC, one of the most frequent requests from shareholder companies was that researchers document not only their successes, but also their failures: paths tried and abandoned. The shareholders did not find it stressful to contemplate the examination of our failures, but we researchers did. We could not bring ourselves to make this effort. Focusing on the positive was far more appealing.

It could be argued that a design capture system provides the best of both worlds, documenting productive as well as abandoned approaches. However, where termination is frequent, a design rationale for an abandoned project will either be discarded quickly, as happens with other forms of documentation, or serve as a persistent reminder of failure. I cannot envision an ongoing practice of recording design rationale in termination-prone environments. Can the effortful documentation of failure be made palatable to those who must do the work to document their own failure?

At issue is the nature of what is possible in these emotionally charged situations, not the benefits that can plausibly be argued to exist. A reviewer of this chapter asked, "If a failed project bequeaths a rationale, then why couldn't the rationale live on?" It could, of course, but who would trust a rationale that supported a failure? And more important, who would help build a rationale if their prior experience suggested it would be a record of failure? Retrospective reconstructions of successful development projects are common; thoughtful accounts of what was learned from failure are almost nonexistent. Ignore human nature at your peril.

6. CUMULATIVE SCIENCE AND ENGINEERING TRADE-OFFS

Many approaches to capturing and using design rationale have as an explicit goal the reduction of irrationality and inefficiency in design decision making. The approaches are developed by researchers with scientific training and incorporate models of rational decision making that resemble a scientific ideal. A full range of alternatives is explored, evidence is accumulated, a tentative decision is reached, and a record of activity is preserved for later reexamination. Whether or not science always proceeds this way, science is cumulative, progress is recorded and preserved, and thoroughness is valued. Engineering values are not in direct opposition to these, but they are not the same. Scientists are not designers. There is a clear danger that in developing tools to support a process, scientists will assume the process resembles that of science more than it does and fail to address the needs of developers.

Scientific exploration is often commended for its own sake. In engineering, exploration is not commended for its own sake. In a world of constant compromise, a point is reached sooner rather than later at which exploration is curtailed to save time or other resources. Those who have designed in areas untouched by standardization are aware that the smallest design decision, if examined closely and imaginatively, can open up a virtually infinite realm of possible solution paths. Setting out to chart a large unexplored territory can be good science; it is usually poor engineering. Scientific territory that is explored and reported augments a body of knowledge, but an open-ended engineering exploration diverts critical resources, misses windows of opportunity, and risks a termination that will leave no traces.

Scientists strive for certainty and replicability, so thoroughness is often commended for its own sake. Engineering trade-offs often force decisions based on much lower probabilities, or even in the face of evidence.

Several of the empirical chapters in this book support this analysis, Buckingham Shum (1995 [chapter 6 in this book]) observed that "most decisions . . . may in fact not be of interest to other domain experts, because they are . . . not contentious given a certain level of expertise." This observation is further reflected in Potts' (1995 [chapter 10 in this book]) distinction between a scientific "bird's-eye view" looking down at design in its entirety and an engineering "turtle's-eye view" rooted in the artifact at hand. Potts finds the latter "much more useful in the practical work of designing." Lewis, Rieman, and Bell (1991 [chapter 5 in this book]) take this a step further by arguing for design rationale that is entirely problem based.

The careful recording of the rationale underlying design decisions can be useful to engineers. But the value is a function of the trade-offs and compromises that constitute engineering practice, which often seem to be ignored by innovators who hope to influence design. If the tools and methods designed by researchers with a scientific background reflect too strongly the values and ideals of science, those tools and methods will be less useful to engineers.

7. CONCLUSIONS

The creation of a record of design decision making, a design rationale, has a strong intuitive appeal. The current level of interest in exploring it is a positive development. For such efforts to enjoy the best prospects for evolving into useful tools and methods, it is important that they be designed with particular use situations in mind. Different development contexts require different approaches.

Several obstacles to the capture and use of design rationale in software development have been identified. It can slow down a project that is under pressure to reach a milestone. It can become a record of failure. It can require a joint effort by people separated by time, geography, platform, and even employer. The effort can be lost in the chaos of a terminated project. It can require commitment to a new, experimental approach. To overcome these obstacles, design rationale systems must provide considerable collective benefit. Those who have to do the work of recording design information should perceive themselves as benefiting from the use of the record. On the positive side, some software development contexts provide relatively promising conditions for such efforts, notably customized software development.

Gruber and Russell (1995 [chapter 11 in this book]) describe the need to look carefully at prospective users of design rationale to see what they need. Adopting this approach would quickly reveal some of the impediments described in this chapter. However, generalizing across cases must be done sparingly; such an analysis should be considered for each project.

This chapter provides suggestions, not definitive answers. These are research issues, to be explored in parallel with the development of tools and techniques for the capture and organization of design rationale.

NOTES

Acknowledgments. General discussions with Gerhard Fischer and the HCC group at the University of Colorado and specific discussions about customized software development with Frank Shipman were helpful. Jack Carroll, Tom Moran, and an anonymous reviewer contributed very helpful comments on an earlier draft.

Support. The preparation and writing of this chapter were supported by the National Science Foundation under Grant IRI-9015441.

Authors' Present Address. Jonathan Grudin, Information and Computer Science Department, University of California, Irvine, Irvine, CA 92717. Email: grudin@ics.uci.edu.

REFERENCES

Boehm, B. (1981). *Software engineering economics.* Englewood Cliffs, NJ: Prentice-Hall.

Boehm, B. (1988). A spiral model of software development and enhancement. *IEEE Computer, 21*(5), 61–72.

Buckingham Shum, S. (1995). Analyzing the usability of a design rationale notation. In T. P. Moran & J. M. Carroll (Eds.), *Design rationale: Concepts, techniques, and use.* Hillsdale, NJ: Lawrence Erlbaum Associates. [Chapter 6 in this book.]

Carroll, J. M., & Rosson, M. B. (1991). Deliberated evolution: Stalking the View Matcher in design space. *Human–Computer Interaction, 6,* 281–318. Also in T. P. Moran & J. M. Carroll (Eds.), *Design rationale: Concepts, techniques, and use.* Hillsdale, NJ: Lawrence Erlbaum Associates, 1996. [Chapter 4 in this book.]

Computer Science Technology Board. (1990). Scaling up: A research agenda for software engineering. *Communications of the ACM, 33,* 281–293.

Conklin, E. J., & Burgess-Yakemovic, KC. (1991). A process-oriented approach to design rationale. *Human–Computer Interaction, 6,* 357–391. Also in T. P. Moran & J. M. Carroll (Eds.), *Design rationale: Concepts, techniques, and use.* Hillsdale, NJ: Lawrence Erlbaum Associates, 1996. [Chapter 14 in this book.]

Conway, M. E. (1968, April). How do committees invent? *Datamation,* pp. 28–31.

Curtis, B., Krasner, H., & Iscoe, N. (1988). A field study of the software design process for large systems. *Communications of the ACM, 31*(11), 1268–1287.

Fischer, G., Grudin, J., Lemke, A. C., McCall, R., Ostwald, J., Reeves, B., & Shipman, F. (1992). Supporting indirect collaborative design with integrated knowledge-based design environments. *Human–Computer Interaction, 7,* 281–314.

Fischer, G., Lemke, A. C., McCall, R., & Morch, A. I. (1991). Making argumentation serve design. *Human–Computer Interaction, 6,* 393–419. Also in T. P. Moran & J. M. Carroll (Eds.), *Design rationale: Concepts, techniques, and use.* Hillsdale, NJ: Lawrence Erlbaum Associates, 1996. [Chapter 9 in this book.]

Gladden, G. (1982). Stop the life-cycle, I want to get off. *Software Engineering Notes, 7*(2), 35–39.

Gruber, T. R., & Russell, D. M. (1995). Generative design rationale: Beyond the record and replay paradigm. In T. P. Moran & J. M. Carroll (Eds.), *Design rationale: Concepts, techniques, and use.* Hillsdale, NJ: Lawrence Erlbaum Associates. [Chapter 11 in this book.]

Grudin, J. (1988). Why CSCW applications fail: Problems in the design and evaluation of organizational interfaces. *Proceedings of the CSCW '88,* 85–93. New York: ACM. (Reprinted in D. Marca & G. Bock (Eds.), 1992, *Groupware: Software for computer-supported cooperative work* (pp. 552–560). Los Alamitos, CA: IEEE Press.

Grudin, J. (1991a). Interactive systems: Bridging the gaps between developers and users. *IEEE Computer, 24*(4), 59–69.

Grudin, J. (1991b). Systematic sources of suboptimal interface design in large product development organizations. *Human–Computer Interaction, 6*(2), 147–196.

Grudin, J. (1994). Groupware and social dynamics: Eight challenges for developers. *Communications of the ACM, 37,* 1, 92–105.

Jarczyk, A. P. J., Loffler, P., & Shipman, F. M. (1992). Design rationale for software engineering: A survey. *Proceedings of the 25th Hawaii International Conference on Systems Sciences,* 577–586. Los Alamitos, CA: IEEE Computer Society Press.

Kidder, T. (1982). *Soul of a new machine.* New York: Avon.

Lewis, C., Rieman, J., & Bell, B. (1991). Problem-centered design for expressiveness and facility in a graphical programming system. *Human–Computer Interaction, 6,* 319–355. Also in T. P. Moran & J. M. Carroll (Eds.), *Design rationale: Concepts, techniques, and use.* Hillsdale, NJ: Lawrence Erlbaum Associates, 1996. [Chapter 5 in this book.]

MacLean, A., Young, R. M., Bellotti, V. M. E., & Moran, T. P. (1991). Questions, options, and criteria: Elements of design space analysis. *Human–Computer Interaction, 6,* 201–250. Also in T. P. Moran & J. M. Carroll (Eds.), *Design rationale: Concepts, techniques, and use.* Hillsdale, NJ: Lawrence Erlbaum Associates, 1995. [Chapter 3 in this book.]

Martin, C. F. (1988). *User-centered requirements analysis.* Englewood Cliffs, NJ: Prentice-Hall.

Potts, C. (1995). Supporting software design: Integrating design methods and design rationale. In T. P. Moran & J. M. Carroll (Eds.), *Design rationale: Concepts, techniques, and use.* Hillsdale, NJ: Lawrence Erlbaum Associates. [Chapter 10 in this book.]

Sharrock, W., & Anderson, R. (1995). Organizational innovation and the articulation of the design space. In T. P. Moran & J. M. Carroll (Eds.), *Design rationale: Concepts, techniques, and use.* Hillsdale, NJ: Lawrence Erlbaum Associates. [Chapter 15 in this book.]

Will it sell? Hard to tell. (1992, August 17). *Business Week,* p. 72 B-E.

Author Index

Reference Index*

---

*Note that chapters written in 1995 for this volume (e.g., Buckingham Shum, 1995) are listed 1995 in text and references. The actual copyright date of this volume is 1996.

Subject Index

485

Empirical Studies
of a copier development project, 434–445
of design explanations, 325–333
of design meetings, 219–234
of design reasoning, 74–86
of Janus, 286, 287
of use of IBIS at NCR, 404–415
of use of PHI notation, 275–277
of use of QOC notation, 187–209

overview, 12, 13

People

Alexander, Christopher, 3
Bacon, Sir Francis, 108, 109, 138, 142
Rittel, Horst, 3, 271, 399
Schön, Donald, 3, 353
Simon, Herbert, 3, 394, 395

Printed and bound by CPI Group (UK) Ltd, Croydon, CR0 4YY

17/10/2024

01775683-0014